Studies in Big Data

Volume 90

Series Editor

Janusz Kacprzyk, Polish Academy of Sciences, Warsaw, Poland

The series "Studies in Big Data" (SBD) publishes new developments and advances in the various areas of Big Data- quickly and with a high quality. The intent is to cover the theory, research, development, and applications of Big Data, as embedded in the fields of engineering, computer science, physics, economics and life sciences. The books of the series refer to the analysis and understanding of large, complex, and/or distributed data sets generated from recent digital sources coming from sensors or other physical instruments as well as simulations, crowd sourcing, social networks or other internet transactions, such as emails or video click streams and other. The series contains monographs, lecture notes and edited volumes in Big Data spanning the areas of computational intelligence including neural networks, evolutionary computation, soft computing, fuzzy systems, as well as artificial intelligence, data mining, modern statistics and Operations research, as well as self-organizing systems. Of particular value to both the contributors and the readership are the short publication timeframe and the world-wide distribution, which enable both wide and rapid dissemination of research output.

The books of this series are reviewed in a single blind peer review process.
Indexed by zbMATH.
All books published in the series are submitted for consideration in Web of Science.

More information about this series at http://www.springer.com/series/11970

Yassine Maleh · Youssef Baddi ·
Mamoun Alazab · Loai Tawalbeh ·
Imed Romdhani
Editors

Artificial Intelligence
and Blockchain for Future
Cybersecurity Applications

 Springer

Editors
Yassine Maleh🆔
Sultan Moulay Slimane University
Beni Mellal, Morocco

Mamoun Alazab🆔
Charles Darwin University
Canberra, Australia

Imed Romdhani
Edinburgh Napier University
Edinburgh, UK

Youssef Baddi🆔
Chouaib Doukkali University
El Jadida, Morocco

Loai Tawalbeh🆔
Texas A&M University
San Antonio, USA

ISSN 2197-6503   ISSN 2197-6511   (electronic)
Studies in Big Data
ISBN 978-3-030-74577-6   ISBN 978-3-030-74575-2   (eBook)
https://doi.org/10.1007/978-3-030-74575-2

This Springer imprint is published by the registered company Springer Nature Switzerland AG
The registered company address is: Gewerbestrasse 11, 6330 Cham, Switzerland

In loving memory of my Mother Fatima

Yassine Maleh

Preface

Cyber threats increase as firms, governments and consumers rely on digital systems for their day-to-day activities. The more they adopt the technologies, the higher the risks they face. Additionally, new solutions to facilitate everyday businesses, such as artificial intelligence for operational systems and enormous IT databases, create complexity. However, these new technologies can also be their most reliable allies! They can provide new protection levels that make a strong shield of protection against hackers if properly designed and integrated. There is growth in IoT use, which increases the risk for organizations and the need for data protection policies. Organizations are not taking enough steps to secure themselves from cyber-attacks; ultimately, there will be an increase in attack size and volume.

AI and blockchain technologies have infiltrated all areas of our lives, from manufacturing to health care and beyond. Cybersecurity is an industry that has been significantly affected by this technology and maybe more so in the future.

Artificial intelligence and blockchain have shown potential in providing various methods for threat detection. Merging artificial intelligence and blockchain will change cybersecurity as we know it and make next-generation solutions more effective.

An open cybersecurity ecosystem, powered by a blockchain, will unlock the enormous opportunity to protect against next-generation threats, eliminate the strain and cost of third-party intermediaries, and ensure a more secure world.

The combination of cyber threat data integrated with artificial intelligence and blockchain is arguably the future of AI-powered cybersecurity.

This book will go in depth, showing how blockchain and artificial intelligence can be used for cybersecurity applications. Merging AI and blockchain can be used to prevent any data breach, identity theft, cyber-attacks or criminal acts in transactions. We accepted 18 submissions. The chapters covered the following three parts:

– Artificial Intelligence and Blockchain for future Cybersecurity Applications: Architectures and Challenges;

- Artificial Intelligence and Blockchain for Cybersecurity: Applications and Case Studies;
- Artificial Intelligence and Blockchain Applications for Smart Cyber Ecosystems.

Each chapter is reviewed at least by two members of the editorial board. Evaluation criteria include correctness, originality, technical strength, significance, quality of presentation, and interest and relevance to the book scope. This book's chapters provide a collection of high-quality research works that address broad challenges in both theoretical and application aspects of artificial intelligence and blockchain for future cybersecurity applications.

We want to take this opportunity and express our thanks to the contributors to this volume and the reviewers for their great efforts by reviewing and providing interesting feedback to the authors of the chapters. The editors would like to thank Dr. Thomas Ditsinger (Springer, Editorial Director, Interdisciplinary Applied Sciences), Professor Janusz Kacprzyk (Series Editor in Chief) and Ms. Rini Christy Xavier Rajasekaran (Springer Project Coordinator), for the editorial assistance and support to produce this important scientific work. Without this collective effort, this book would not have been possible to be completed.

Beni Mellal, Morocco   Yassine Maleh
El Jadida, Morocco   Youssef Baddi
Canberra, Australia   Mamoun Alazab
San Antonio, USA   Loai Tawalbeh
Edinburgh, UK   Imed Romdhani

Contents

Artificial Intelligence and Blockchain Applications for Smart
Cyber Ecosystems

About the Editors

Prof. Yassine Maleh is Associate Professor at the National School of Applied Sciences at Sultan Moulay Slimane University, Morocco. He received his PhD degree in computer science from Hassan 1st University, Morocco. He is a cybersecurity and information technology researcher and practitioner with industry and academic experience. He worked for the National Ports Agency in Morocco as IT manager from 2012 to 2019. He is Senior Member of IEEE and Member of the International Association of Engineers (IAENG) and the Machine Intelligence Research Labs. He has made contributions in information security and privacy, Internet of things security, and wireless and constrained networks security. His research interests include information security and privacy, Internet of things, networks security, information system and IT governance. He has published over 50 papers (chapters, international journals and conferences/workshops), 7 edited books and 3 authored books. He is Editor in Chief of the International Journal of Smart Security Technologies (IJSST). He serves as Associate Editor for IEEE Access (2019 Impact Factor 4.098), the International Journal of Digital Crime and Forensics (IJDCF) and the International Journal of Information Security and Privacy (IJISP). He was also Guest Editor of a special issue on Recent Advances on Cyber Security and Privacy for Cloud-of-Things of the International Journal of Digital Crime and Forensics (IJDCF), Volume 10, Issue 3, July–September 2019. He has served and continues to serve on executive and technical program committees and as a reviewer of numerous international conferences and journals such as Elsevier Ad Hoc Networks, IEEE Network Magazine, IEEE Sensor Journal, ICT Express and Springer Cluster Computing. He was Publicity Chair of BCCA 2019 and General Chair of the MLBDACP 19 and MLBDACP 21 symposiums.

Prof. Youssef Baddi is full-time Assistant Professor at Chouaïb Doukkali University (UCD), El Jadida, Morocco. He obtained his PhD thesis degree in computer science from ENSIAS School, University Mohammed V Souissi of Rabat, Morocco, since 2016. He also holds a research master degree in networking obtained in 2010 from the High National School for Computer Science and Systems Analysis, ENSIAS, Rabat, Morocco. He is Member of Laboratory of Information and Communication Sciences and Technologies STIC Lab, since 2017. He is Guest Member of Information Security Research Team (ISeRT) and Innovation on Digital and Enterprise Architectures Team, ENSIAS, Rabat, Morocco. He was awarded as the best PhD student at the University Mohammed V Souissi of Rabat in 2013. He has made contributions in group communications and protocols, information security and privacy, software-defined network, the Internet of things, mobile and wireless networks security, and mobile IPv6. His research interests include information security and privacy, the Internet of things, networks security, software-defined network, software-defined security, IPv6 and mobile IP. He has served and continues to serve on executive and technical program committees and as a reviewer of numerous international conferences and journals such as Elsevier Pervasive and Mobile Computing PMC, International Journal of Electronics and Communications (AEUE) and Journal of King Saud University: Computer and Information Sciences. He was General Chair of IWENC 2019 Workshop and Secretary Member of the ICACIN 2020 Conference.

Prof. Mamoun Alazab is Associate Professor in the College of Engineering, IT and Environment at Charles Darwin University, Australia. He received his PhD degree in computer science from the Federation University Australia, School of Science, Information Technology and Engineering. He is a cybersecurity researcher and practitioner with industry and academic experience. His research is multidisciplinary that focuses on cybersecurity and digital forensics of computer systems including current and emerging issues in the cyber environment like cyber-physical systems and the Internet of things, by considering the unique challenges present in these environments, with a focus on cybercrime detection and prevention. He looks into the intersection of machine learning as an essential tool for cybersecurity, for example, for detecting attacks, analyzing malicious code or uncovering vulnerabilities in software. He has more than 100 research papers. He is the recipient of a short fellowship from Japan Society for the Promotion of Science (JSPS) based on his nomination from the Australian Academy of Science. He delivered many invited and keynote speeches, 27 events in 2019 alone. He convened and chaired more than 50 conferences and workshops. He is Founding Chair of the IEEE Northern Territory Subsection (February 2019–current). He is Senior Member of IEEE, Cybersecurity Academic Ambassador for Oman's Information Technology Authority (ITA) and Member of the IEEE Computer Society's Technical Committee on Security and Privacy (TCSP). He has worked closely with government and industry on many projects, including IBM, Trend Micro, the Australian Federal Police (AFP), the Australian Communications and Media Authority (ACMA), Westpac, UNODC and the Attorney General's Department.

Prof. Loai Tawalbeh completed his PhD degree in electrical & computer engineering from Oregon State University in 2004 and MSc in 2002 from the same university with GPA 4/4. He is currently Associate professor at the Department of Computing and Cyber Security at Texas A&M University-San Antonio. Before that, he was a visiting researcher at the University of California, Santa Barbara.

Since 2005, he taught/developed more than 25 courses in different computer engineering disciplines and science with a focus on cybersecurity for the undergraduate/graduate programs at New York Institute of Technology (NYIT), DePaul University and Jordan University of Science and Technology. He won many research grants and awards with over 2 million USD. He has over 80 research publications in refereed international journals and conferences.

Prof. Imed Romdhani is full-time Associate Professor in networking at Edinburgh Napier University since June 2005. He was awarded his PhD from the University of Technology of Compiegne (UTC), France, in May 2005. He also holds engineering and a master degree in networking obtained, respectively, in 1998 and 2001 from the National School of Computing (ENSI, Tunisia) and Louis Pasteur University (ULP, France). He worked extensively with Motorola Research Labs in Paris and authored 4 patents.

Artificial Intelligence and Blockchain for Future Cybersecurity Applications: Architectures and Challenges

Artificial Intelligence and Blockchain for Cybersecurity Applications

Fadi Muheidat and Lo'ai Tawalbeh

Abstract The convergence of Artificial Intelligence and Blockchain is growing very fast in everyday applications and industry. In centralized systems and applications like healthcare, data access and processing in real-time among various information systems is a bottleneck. Blockchain's decentralized database architecture, secure storage, authentication, and data sharing would offer a solution to this problem. Besides, Artificial Intelligence can live at the top of Blockchain and generate insights from the generated shared data used to make predictions. Blockchain is a high-level cybersecurity technology that forms chains that connect the existing blocks stored in nodes and the new block chronologically by mutual agreements between nodes. Technology convergence accelerates various industries' growth, such as banking, insurance, cybersecurity, forecasting, medical services, cryptocurrency, etc... The more digital systems are adopted and services provided by these industries, the greater the risk of hacking these systems. Combining blockchain power and artificial intelligence can provide a strong shield against these attacks and security threats. In this chapter, we will study the convergence of AI and Blockchain in cybersecurity. We will expand on their role in securing cyber-physical systems.

Keywords Artificial intelligence · Blockchain · Cybersecurity · Authentication · Data integrity · Encryption · Cyber-physical systems · IoT · Security

1 Introduction

The world we live in today is highly revolutionized, whereby we have to depend on technology to move forward. Unlike a decade ago, everything is interconnected. When discussing history, we talk about the industrial revolutions that changed the face of the industrial world. There was a time that technology was so minimal that

F. Muheidat
California State University, San Bernardino, San Bernadine, CA, USA

L. Tawalbeh (✉)
Texas A&M University, San Antonio, TX, USA
e-mail: ltawalbeh@tamusa.edu

© The Author(s), under exclusive license to Springer Nature Switzerland AG 2021
Y. Maleh et al. (eds.), *Artificial Intelligence and Blockchain for Future Cybersecurity Applications*, Studies in Big Data 90,
https://doi.org/10.1007/978-3-030-74575-2_1

there barely any planning was needed for it. For example, look at how the telephone started. Once it was invented, it was only available to the very elite, which means they could only speak amongst themselves. Fast forward to the invention of the computer. The machines were so big they could barely be transported. Today, we have very sleek computers that are desktops and even laptops, which are very portable.

On the other hand, old telephones have evolved so much that now we have foldable mobile phones. The globe has turned into a village with all the technological advancements that have been made, and now it is easy to communicate with anyone as long as they are in a place where they can access the Internet. The advancements have not been solely within the phone and computer sectors, neither. In the era we live in, our existence's whole dynamic has changed [1]. We have machines in place to do almost every chore the human being can take up. Companies have turned into a digital era in which machines can now carry out some workers' duties, which means the workers are becoming very expandable. Before the emergence of HTML, WWW, among other things within the Internet, computers and their connectivity were very much limited and on a local basis.

Another significant advancement within the Internet and technology is the Internet of Things (IoT)/Cyber-Physical Systems (CPS). While technology has significantly advanced, devices have also been on the increase. That is why most companies have taken up establishing their devices to be connected with other devices. We live in a world where all devices we own can be interconnected so they can, in a way, communicate among themselves. This is meant to increase the accessibility of the devices, as well as connectivity. With devices such as Alexa, which can even communicate with us, is a significant change and development in the technology world. The beginning was how mobile phones could be connected to laptops that we had, but now, most electronic devices we own at home can interconnect [2]. Using a more comprehensive look at the developments is very welcomed and will help bring many advancements to how people live. Having a fridge whereby, due to interconnectivity, it can identify what is depleting and needs to be bought, among other things, is a considerable advancement. Every aspect of our lives is filled with technology and its advances. Think of vehicles; some, like the Tesla, is driverless, meaning they can interconnect with our phones, among other technologies.

While we can have the interconnectivity of devices, the different technological devices existing individually makes it simple for them to be accessed by outsiders if the right security steps are not taken. Cybersecurity comes hand in hand with every development that is made within technology. Cybersecurity refers to protecting and defending computers, electronics, and their technological devices from external attacks. The data and information stored within the devices are essential, making them vulnerable to malicious attacks if not well protected. While the technology gurus continue advancing their innovations, the cybersecurity threats metastasize. That is why cybersecurity has to move in sync with every advancement that is made. In technology, most advancements include updates on devices already in existence, which can cause the cracks that the threats are waiting to take advantage of if not well overwritten and covered. Therefore, as much importance that we give to the developments, we should give the same or more attention to the cybersecurity sector.

We need advancements connected to existing or new technology every day, but they all deal with data and access some aspects that require privacy. Simultaneously, almost every sector starting from the companies we work for, the banking system, and even the health sector, uses advanced technology. They require data, and if it falls into the wrong hands, it can cause so much harm to people. That is why cybersecurity is necessary for every aspect that has any connection to using any technology within their running.

Blockchain is one of the most exciting technologies, gaining tremendous popularity as a horizontal technology that is commonly used in different areas [3–5]. The need for a central authority to monitor and validate communications and transfers between many participants can be highly economically removed by Blockchain. Each transaction in Blockchain shall be secured and checked by all mining nodes comprising a duplicate of the entire ledger containing chains across all transactions. This offers stable and coordinated information that cannot be manipulated and exchanged [6].

Artificial Intelligence, another influential aspect that gains immense momentum, enables a device to understand, conclude, and adjust cognitive capacities based on data it gathers. According to recent market reports (Economics of Artificial Intelligence), AI is projected to rise by 2030 to $13 trillion [8]. Artificial Intelligence has evolved due to large-scale development and knowledge generation through sensing systems, IoT devices, social media, and web applications. Various machine learning techniques can use such data. The majority of AI's machine learning techniques depend on a centralized training model using clusters or cloud services provided by companies such as Google, Amazon, Oracle …etc. [9, 10]. A point to keep in mind that the sensed data is error and security prone. Blockchain decentralized architecture, security, and authenticity can be of great support to AI. Intelligent Algorithms run on shared data but are secured, trusted, and authentic [11]. Blockchain-based Artificial Intelligence techniques use decentralized learning to help ensure the trust and exchange of information and decision-making by many agents who can participate, collaborate on making decisions [12, 13].

Blockchain and Artificial Intelligence present different characteristics according to their nature. AI solutions can be applied to produce a learning security behaviour capable of detecting and eliminating threats, just like humans do, but thousands of times faster. On the other hand, Blockchain leverages a secure and highly encrypted digital ledger platform, only accessible by authorized peers [7].

We will study Cybersecurity, Blockchain, Artificial Intelligence, and their convergence in more detail in the following sections.

2   Cybersecurity and Applications

Over time, there has been much advancement in technology; thus, people have several research on cybersecurity to understand it better and even give a breakdown of how the issue can be rectified. We are moving to an era where technology, and then the

Internet, will be in every aspect of our lives if it already is not. For example, take a quick look at social media platforms where nowadays, everyone spends their time. That is why, among other sectors, it is the area that manages to bring people together and contains so much data, some of which is private and can be used to commit a crime if it falls into the wrong hands. To better understand cybersecurity and its application, we have to have a clear comprehension of the threats that exist; the only way that we can manage to control and prevent a problem is by first understanding how it works. We also need to understand the vulnerable areas whereby we can come up with the most suitable actions. Cyber threats are also driven by various reasons and knowing them gives us an advantage as we can work towards the crime's intentions to ensure prevention via security measures.

When it comes to cyber threats, they are divided into three folds, which help identify how best to respond and what security measures to apply. First, we have *cyber terrorism*, which is on broader coverage. It aims to cause panic and fear of technology in the masses. It is used by radicals who are against the idea of technological developments. They rely on panic and fear to disrupt the running of society. We have *cyber-attacks* mainly aimed at attacks on political grounds to share information that is not meant for the public. Lastly, we have *cyber-crime*, which is driven by financial gain and disruptions, which affects the various systems. There are already existing cyber threats, but as technology increases and advances, more threats appear.

Developments in cyber threats accompany every advancement within technology. Among the common cyber threats, we have phishing, malware, ransomware, and social engineering. *Social engineering* relies on interactions between humans, where the attacker befriends or uses the person to override the security measures to protect information [18]. In such cases, the person who falls victim is left to bear the effects of the attack. Such attacks take planning, as most security measures also require physical cases to the devices or the building where the information is being keyed in.

Additionally, *phishing* involves the use of fraud, whereby scammers send fraudulent emails, and they use them to gather the information they are aiming to acquire. The emails sent resemble emails from reliable sources used as a dangling carrot for the victims. *Ransomware* is given the name due to the ransom asked for once the attacker controls the information or system files. Ransomware involves the feeding of malware into the computer system. *Malware* involves introducing malicious software, which causes harm to the system file or grants access to the information within the system files. Cybercriminals use a structured language query to insert malicious ware or viruses to gain access to the information. SQL submits a malicious SQL statement, which gives criminals access to the database being targeted. There are others like a *man-in-the-middle* attack or the *denial of service*, among others. A look at past cyber-attacks will help provide a better view of how dangerous they can be.

In December 2019, an organized cyber-crime unit leader was charged with the Dridex malware attack by the United States Department of Justice [19]. The government, the public, and infrastructure worldwide were affected by the malware. The thing about malware and the state of connectivity that we have today is that attackers can simultaneously affect many users. Currently, there is a surge in zero-days, which are threats that emerge and do not carry any detectable digital signatures [20]. When attacks happen, the experts rely highly on the digital signature the hacker left to triangulate and gather information about them to get to the hacker. With a threat that has no detectable digital signature, it means it is hard to protect from. Notably, when dealing with cyber threats and enhancing cybersecurity, they triangulate and check the vulnerabilities the hacker used in cases of existing threats. The use of digital signatures sheds light on how the hackers got access to the network or system files. Within the information, they can now work on preventing such attacks by eradicating the vulnerabilities and upping the systems' security. Cyber-attacks can come from hackers or even people looking to hurt the organization. It can be even those that we do not and could never suspect. That is why the issue becomes critical because, in such cases, they know the security protocols and applications in place. That is which security put in place should be very legitimate and strong to avoid such attacks. Unfortunately, for every safety measure, there is an override put in place for risk management.

Cybersecurity falls in so many sectors within the economy, and that is why it is divided into different sections to fit the other sectors. The various elements of cybersecurity are as follows:

- *Information Security*

In every context, it covers different sectors and departments within the sector, which makes it easily covered by using the distinct divisions set. Information is one essential thing in the technology world. In every device we purchase, any technology that we have to use, there is a requirement to give a certain amount of information. At times, the information is the basic protocol, while other times, it involves private information that helps identify the owner of the product. In such cases, the data had to be well stored, so it does not fall into unauthorized hands, and that is where the cybersecurity necessity comes in.

- *Network Security*

Network security refers to protecting the internet network, be it within the computer usage, phone usage, or any other device. Network security does not mean prevention from authorized members, but malware can corrupt the device's information or cause the information to be lost entirely. With network security, we are covering the safeguard of the network infrastructure. Under network security, we can take several cybersecurity measures. We can compromise network security in various ways, and getting to know them gives us a way of handling them. One of the pernicious attacks is the denial of service, which is very popular within network security. Denial of service involves an attacker that is a hacker making the access of a network unavailable to the intended users for some time. It is very complicated to achieve, as in most cases, it involves compromising a network, which is a platform that is accessed by so many users a minute.

A good example was in 2016, an attack of a company by the Dyn, and they ended up denying access to Twitter for so many users even though it was for a short time. A platform like Twitter is always busy flowing with people interacting, communicating, or using the app. It connects people globally; therefore, one cannot understand how many users it has per second globally. So imagine what a denial of service attack can do if it lasts for five minutes or more. Dyn company provides DNS services, and with the network attack, the intruders took control of over 600 000 devices [21]. In 2016, during a DDoS attack, it was discovered that they had used the Mirai IoT Botnet. Attacks were mainly on IP cameras and Routers. Such attacks leave the question of how we can up network security so that we can achieve cybersecurity.

Even then, the network DDoS attacks have declined since then, but that does not mean that they are not an issue to grapple with. Trends show that when there is a new vulnerability, the attacks increase for a time before they are mitigated, which leads them to slow down. Figure 1 below shows the trends of the DDoS attacks 2017–2018. The Y-axis represents the attack size, while the darker the dots get, the more prevalent the attacks on that size. From 2017 to 2018 in January, there was an increase in attacks by density from 560 to 738 Mbps. That shows a tremendous increase in the attacks, and the move now is ensuring that we can give out workable security measures before the attacks. The attacks spur for a moment until they are discovered and then solved, but then the damage has already been done. Although it can be challenging, having measures in place that prevent attacks at all and risk management can help network security. The more the attacker stays within the network undetected, the more information they gather and the better chance of destroying the enterprise. They get to learn the company's trends and can cause damages slowly because they do not arouse suspicion, and by the time they are done, they have done so much it is impossible to regain the previous state of the network. That is why cybersecurity should be applied before the attacks.

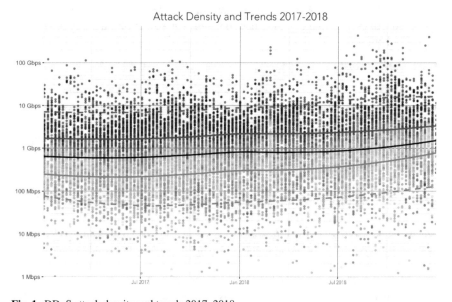

Fig. 1 DDoS attack density and trends 2017–2018

- *Application Security*

Observably, within the devices we have, we use different applications for various reasons. Therefore, when we get a device, say a mobile phone, the first step involves downloading the various applications that one finds essential. We have things like social media applications, workout apps, and games. In each of the apps, there is a need to give personal information to get accessibility and identify you as the user. Some apps are more critical than others, and they require more factual information. For instance, apps connected to the banks require specific keys, as the information is crucial. That is why everyone must ensure that they have installed the proper security measure for the various apps. Some apps require little security like general apps like the workout, a bible app, among others [1]. However, apps that are dealing with, say, the security system of your home. Then it has to be well secured.

In the past, some intruders have relied on the apps on their target devices to get the information that they need so they can exploit them. In most cases, it provides such programs with security protocols before they are even deployed to the mass. Even then, going an extra step can never hurt anyone. Cybersecurity is diverse, and every aspect ah to be dealt with at its capacity. Such threats are what most people deal with, especially with the Internet of things that are spreading swiftly. Now, if a person can access the app that creates the interconnectivity between your device, and they can access every device that one owns. Therefore, taking care of the security measure on the applications is essential.

2.1 The Scale of Cybersecurity Threat

A study done at the University of Maryland [22] shows that every 39 s, there is a cybersecurity attack within the united states. Every day, several people are victims of the cyber threat and have to bearer the consequences. Research has shown that most of the attacks are made possible because of the insecure usernames and passwords we use. For creating a username and password, people tend to make the simplest so they can remember them. For anyone not within the IT department, they do not take time devising the most secure username of the password as long as they can remember them. For hackers, this is a fantastic opportunity for them, and they take advantage of that. One, there is the fact that the hacker is well versed with the workings of the technology and the networks; therefore, having a less complicated security measure makes things easier for them. The thing with cyber threats is that they do not necessarily target one sector. As long as the sector ash technology and networks, then they can easily fall victim to cyber-attacks. Every sector today has turned to digitalization, even the health care systems and banking, and without the suitable security measures, they fall under the category of the attackers.

By September 2019, there had been around cases of data exposed due to breaches in the various sectors. It was a substantial increase compared to the previous year. Public entities and medical care centers are among the most hit by cybercrimes within the United States. The heath institutions, they record vital information worth so much, especially in the black markets. By the time a breach is identified and responded to, the hackers have the chance even to sell the information that they got and move on to other areas that they can exploit. The health care system has shown a significant increase in the technology's adoption, and networks are the latest models [23].

It drives health care towards improving their health care services, and what better way to do that than assimilate the use of technology. Unfortunately, they also report being the one sector that spends the least of their budgeting's on cybersecurity. The United States federal government allocated $18 billion to cybersecurity, which increased from what they had allocated in the previous years. They spent the least amount of their budgets on cybersecurity with health care institutions, which amounted to 5% [24].

One thing that happens with the technological devices in cybersecurity is that the manufacturers are more driven to release the products to the market. Cybersecurity comes as an afterthought when it comes to devices and networks out there. That is immediately when a product is complete. It is released into the market so that the consumers can have access, and the company can be on the trend of making money. Later on, when the devices and technology are within the market, they offer cybersecurity applications options. Some people have already suffered victims of cyber-attacks. As mentioned before, cyber-attacks and threats are evolving just as fast as cyberspace is developing. A modern technology is devised every day, so it is a cyber threat that develops [25]. The same minds that come up with the updates for technology and the innovations are the same as those running the black and grey hackers. Therefore, when a new device or system is released into the world, they quickly identify the vulnerabilities and exploit them to their benefit before they are identified. That is why, with some cyber-attacks, it is always a case of cat and mouse. With the black hackers, they are still out before they can be tracked done, which is they are always a step ahead of the white hackers and other cyber experts who are working towards increasing and enacting cybersecurity. Most of the time, cybersecurity measures are put to prevent an attack that has been previously done. Most of the cybersecurity applications are reactions and responses to an action already taken.

Modern problems require modern solutions, which is what cybersecurity application is all about now. The world is moving to a place that will be technology-driven. We have to be prepared for what we will do and how we will be handling cyber threats. We have to establish a working cybersecurity.

According to Parrend et al. [26], zero-day and multi-step attacks are on the rise in cyber-attacks. The two attacks are hard to identify or even rectify before any damage is done. With such an attack, the hackers or exploiters identify the weak spots of the vulnerabilities. They then send coded messages or notifications to groups of people within their line of work. Therefore, the end-user has no idea they have a vulnerability within their system while groups out there are very well aware. They all can tap into the vulnerability and collect the information they need or want for their purposes. That is why, within their research and study, they suggested the use of two approaches so they can ensure that cybersecurity is applied. The two approaches included the use of artificial intelligence and the use of statistics and machine learning. With artificial intelligence, it is possible to identify the vulnerabilities before the black hackers. If not, the machine and statistics learning will immediately recognize a new entry into the system. Then the best ideology here is artificial intelligence, which is identifying the vulnerabilities. The access of new users or unauthorized users can be hard as the systems get new users, especially in platforms like websites and social media platforms. This system will detect any behavioural abnormality within the systems and even track the event sequences. With such systems, it becomes easier to maintain cybersecurity and even put it into action. In a world where technology is taking over, we have to be very ready to handle the distinct threat that is waiting to happen. The best way to do that is by enacting the use of more reliable cybersecurity measures.

2.1.1 Importance of Cybersecurity

Considerably, there are so many issues that arise within cyber-attacks and lack of cybersecurity matters. There is also how some people do not take the matter seriously and rarely take any measures in cybersecurity. That is why they must get a better understanding of the advantages of cybersecurity applications. The protection of data within a system that is on an individual or company level is significant. The data stored within the systems or drives is both essential and confidential. In terms of companies, their competitors can use it to destroy or even run the company into the ground. It can give personal data like banking data, which can bankrupt a person with individual matters. Either way, any data that is not meant for public access should be well secured, which can only be done with cybersecurity applications.

With cybersecurity, the speed is maintained at a high level. Where the system is bugged with malware and virus, it gets slower, which impacts the effectiveness. That is why cyber securities application is essential as the cyber speed is left at a high. Slow systems can cause a lot of damage and losses, especially if the billing department of a company is dependent on the speed of the system. In the past, so much has been used in restoring cybersecurity after attacks [27]. Cyber-attacks are not helping to handle, and ensuring one has cybersecurity beforehand saves on cost significantly. With big companies, attacks on cyberspace can lead to loss of customers and popularity. One aspect of the cyber is the people, and when they engage with a company, they get to share a piece of private information. They rely upon that confidentiality will be observed. Therefore, when there are cyber-attacks, their information and data are taken, they lose trust in the company, affecting their popularity within society. Companies rely heavily on how the population sees them. Image is essential for companies to survive, and if it is tainted, they are bound to lose customers. Like other sectors, even in cybersecurity, prevention is better than cure; thus, it is better to secure our systems than grapple to find balance again after someone has hacked the systems. In such instances, it takes a while before the stolen information can be identified [27].

A good example is that companies have many data within the systems. While breaches and hacks are searching for meaningful information, such attacks target the minor and benign data. In such cases, it takes time before they can identify the actual damage, and in such cases, the hacker is already in the wind, and the damage has already been done.

2.1.2   Findings of Cybersecurity

Today, the world is getting smaller and smaller in terms of communication, interactions, and especially interconnectivity. The Internet of Things is really on the rise, whereby everything is now getting connected via the provided networks. Different companies are working together so they can establish connectivity ranging from various sectors. A special report by Steve Morgan EIC of Cybercrime Magazine showed that by 2021, around $6 trillion would have been spent on cybersecurity. Such statistics show companies and individuals should take it upon themselves to ensure the security measures within their systems are strong to prevent attacks and spare the spending on handling a breach that has already occurred. With cyber-attacks, some are unpredictable and only come to light after it has happened. This is one weakness that cybersecurity applications face.

Because so many people use the Internet daily, they lack knowledge about the importance of cybersecurity. Most only do it as it is necessary, especially with the accounts they have to create within their applications. Even then, they do not make them as strong as possible. The communication of information makes it possible for there to be a physical infrastructure within cyberspace. If one does not know the importance of cybersecurity applications, they do not put enough effort into securing their devices and systems. Today, we store all our information in the cloud, so if someone can access the cloud, they can paralyze someone technology-wise. Such events lead to identity theft, financial losses, among other losses. Creating awareness so the masses can understand how essential cybersecurity is something that needs to be done. The people within IT departments have the advantage of fully understanding the dangers of cyber threats and taking the required precautions. Even then, that is not enough within a company as the black hackers will look for the weak link and exploit that to their benefit. According to [28], so many people ride on the idea that everything within the cyber-physical society will work things out by themselves. They forget the damage that can be inflicted can affect them too first hand. Ignorance is bliss, and we have to ensure that it is eradicated by creating awareness on the importance of cybersecurity and the best way to go about it.

In line with this, end-user protection is very crucial. Many are times that individuals upload malware to their system without even knowing. Only the professionals in this sector can quickly identify the malware. That is why awareness can also help. In a company set up, there can be training pieces that will shed light on the employees who can identify malware and how best to improve their security. There are end-user security protocols and applications that can be used. Cybersecurity is a never-ending issue, so when we get away from handling one threat, there will always be new emerging threats. That is why we have to be alert with cybersecurity [29]. The world of technology keeps changing, and we have to change with it, so we have a fighting chance against cyber threats. Adversaries keep on evolving, which means that within cyberspace, we have to keep evolving too.

Cybersecurity applications have to be focused on two aspects: Using what we know effectively in coming up with strategies and techniques to prevent cyberspace threats. That will involve the reliance on past threats and attacks and how we get out of them. Here, the tactics put in place will aim to avoid having the same vulnerabilities as before. That is known as prevention from the known rather than the unknown. It should be inclusive of the factors we know or assume as a basis for cybersecurity, yet we ignore it. Second, cybersecurity requires constant alerts on new vulnerabilities. That is, preparing for the unknown. While there are the known threats, there those that emerge with time, and experts' have to be well prepared for them [30]. This could involve using white hackers to find vulnerabilities within the systems now and then so they can in them and handle them before they fall prey to the wrong hands. It uses the traditional cybersecurity systems and techniques to come up with new ones that are well upgraded and can handle the new set of attacks that are expected to emerge. The first section relies on existing knowledge to come up with security techniques. In contrast, the second part requires new techniques as they are dealing with issues they expect to be possible with the developments within the systems.

2.1.3 Challenges in Cybersecurity Applications

Unfortunately, cybercriminals have turned cybercrime into business opportunities. Professional black hackers are selling the tools they use, especially for zero-day attacks, to other people within the black markets. With so many people having access to such tools, they can cause many problems within the systems and cyberspace. Over time, people have taken up the use of mobile phones in almost everything. Now, every aspect of our lives has an application that we can use to communicate with, such as within the bank's systems, interactions at work, etc. [31]. Add this to how affordable mobile phones have become, as everyone is now using them. It becomes effortless for hackers to access the mobile phone systems and manipulate them to their liking. In such matters, the systems have to use extreme measures like voice and facial recognition to increase security.

We have proficient use of the Internet of Things around the globe. We have made quite some progress towards technological development. Today, the most growing part is the Internet of Things, whereby all devices are interconnected to help increase management and control. While this is great for individuals, accessing and controlling all your devices from just one place is also beneficial to hackers. With their prowess in attacking the systems, now they can do it by accessing the one device with control over the rest of the device within a person's life. We cannot eradicate such a challenge as there has to progress in cyberspace and technology [32]. We have to come up with more robust and more effective cybersecurity techniques and strategies so that the end consumers can enjoy the technological advancements. Estimates show in the next ten years, the number of devices that will be under the connectivity of the Internet of things will be around 125 billion. Therefore, the challenge will continue [27].

As discussed, cybersecurity issues are continuous and never stop evolving. What changes is how they handle it, which is why the techniques to deal with the attacks and the rest should also keep changing and updating to better and more reliable versions. Another challenge is at the time, and the hackers administer their attacks by attacking third-party vendors. That is people who are not under the wing of the company. For instance, within a company, they can treat a person who does delivery to that company. Prevention of such breaches can be hard to manage as the most secured areas are the company's workers.

Notably, there is a significant disparity between cybersecurity areas and readiness to address them. Many companies know of the cyber-attacks and security issues they can face, yet their response to it is not up to standard or takes too long when faced with the problem. For detecting and responding to attacks, they are very slow and always seem unprepared. Figure 2 shows the response time to attacks as an example within the global financial sectors. Attack success covers the time between when an attack is initiated and how long it takes to be successful, while discovery success is the time to realize there has been a breach into the system [33]. Last, there is the time between the discovery of a breach, while the time to manage and contain it is the recovery time. From the data below, we can see the attack time is very swift and happens in a matter of seconds. Still, the breach's discovery, which takes some time, shows how slow it takes to determine or detect an intrusion within the systems, especially if it is done swiftly and by triggering no red flags. Cleaning up the mess

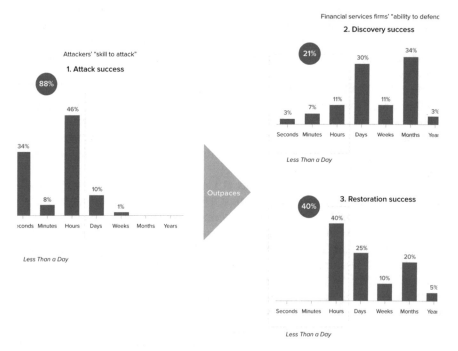

Fig. 2 Response time to attacks within global financial sectors

after a breach takes a lot of time in which a company can have lost popularity and customers, and the information or data that was taken [41].

3 Blockchain

In recent years Blockchain, the core component of Bitcoin, has evolved extraordinarily quickly, and its use is no longer limited to digital currencies. A Blockchain is a distributed public database that keeps a permanent record of digital transactions. The distributed ledger records the transactions of Blockchain blocks, and every block is related with a hash function to preserve the chain with its previous block [14]. The network elements/nodes will receive a pair of the public key and private key upon registering to the network. Public key works as a unique identifier for each element. Private key also helps to sign transactions in the network and is used for encryption and decryption. The transactions are received by all the nodes and are validated. They are grouped into a timestamped block by few nodes designated as miners [14]. Blockchain is a "*No Central Authority,*" consensus algorithm used to select a block, among the number of blocks created by the miners, added to the Blockchain network. For making any changes to the existing block of data, all the nodes present in the network run algorithms to evaluate, verify, and match the transaction information with Blockchain history. If the majority of the nodes agree in favor of the transaction, then it is approved, and a new block gets added to the existing chain.

Implementing Blockchain comes with benefits such as securing data, reducing errors, ensure reliability, and improve integrity and effectiveness [44]. Figure 3 below shows an overview of how Blockchain works. Today we have three options to manage the buying transaction between the buyer (left) and the seller (right):

Blockchain - Process

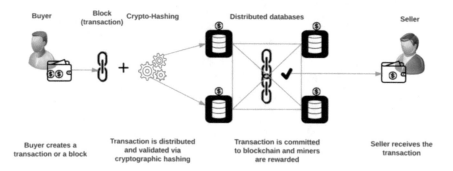

Fig. 3 How Blockchain works

1. Both the buyer and the seller trust each other, and they can manage it. Depending on their relationship, if they are friends, they will finish the deal; if no, the buyer can quickly not pay.
2. The buyer and the seller sign a contract; if the buyer did not pay, the court would be involved and more time to settle.
3. Introduce an intermediary who can manage the process, but this third party might take the money and not pay the seller; we are back to the first two options.

With Blockchain, we can offer the intermediary, but it is guaranteed, secure, quick, and cheap. As we can see from the figure, the transaction is secure (hashed) and stored in distributed databases (data sources). At any given time, both the buyer and seller can check the status of the transaction. Figure 4 shows how the blockchains work. The blockchain network orders transactions by grouping them into blocks; each block contains a definite number of transactions and links to the previous block. Blocks are organized into a time-related chain [43].

Blockchain has four important characteristics: distributed database architecture, almost real-time transactions, irreversibility, and censorship resistance. The strength of the consensus ensures fraud less transaction. It is improbable to have all or the majority of the nodes to be complicit [45]. We are not aiming to provide all the Blockchain details. Still, we are interested in its role in privacy and cybersecurity applications and systems such as IoT, Electrical grid, banking industry, food supply chains, healthcare, and more [46–49].

One of the main characteristics of blockchain technology is that it is its security. Using its distributed ledger, we can securely store millions of data within its platform, leveraging a series of architecture tweaks. It is conceived that all users' modifications and changes have to be approved through its proof of work protocol. Likewise, this system allows for a trustless principle where all transactions are anonymous, but they

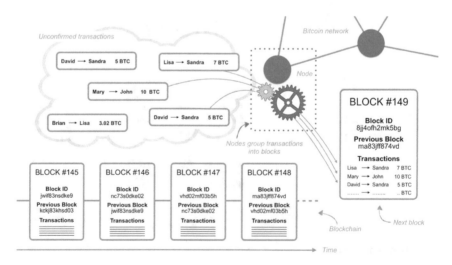

Fig. 4 Blockchains sequence and hashing

stay recorded in the chain. All of these measures go towards keeping the integrity of data. This is one of the issues or limitations of Blockchain. It needs to develop a way through which the identity may be revealed if a transaction is suspected to be fraudulent to counter fraud.

A blockchain platform is formed out of infinite blocks; every time a block is solved, the platform itself will create a new block carrying a "hash" or fingerprint from the previous block. There is never a duplicate recording of the same transaction. As such, the need for a central intermediary is not there anymore.

Also, blockchain platforms use two types of consensus among all members called *proof of stake* (PoS) and *proof of work* (PoW) to validate any changes make on any given block amid asking users to prove ownership. Proof of stake (PoS) is a method by which a blockchain network aims to achieve distributed consensus. The proof of work (PoW) method, on the other hand, asks users to repeatedly run hashing algorithms or different client puzzles to validate electronic transactions. Both ways are thought upon leveraging that any change made on the Blockchain is generally validated among all users, making the blockchain intruders-free as any external change will be watched by hundreds if not thousands of users [7].

According to Ed Powers, Deloitte's US Cyber Risk Lead, "*while still nascent, there is a promising innovation in blockchain towards helping enterprises tackle immutable Cyber Risk challenges such as digital identities and maintaining data integrity.*" Blockchains could help improve cyber defense as the platform can secure, prevent fraudulent activities through consensus mechanisms, and detect data tampering based on its underlying characteristics of immutability, transparency, auditability, data encryption & operational resilience (including no single point of failure). However, as Cillian Leonowicz, Senior Manager at Deloitte Ireland, opines, "*blockchain's characteristics do not provide an impenetrable panacea to all cyber ills, to think the same would be naïve at best, instead of as with other technologies blockchain implementations and roll-outs must include typical system and network cybersecurity controls, due diligence, practice, and procedures*".

Blockchain can enhance cybersecurity and solve issues related to users' malpractice, incautious users. Nevertheless, we need to be vigilant to the possibility of breaking the hash code by enhancement in the computing power and decryption algorithms.

4 Artificial Intelligence

Artificial Intelligence is growing and expected to grow father than before. In the last four years, the number of businesses using artificial intelligence has risen by 270%. It has tripled from 25% in 2018 to 37% in 2019 [50]. AI research is characterized as the study of intelligent agents, i.e., any system perceiving its environmental state and taking action to increase its chances for success [51]. The fields of AI include machine learning, deep learning, natural language processing, robot, etc. AI can be made more powerful and efficient by ensuring data sharing that is scattered across different stakeholders [52]. AI is being seen in many places and is being taken advantage of by a variety of individuals. However, AI can also steal private data, allowing a number of illegal users to launch safety attacks [53]. Artificial intelligence is well covered in the literature as it has been around for almost 50 years. We are looking into how we can apply AI to strengthen cybersecurity applications.

Artificial Intelligence supports security solutions through protocols, software, or even raw code. With its ability to learn, adapt, and act, artificial intelligence adds layers of security capable of learning from threats, security breaches, and other data collected through their mechanisms. Hence they know from security breaches to avoid them in the future. Consequently, the more attacks a system is hit with, the more reliable the security will be to defend itself in the future. Zeadlally et al. discussed the role of AI in cybersecurity in three different domains: Internet, Internet of Things (IoT), and Critical infrastructure. Figure 5 list the role of AI in these three areas [54]. The role of AI in cybersecurity is expanding as the Internet grows. AI methods are used in national-security and human well-being-critical systems. AI methods are used to rationally solve problems and make computers think and function like humans.

In internet domain area, Human suffers fishing attacks. Hence AI solves this by automatic phishing detection. Network and Application layers suffer Denial of services and change the semantics of the messages, respectively. AI handles this by learning patterns and adapt and build smarter classifiers. IoT domain suffers impersonation, and insecure data collections and sharing, AI added layers of cloud and distributed environments security. In Critical infrastructure, where all attacks are in the cyberattack category, AI used a logic-based framework and policies.

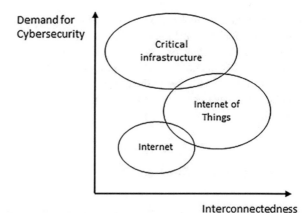

Fig. 5 Role of AI in assisting cybersecurity. AI applications grow from two main drivers: the degree of interconnectedness and the demand for having security systems

5 Blockchain and Artificial Intelligence Convergence

When we mention Blockchain, its decentralized architecture, what gets our attention first. When we bring artificial intelligence and Blockchain and put them together, we get a decentralized artificial intelligence. Machine learning Artificial Intelligence (AI) has been around since the 50s. It's not a new technology. It's a pretty old technology, but basically, the idea has always been the same. We have some input data set. We have some objective we have something that we want the AI to do in this case, classify an object (say a pine tree) as a pine tree or is it not a pine tree, so what we do is we feed this AI a training set of images of different types of trees. Hence, the idea is that the AI learns the mapping the relationship between the input data and output data (labels). The idea is that the Human would have to extract all the features of what it means to be a pine tree; the shape of a tree, the leaves, the length of the stem …etc. This was a very tiring process, but this is what humans by hand, and so what happened is once we've extracted those features, we would feed it to some kind of machine learning model. There's a lot out there; eventually, it would learn the mapping and then give it a new tree picture, and it would know this is a pine. In recent years, a neural network which is machine learning model, we feed it with the dataset and massive computing power and add a lot of layers deep layers aka deep learning that's what we call it what happened were when we did this it started outperforming every other machine learning model. Now deep learning is outperforming everything else self-driving cars, drug discovery everywhere AI is applicable.

Satoshi Nakamoto who released a paper on a cryptography mailing list [15] detailing a system called Bitcoin that allows two people to transmit value online without needing a third party, namely a bank. So, what happens is instead of a bank being the third party, there are a group of people called miners, and anybody can become a miner; you just need a computing device right, and the idea is that when someone (let say it is me) transmit value to (you), these miners have to approve this transaction and say okay. Let me check this list of transactions, so every miner has a copy of every transaction that has occurred in the network, and they have to approve whether or not this transaction is valid or not. You might be thinking that someone might fake a bunch of accounts approves a transaction as if there were different miners! Because of such a scenario, Satoshi Nakamoto said every single miner has to prove that they have solved some random mathematical problem; it's called the proof of work algorithm (PoW) [16], and that means that you have to have more computing power than the fastest supercomputers in the world combined with having the majority of the computing power in the Bitcoin network. Because no one has that much computing power, no one's been able to hack Bitcoin, and that's why it has over 350 billion market cap as of this writing [17] and no one's been able to tack it; it's been around for a decade it's a really powerful technology.

These two technologies go very well together, but like oil and water, they do not mix. The unique structure of Blockchain, the immutable and unalterable data structure, and the proof of work algorithm make the blockchains deterministic.

At any given time, we know exactly what is happening, and they are unchangeable. Artificial Intelligence is based on prediction and probability models. What we can do is to combine Blockchain and AI together. We can use this immutable ledger and have an AI speak to the Blockchain, and data is being pointed to some kind of decentralized storage source. Let us imagine that we can allow AI to live on the Blockchain that means it has full access to the power of immutable decentralized architecture and a very secure signature. If AI lives there and no human intermediary is available, we are giving the AI the full power to learn, adapt, and expand and then gain control! (scary right?) Knowing the PoW algorithm and the need for supercomputers to hack and attack, it is not easy to shut down such a huge monster trying to get control. This convergence of AI and Blockchain needs to be normalized and regulated so that we can keep control. Recently, a model called *SecNet,* which incorporates Blockchain technologies and AI to provide protections for the whole system, has been proposed in [42]. This architecture guarantees that the safety and protection of data exchanged by the various system members are maintained. The authors also carried out a security review of the architecture, which resisted a Distributed Denial-of-Service (DDoS) threat.

We can develop decentralized AI applications and algorithms with access to an identical view of a secure, trusted, the shared platform of data, logs, knowledge, and decisions by integrating Artificial Intelligence and blockchain technologies [55]. Blockchain decentralized structure can overcome centralized AI structure and hence enhance data security. Blockchain Deterministic feature overcomes AI probabilistic (changing), therefore, improved trust. Blockchain data integrity helps AI in Decentralized intelligence. All in all, Blockchain allows AI to make a transparent, trustworthy, and explainable decision. As we know, A blockchain's architecture and operation entail thousands of criteria and compromises between protection, efficiency, decentralization, and many more. AI can quickly ease those decisions and optimize Blockchain's efficiency. Besides, AI has a crucial role in maintaining consumer anonymity and protection because all Blockchain data are freely accessible.

5.1 Proposed Model

Individually, each of Blockchain and AI has their strengths and weaknesses. In [56], Salah et al. provide an excellent review of AI and blockchain integration challenges. In some cases, AI can support Blockchain operations, and blockchain characteristics can support AI. The goal is to build new digital systems utilizing both AI and Blockchain's power to provide faster solutions with transparency and trustworthiness as possible. With many of the developed systems on the market [60]. Each of these systems is specialized in specific applications or domains such as healthcare, finance, banking, energy…etc. There is always a need for a generic model that provides that convergence utilizing and harnessing the power of shared data sets and artificial intelligence (machine learning) model to help end-users infer accurate decisions. Think about having all these giant companies with their hidden (private) huge data sets and

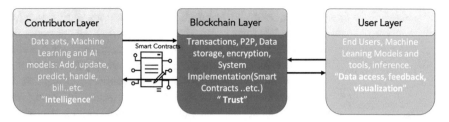

Fig. 6 Proposed AI and Blockchain convergence model

machine learning models, collaborating, and building a shared dataset and models publicly accessible by end-users (could be free or with paid incentives).

Artificial Intelligence algorithms can perform better if it is being trained on more extensive and diverse datasets. So, there is a need for cross-organization data sets and machine learning models without hindering the privacy and secrecies of these organizations and their clients. Blockchain can be that trustworthy storage of the data, and it provides encryption, trustworthiness, transparency, and validation. Contributors to the dataset and AI models can utilize the power of smart contracts to perform operations such as adding data, updating data, and data analytics safely and securely. As an end-user of Blockchain, they have access to shared data and trained models. We are looking to develop a convergence model that can use Blockchain as data exchange and storage for training data and models and access to ownership of data and incentivizing systems. At the top of that is a mechanism to ensure the bad players are not spamming the datasets with invalid data, hence affecting the training models' performance. The model is proposed around the collaborative dataset's idea to harness AI's power on decentralized, secure storage provided by Blockchain. This is proposed model, not well verified. Future work to verify the model using Ethereum blockchain systems [57]. Ethereum is a distributed blockchain network that uses the idea of Blockchain that was previously used in the popular cryptocurrency Bitcoin.

In Fig. 6 above, we show our proposed model for AI and Blockchain convergence. The system consists of the followings three layers:

1. ***Data set and AI models Contributors Layer:*** In this layer, the organization add their well-trained datasets and their AI/ML models. It consists of multiple functionalities and services; adding data, updating data, validating data, incentive mechanism, develop prediction or data, analytics model. We can call it the admin user layer depending on its role and the application under study. For example, in healthcare systems: The actors of this layer could be different hospital systems. Each design can share its reliable and well-trained dataset with a high accuracy prediction model. Each type of data and its combinations have significant connotations for specific disease conditions depending on medical record quality and its biological inference. It will be easy and useful to infer if a particular patient is most likely to have prostate cancer based on similar patient records.

2. ***Blockchain Layer*** keeps tracks of all transaction records, communications, p2p, encryption, censorship, policy roles …etc. Users and Contributors perceive this layer as Blockchain As A Service (BaaS). They do not need to worry about the underlying storage, network, or computational structure. Operations of the Blockchain layer varies mainly: perform the transaction, confirm the transaction, display result, communicate with a lower level of system implementation (this implementation is application dependent). For example, in a healthcare system, in system implementation, we can have roles creating and validation, access policies, …etc.

3. ***End-User Layer:*** it represents various users interacting with the system. These users can be of different roles and levels. For example, in the healthcare system, we can think of users as doctors, patients, staff, administrators, each with different tasks and roles.

As stated above, there are few areas to consider in our future work; implementing the model using open-source frameworks considers different types of AI models: supervised, unsupervised, clustering, vetting the spammers, ranking, and incentive mechanisms. The decentralized nature of Blockchain can be a bottleneck, especially dealing with big data and the need to store all the database on every node. Security and privacy concerns; do we need to share data publicly or just update the data and model. In AI and Blockchain convergence, we look for: continual data updates, train, and test data to make the inference, storage, and how to spill out the spammers and cancel or reject their effects (data sets).

5.2   Use Cases for AI and Blockchain Convergence

As we stated in previous section, each system comes with different needs and requirements, yet they fit within the proposed model.

5.2.1   Energy Grids

The driving forces behind many developments in the energy sector have been renewable energy sources and the growing interest in green energy, such as how utility companies communicate with their customers and vice versa. In what is effectively a merger of the conventional energy grid with the IT market, smart grids' implementation is one of those developments. It suffers an increased cyberattack. AI and Blockchain can help alleviate these attacks [58]. Figure 7 below shows the network architecture of the Smart Grid. Due to the openness of wireless communications and the distributed nature of the Advanced Metering Infrastructure (AMI), they are vulnerable to cyberattacks.

We can see the role of AI and Blockchain through our model, as shown in Fig. 8 below.

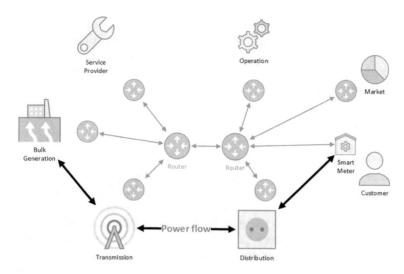

Fig. 7 Smart grid network architecture

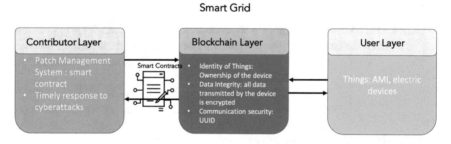

Fig. 8 Smart grid AI and Blockchain model

- The **user layer** is simply the infrastructure devices. However,
- The **Blockchain layer** can do multiple functionalities: 1) *Identity of Things*: The ownership of a device can change/be revoked during its lifetime. Each device has attributes such as manufacturer, type development GPS coordinates. Blockchain will be able to register and provide identity to the connected devices with their attributes and store them in the distributed ledger. 2) *Data Integrity*: Every transaction is encrypted, so all data transmitted by the devices will cryptographically be signed and proofread by the send. Each node will have it si Own unique private and public key and ensure the data's integrity by timestamping each data transaction. 3) *Secure communication*: In Blockchain has a unique universal identifier UUID and creates an asymmetric key pair once the node joins the networks. This will create a faster handshake compared to the PKI certificates.
- In the **Contributor layer**, different vendors can utilize smart contracts to validate the transportation of the correct patch (in patch management) to the meters.

The contract will operate on the basis of device-specif information, model, and firmware version. The smart contract will decide whether to update the device and instruct the device to do so. We might have a compromised device, so if the device refused the updates, then the smart contract will give a lower rank to that device and notify the energy provider (vendor).

5.2.2 Voting System

Elections are seen as a celebration of democracy, in which every person has the opportunity to exercise his or her rights in selecting the right representative to take office. It is also important that the elections are held in an apolitical manner, without the gain of any candidate or party [59]. The majority of elections take place offline in polling booths. The ability to vote on your phone or computer and make our life more comfortable, and speedup the results process. By doing so, we can reduce the time taken to announce election results. The speedup of reporting results is a significant problem in the current voting systems. The government can develop a secure mobile application that can be easy to use and help citizens vote. Each voter will have a private key (biometric signature) as evidence of presence during the voting process. Figure 9 shows the approach. Each layer is described below:

- **User Layer:** Voters use a mobile application to cast their vote. This application also needs to be able to capture the photo ID or any biometric identifier. Election officials can have a dashboard to monitor overall voter turnout and also to handle frauds.
- **Blockchain Layer:** The blockchain network will provide wallets for every single voter. Votes are cast in a smart contract where mapping exists to count the votes made for each candidate. The Blockchain can also provide conflict management if the AI identifies or warns of false proof provided by a voter.
- **Contributor Layer:** Use the Machine Learning or AI-enabled Classifier and tools to vet the proofs shared with the authorities.

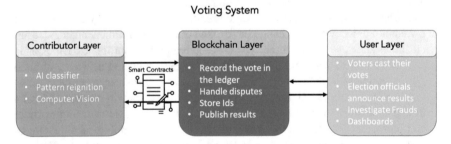

Fig. 9 Voting system AI and Blockchain model

5.3 Recommendations for Cybersecurity Applications

Remarkably, we start with the basic requirement, like ensuring every enterprise has a multi-layered defense mechanism against any attempts to attack or enter by authorized members. That is using encryptions with over two authentication methods for access. Such a defense mechanism can work very well with the mobile technology we have taken up in our daily proceedings. In the same line, handle the third-party vendors here. We can ensure they adhere to the security policies of the company. This can be achieved by offering the service level agreement, so they can fully understand and concur with terms, especially as they do not fall under the company's staff, but they associate with the company regularly. It could also be a go-ahead to include Identity and Access Management (IAM). It provides for who will access what within an enterprise under both the employees' docket and the customers. With identity and access management, it is easier to monitor employees' activities in terms of how often and what information and systems they are accessing. It can help in identifying when they have accessed a system that they are not supposed to. With a breach, it is easier to determine where the vulnerability was and work to rectify it as soon as possible.

Regular audit of the information technology sectors and technological sectors on that the company can be sure they are well aware of how things are running. While auditing will help bring up to speed how their systems are, that is not enough. They should bring in a white hacker who will try to hack into the systems in all ways possible that he or she can think of. This exercise identifies the vulnerabilities the system might have so they can override it. As long as they have real-time intelligence, a company or an individual can avoid any threats getting in. Imagine being in a situation where you know when you are about to get into danger. That would give you ample time to work on the way to avoid the threat coming your way. The same works with the audits; they give real-time information the black hackers could have otherwise used. With the information, they can rectify their weak links and put effective measures in place. Information and data stored within systems are crucial, and in the wrong hands, it can cause so much harm [34]. On personal levels, this can be substituted by the regular updates on the security systems that have been input into the devices we are using. With regular system updates, they can get to override any viruses or malware found within the device system.

Artificial Intelligence and machine learning have been considered as great ways to manage cybersecurity applications. One of the best ways to manage cyber-attacks and threats is to understand how the attacker thinks and maneuvers within the system. An overlook of how the defense actions and the competing attacks react and relate with each other is a revelation on so many levels. The cybersecurity issue is that there is a new emerging threat every day that was not there before [35]. Therefore, tactics used a month ago may not apply to a recent threat now, which is why the cybersecurity applications team should always be alert. Defensive designs are created to tackle unforeseen attacks that have not happened before that predict and work. If they do appear, they can handle them and protect the data, information, or network in question. There is a need to stimulate such a scenario for this kind of understanding and apply multiple attackers with different behaviors and environmental enhancements [36]. Such a setup is difficult to develop, but with genetic programming, it is made possible.

Genetic programming combined with adversarial evolutionary algorithms shed great light on how the attacks can happen and the best ways to subdue them, if not prevent them from happening at all [37]. For the success of genetic programming, there is a need to conduct some cybersecurity research, which will help alienate the data needed to identify the best defense to use against attacks. In such cases, it is used to determine the vulnerabilities or attacks that are yet to happen to cultivate a defense mechanism that will help in doing so [38]. Below are examples of applications it associates with to eliminate the areas that are not likely to be attacked so that by the end of the process, it can identify weak areas. Network defense investigation deals with the isolation defenses, distributed denial of service, and others, within cybersecurity. The defense of the physical infrastructure is also essential to determine the systems are interacting. Then there is vulnerability testing and the detection of anomalies, which will help identify areas that can be hot spots for the hackers to access.

Additionally, risk management is very crucial in cybersecurity applications. As mentioned before, the threats are new to the systems. Organizations have to take the risk by taking the security measures they believe are best suited before getting any black hackers' intrusion. Before then, they have to administer robust identification access management, the nature of the vulnerability, and the data that can be accessed if they exploit the vulnerability [39]. Taking risk management approaches is not that simple, and they have to make sure that with the known variable, they have the correct data on a piece of information to act on it. Lack of awareness is one challenge that is mainly faced by cybersecurity applications. That is why, within organizations, they should conduct training sessions for the employees. That way, they will know how to go about cybersecurity, and it becomes easier to identify or detect an intrusion if they know what to observe. With training and awareness, there can be a reduction in recovery and discovery times when dealing with cyber-attacks.

Most important of all is reaching the masses. Everyone out here is using technology and the Internet daily. That is why everyone must take responsibility for ensuring that they maintain cybersecurity. It starts by ensuring that our devices have strong passwords; information is well secured within our devices. Avoid sharing so much information on public platforms. As the globe turns into the Internet of things connectivity, it will be paramount that the masses observe privacy with their information [40]. With a web of connectivity, a slipup in one sector can help create a vulnerability that can help hackers attack the rest of the system. Today, even social media platforms are sectors of cyber threats, and we have to maintain security at all costs. Insignificant details like the use of public Wi-Fi can go a long way in promoting the application of cybersecurity.

6 Summary

Cybersecurity affects everyone subject to the use of technology and the Internet. That is why, because the organizations have to develop techniques to prevent cyber-attacks, it is essential we also play our parts. The unpredictability of cyber-attacks as technology advances means they are always running and open for discoveries. Therefore, younger generations should be encouraged to engage in cyber-related professions to have better control over the issue. Research has shown that there is little understanding of the dynamics of information technology, which includes cybersecurity. Artificial Intelligence and Blockchain are born out of human behaviors that have worked in the past, prediction, responsiveness, and validation, making them two of the most promising security measures for the near future. Potential Applications in Information Security may allow Blockchains to manage digital identities, protect large amounts of data, and secure edge devices. A few critical elements make Blockchain one of the most secure suppliers of cyber defense. It is capable of supplying protection for financial activities such as cryptocurrency transactions. Some of the characteristics that make Blockchain very secure are decentralization, immutability, and accountability. Blockchain implements numerous cryptographic protocols, most of which are difficult to decipher. The integration of Blockchain and AI technologies can empower AI by providing a secure and trustful environment.

Acknowledgments This research is supported by the Texas A&M University Chancellor Research Initiative (CRI) grant awarded to Texas A&M University-San Antonio, TX, USA.

References

1. Thames, L., Schaefer, D.: Industry 4.0: an overview of key benefits, technologies, and challenges. In: Cybersecurity for Industry 4.0, pp. 1–33. Springer, Cham (2017)
2. Zhao, K., Ge, L.: A survey on the Internet of Things security. In: 2013 Ninth International Conference on Computational Intelligence and Security, pp. 663–667. IEEE, December 2013
3. Hassani, H., Huang, X., Silva, E.: Big-crypto: big data, blockchain, and cryptocurrency. Big Data Cogn. Comput. **2**(4), 34 (2018)
4. Maxmen, A.: Ai researchers embrace Bitcoin technology to share medical data. Nature **555**, 293–294 (2018)
5. Baynham-Herd, Z.: Enlist blockchain to boost conservation. Nature **548**(7669), 523 (2017)
6. Swan, M.: Blockchain: Blueprint for a New Economy. O'Reilly Media, Newton (2015)
7. Aghiath, C.: How AI and Blockchain will be the Future of Cybersecurity (2020). https://www.intelligenthq.com/how-ai-and-blockchain-will-be-the-future-of-cybersecurity/. Accessed Dec 2020
8. Sirui, Z.: Economics of Artificial Intelligence (2019). https://equalocean.com/analysis/2019100712033. Accessed Dec 2020
9. Koch, M.: Artificial intelligence is becoming natural. Cell **173**(3), 531–533 (2018)
10. Nebula AI (NBAI): Decentralized AI blockchain whitepaper. Nebula AI Team, Montreal (2018)
11. Dinh, T.N., Thai, M.T.: AI and blockchain: a disruptive integration. Computer **51**(9), 48–53 (2018)
12. Wang, S., Yuan, Y., Wang, X., Li, J., Qin, R., Wang, F.-Y.: An overview of smart contract: architecture, applications, and future trends. In: Proceedings of the IEEE Intelligent Vehicles Symposium (IV), pp. 108–113, June 2018

13. Panda, S., Jena, D.: Decentralizing AI using blockchain technology for secure decision making (2021). https://doi.org/10.1007/978-981-15-5243-4_65
14. Mohanta, B.K., Jena, D., Panda, S.S., Sobhanayak, S.: Blockchain technology: a survey on applications and security privacy challenges. Internet Things **8**, 100107 (2019)
15. Nakamoto, S.: Bitcoin: a peer-to-peer electronic cash system (2009)
16. Wood, G.: Ethereum: a secure decentralized generalized transaction ledger. Ethereum Project Yellow Paper, vol. 151, pp. 1–32, April 2014
17. Market capitalization of Bitcoin from October 2013 to 13 December 2020. https://www.sta tista.com/statistics/377382/bitcoin-market-capitalization/. Accessed Dec 2020
18. NIST (2020). https://csrc.nist.gov/glossary/term/social_engineering. Accessed Dec 2020
19. Department of Justice (2019). https://www.justice.gov/opa/pr/russian-national-charged-dec ade-long-series-hacking-and-bank-fraud-offenses-resulting-tens. Accessed Dec 2020
20. Al-Rushdan, H., Shurman, M., Alnabelsi, S.H., Althebyan, Q.: Zero-day attack detection and prevention in software-defined networks. In: 2019 International Arab Conference on Information Technology (ACIT), Al Ain, United Arab Emirates, pp. 278–282 (2019). https://doi.org/10.1109/ACIT47987.2019.8991124
21. Wikipedia (2016 Dyn cyberattack). https://en.wikipedia.org/wiki/2016_Dyn_cyberattack. Accessed Dec 2020
22. Michel, C.: Study: Hackers Attack Every 39 Seconds (2007). https://eng.umd.edu/news/story/study-hackers-attack-every-39-seconds. Accessed Dec 2020
23. Yao, M.: Your electronic medical records could be worth $1000 to hackers, Forbes (2017). https://www.Forbes.com/sites/mariyayao/2017/04/14/your-electronic-medical-records-can-be-worth-1000-to-hackers. Accessed Dec 2020
24. Purplesec: 2020 Cyber Secuirty Statistics the Ultimate List of Stats Data & Trends (2020). https://purplesec.us/resources/cyber-security-statistics/. Accessed Dec 2020
25. Singer, P.W., Friedman, A.: Cybersecurity: what everyone needs to know. OUP, USA (2014)
26. Parrend, P., Navarro, J., Guigou, F., Deruyver, A., Collet, P.: Foundations and applications of artificial Intelligence for zero-day and multi-step attack detection. EURASIP J. Inf. Secur. **2018**(1), 4 (2018)
27. Raj, R.K., Anand, V., Gibson, D., Kaza, S., Phillips, A.: Cybersecurity program accreditation: benefits and challenges. In: Proceedings of the 50th ACM Technical Symposium on Computer Science Education, pp. 173–174, February 2019
28. De Bruijn, H., Janssen, M.: Building cybersecurity awareness: the need for evidence-based framing strategies. Gov. Inf. Q. **34**(1), 1–7 (2017)
29. Maalem Lahcen, R.A., Caulkins, B., Mohapatra, R., Kumar, M.: Review and insight on the behavioral aspects of cybersecurity. Cybersecurity **3**, 1–18 (2020)
30. Patterson, I., Nutaro, J., Allgood, G., Kuruganti, T., Fugate, D.: Optimizing investments in cyber-security for critical infrastructure. In: Proceedings of the Eighth Annual Cyber Security and Information Intelligence Research Workshop, pp. 1–4, January 2013
31. Jones, S.L., Collins, E.I., Levordashka, A., Muir, K., Joinson, A.: What is 'cyber security'? Differential language of cyber security across the lifespan. In: Extended Abstracts of the 2019 CHI Conference on Human Factors in Computing Systems, pp. 1–6, May 2019
32. Romero-Mariona, J., Ziv, H., Richardson, D.J., Bystritsky, D.: Towards usable cyber security requirements. In: Proceedings of the 5th Annual Workshop on Cyber Security and Information Intelligence Research: Cyber Security and Information Intelligence Challenges and Strategies, pp. 1–4, April 2009
33. Yavanoglu, O., Aydos, M.: A review of cybersecurity datasets for machine learning algorithms. In: 2017 IEEE International Conference on Big Data (Big Data), pp. 2186–2193. IEEE, December 2017
34. Yampolskiy, R.V., Spellchecker, M.S.: Artificial intelligence safety and cybersecurity: a timeline of AI failures. arXiv preprint arXiv:1610.07997 (2016)
35. Dilek, S., Çakır, H., Aydın, M.: Applications of artificial intelligence techniques to combating cyber crimes: a review. arXiv preprint arXiv:1502.03552 (2015)
36. Li, J.H.: Cybersecurity meets artificial Intelligence: a survey. Front. Inf. Technol. Electron. Eng. **19**(12), 1462–1474 (2018)

37. O'Reilly, U.M., Toutouh, J., Pertierra, M., Sanchez, D.P., Garcia, D., Luogo, A.E., Hemberg, E.: Adversarial genetic programming for cybersecurity: a rising application domain where GP matters. Genet. Program. Evol. Mach. **21**, 219–250 (2020)
38. Banković, Z., Stepanović, D., Bojanić, S., Nieto-Taladriz, O.: Improving network security using a genetic algorithm approach. Comput. Electr. Eng. **33**(5–6), 438–451 (2007)
39. Willard, G.N.: Understanding the co-evolution of cyber defenses and attacks to achieve enhanced cybersecurity. J. Inf. Warfare **14**(2), 16–30 (2015)
40. Albladi, S.M., Weir, G.R.: Predicting individuals' vulnerability to social engineering in social networks. Cybersecurity **3**(1), 1–19 (2020)
41. Melissa, L.: Cybersecurity: What Every CEO and CFO Should Know (2016). https://www.top tal.com/finance/finance-directors/cyber-security. Accessed Dec 2020
42. Wang, K., Dong, J., Wang, Y., Yin, H.: Securing data with blockchain and AI. IEEE Access **7**, 77981–77989 (2019)
43. Michele, D.: How Does the Blockchain Work? (2016). https://onezero.medium.com/how-does-the-blockchain-work-98c8cd01d2ae. Accessed Dec 2020
44. Smith, S.S.: Blockchain: what you need to know. Account. Today **31**(11), 42 (2017)
45. Mahbod, R., Hinton, D.: Blockchain: the future of the auditing and assurance profession. Armed Forces Comptrol. **64**(1), 23–27 (2019)
46. Dorri, A., Kanhere, S.S., Jurdak, R., Gauravaram, P.: Blockchain for IoT security and privacy: the case study of a smart home. In: 2017 IEEE International Conference on Pervasive Computing and Communications Workshops (PerCom Work 2017), pp. 618–623 (2017). https://doi.org/10.1109/PERCOMW.2017.7917634
47. Sikorski, J.J., Haughton, J., Kraft, M.: Blockchain technology in the chemical industry: machine-to-machine electricity market. Appl. Energy **195**, 234–246 (2017). https://doi.org/10.1016/j.apenergy.2017.03.039
48. Liang, X., Zhao, J., Shetty, S., Liu, J., Li, D.: Integrating blockchain for data sharing and collaboration in mobile healthcare applications. In: IEEE International Symposium on Personal, Indoor, and Mobile Radio Communications, PIMRC 2018 (2018). https://doi.org/10.1109/PIMRC.2017.8292361
49. Guo, Y., Liang, C.: Blockchain application and outlook in the banking industry. Financ. Innov. **2** (2016). https://doi.org/10.1186/s40854-016-0034-9
50. Gartner: Gartner Survey Shows 37 Percent of Organizations Have Implemented AI in Some Form, Gartner (2019). https://marketbusinessnews.com/ai-gartner-survey/194856/. Accessed Dec 2020
51. Marr, D.: Artificial intelligence-a personal view. Artif. Intell. **9**(1), 37–48 (1977)
52. Harini, B.N., Rao, T.: An extensive review on recent emerging applications of artificial intelligence. Asia-Pacific J. Converg. Res. Interchange **5**(2), 79–88 (2019)
53. Calderon, R.: The benefits of artificial intelligence in cybersecurity (2019)
54. Zeadally, S., Adi, E., Baig, Z., Khan, I.A.: Harnessing artificial intelligence capabilities to improve cybersecurity. IEEE Access **8**, 23817–23837 (2020). https://doi.org/10.1109/ACCESS.2020.2968045
55. Panarello, A., Tapas, N., Merlino, G., Longo, F., Puliafito, A.: Blockchain and IoT integration: a systematic survey. Sensors **18**(8), E2575 (2018)
56. Salah, K., Rehman, M.H.U., Nizamuddin, N., Al-Fuqaha, A.: Blockchain for AI: review and open research challenges. IEEE Access **7**, 10127–10149 (2019). https://doi.org/10.1109/ACCESS.2018.2890507
57. The foundation for our digital future. https://ethereum.org/en/what-is-ethereum/. Accessed Feb 2021
58. Mengidis, N., Tsikrika, T., Vrochidis, S., Kompatsiaris, I.: Blockchain and AI for the next generation energy grids: cybersecurity challenges and opportunities. Inf. Secur. Int. J. **43**(1), 21–33 (2019). https://doi.org/10.11610/isij.4302
59. Kumble, G.P.: Practical Artificial Intelligence and Blockchain: A Guide to Converging Blockchain and AI to Build Smar Applications for New Economies (2020)
60. BlockchianInsight: Top Blockchain Technology Companies 2021 (2021). https://www.leeway hertz.com/blockchain-technology-companies-2021/. Accessed Feb 2021

Securing Vehicular Network Using AI and Blockchain-Based Approaches

Farhat Tasnim Progga, Hossain Shahriar, Chi Zhang, and Maria Valero

Abstract Intelligent vehicles have become a common phenomenon whereas establishing secure communication between those vehicles through multiple networks has become a universal solicitude. Vehicular communication aims to provide secure communication and reduce the cost of traffic congestion by processing real-time data. This proliferation paradigm of vehicular systems represents several options for communication such as message sharing and data transmission, and thus it becomes vulnerable in terms of security and privacy. However, artificial intelligence has encountered an undeniable development in every research field including healthcare, transportation management, academia, and genetic engineering. Consequently, blockchain technology has brought plausible accomplishments in those fields where maintaining security is the first precedence. Considering the recent establishments of both artificial intelligence (AI) and blockchain technology, researchers have solved vehicular network-related security problems using those technologies separately and combinedly as well. The common security concerns include Sybil attacks, Denial-of-service (DoS) attacks, man-in-the-middle (MITM) attacks, malicious attacks, which cause data manipulation, data outflow, message delay, and traffic congestion. In this paper, we reviewed recent developments based on research works addressing the issues related to vehicular ad hoc networks and vehicular social networks. We have highlighted the proposed solutions relying on AI and blockchain technologies while identifying new research directions.

F. T. Progga
Department of Electrical and Computer Engineering, North South University, Dhaka, Bangladesh
e-mail: farhat.progga@northsouth.edu

H. Shahriar (✉) · C. Zhang · M. Valero
Department of Information Technology, Kennesaw State University, Marietta, Georgia, USA
e-mail: hshahria@kennesaw.edu

C. Zhang
e-mail: czhang4@kennesaw.edu

M. Valero
e-mail: mvalero2@kennesaw.edu

© The Author(s), under exclusive license to Springer Nature Switzerland AG 2021   31
Y. Maleh et al. (eds.), *Artificial Intelligence and Blockchain for Future Cybersecurity Applications*, Studies in Big Data 90,
https://doi.org/10.1007/978-3-030-74575-2_2

Keywords Vehicular network · Security and privacy · Intelligent vehicle ·
Artificial intelligence · Blockchain technology

1 Introduction

Vehicular networks are considered one of the most demanding yet challenging classes
of mobile networks that provide the vehicles with the media to communicate with
other vehicles, infrastructures, and devices. Consequently, they play a vital role in
maintaining a rigorous traffic system. The expansion of Internet-of-things (IoT) in
vehicular networks leads to building [1] vehicle-to-vehicle (V2V), vehicle-to-human
(V2H), vehicle-to-infrastructure (V2I), and vehicle-to-pedestrian (V2P) communications, and thus summarizes vehicle-to-everything (V2X) communication. Over
the last few years, intelligent transportation systems have become a significant yet
prominent research field both in industry and academia. Within a few decades,
Intelligent Transportation Systems (ITS) will become a full-fledged reality due to
the advancements of intelligent vehicles. Rapid modifications in technology have
brought new features amplified in both vehicular ad hoc networks (VANET) and
Vehicular Social Networks (VSN) systems. Therefore, vehicular network systems
have become dynamic and heterogeneous as well. Also, these modern vehicles have
several applications embedded in their systems, most of which require intelligent
decision-making [2]. Indeed, these applications have made today's vehicles more
efficient and modernized. These applications can be categorized based on target use,
transmission method, communication method, vehicle technology, delay tolerance
mode, and radio technology. For instance, real-time navigation features using GPS,
street surveillance features, autonomous driving features, and so on [1]. Although
VANET and VSN connections have several significances in vehicle communication,
they are vulnerable [3] in terms of security, privacy, and trust issues.

Security is considered as one of the elementary concerns of today's vehicular
network systems due to their increased vulnerability and complexity. As a vehicular
security system, both VANET and VSN should ensure that not only the transmitted
data comes from a trusted origin but also that the transmitted data should not be
encountered by any other sources. Also, these networks should maintain security and
privacy at the same time, which sometimes happens to be contradictory. Accordingly,
privacy and authentication are also considered fundamental concerns of vehicular
networks. Privacy allows an individual to be the controller of his or her data sharing
with others, and the creation of anonymous authentication helps to protect privacy
in general. However, security and privacy often are compromised through several
denial-of-service (DoS) attacks which result in inaccurate data transmission and
data outflow. Such attacks include side-channel attack, message injection, spoofing,
packet sniffing, bogus information transmission, in-vehicle spoofing, fuzzy attack,
and malicious attack [4]. Vehicle networks have been interconnected with cellular
and IT mechanisms which have made them vulnerable in various conditions. Nevertheless, considering the vulnerabilities of vehicular networks, nowadays researchers

have implemented various artificial intelligence and blockchain technology-based solutions to prevent security threats.

Through time, artificial intelligence (AI) has enabled various opportunities for improvements in almost every sector of research, especially in transportation management systems for upgrading traffic systems in terms of security. AI panoramically has the ability to rationalize human intelligence through simulations and then can take the necessary actions that have the best chance to attain a particular goal. AI pushes such systems to get smarter and more user-friendly. Applications of AI can be achieved through the implementations of different Machine Learning (ML) based algorithms such as supervised learning, unsupervised learning, and reinforcement learning. Nowadays, vehicles [5, 6] can also perform human-like or even superhuman behaviors accordingly due to the applications of AI techniques such as the automated vehicles of Cruise, Waymo, and Google. The advancements of AI for Vehicles (AIV) have enlightened the possibility of autonomous driving, which depended upon feature extraction of human driving behavior by mitigating human faults. To date, AI has been able to acknowledge [7] V2X communication through its implementation by dint of swarm intelligence, machine learning, deep learning, and optimization techniques. ML as a part of AI has led several researchers to utilize supervised, unsupervised, reinforcement learning upon both VANET and VSN to secure privacy, proper data transmission [8]. In this chapter, we have deliberately focused on security and data related issues of vehicular networks and their AI and blockchain-based solutions.

Blockchain technology is commonly defined as the decentralized, distributed ledger that allows storing the sources of digital assets/information [9]. In recent times, blockchain has become an uprising technology for data privacy due to its transparency in accessible ways to reduce fraud risk. Bitcoin, the implementation of blockchain, is supposedly the main reason behind the development of blockchain technology. Blockchain technology is considered as the solution to those security and privacy concerns where data integrity and secured data sharing are the foremost concerns. Working for [10] cryptocurrency and bitcoin, blockchain technology has made its way to secure intelligent vehicle's privacy while sharing data. Blockchain-based solutions [11] have ensured safety, data privacy, and resilience of telecommunication-based transportation systems. Blockchain supports programmable smart contacts and has the ability to automate the generation of structured events, actions, and payments as well. Blockchain technology comprises four major modules [12]: cryptography, structure, census, and smart contract. The vision behind the census mechanism is to make suspicious entities agree on a version of a valid block so that the mechanism can provide verifiable data over the network. Besides, cryptographic hash functions help the vehicular networks to attain security, privacy, and anonymity.

Both the AI and blockchain-based solutions have separately been able to reduce the issues of vehicle networks such as data privacy, confidentiality, outflow, and delay. However, convergence and combination of both AI and blockchain [13] have been comprehensively studied for sustainable future generations. Deploying both AI and blockchain technology simultaneously in an intelligent vehicle system raises many queries due to high resource demands and computational complexity. Until now, very few studies have taken place regarding this issue because of the feasibility questions that arise. Apart from the conditions, AI's learning algorithms on a scalable blockchain technique embedded in the same architecture would portray a smart, secure, and efficient vehicular network system.

In this chapter, we have focused on vehicle network security-based issues addressed by the recent research studies. Also, the techniques they followed to solve those problems have been elaborated here for a better perception of this research field. Moreover, we have also enlightened the limitations of those researches and elaborated those limitations on the open scopes section of the paper.

The rest of the paper is organized as follows: Sect. 2 presents the methodology of the study; Sect. 3 discusses the detailed findings of the study, including the addressed problems and their solution methods; Sect. 4 illustrates the open challenges and limitations of those research studies; and finally, Sect. 5 draws the conclusion of our study.

2 Methodology

All the papers we have been discussing for this literature review-based project were listed from Google Scholar [14]. Intending to execute a thematic search focusing on vehicle network attack prevention based on AI and Blockchain technology, we prioritized two types of attacks of vehicle networks: application-based and network-based attacks. Application-based attacks refer to the attacks occurring upon the features of vehicular systems. Application-based attacks include inaccurate data transmission, VANET message injection, outflow of transmitted data, secure message violation, message delivery delay whereas network-based attacks include Sybil attacks, DoS (Denial of service) attacks, Man-in-the-Middle (MITM) attack, traffic congestion, abnormal behaviors of local sub-networks, etc. These existing problems were resolved by the researchers using AI and Blockchain technology [15–31]. We gathered a total of seventeen recent papers (published between 2017 to 2020) for our research. Out of those seventeen papers, four papers [15–18] analyzed the DoS attacks and Sybil attacks of VANET and proposed solutions accordingly using ML techniques whereas other papers described different application and network-based problems of VANET and provided AI and blockchain technology-based solutions.

3 Findings and Discussion

Using AI and blockchain technologies on learning-based algorithms and detection methods can help researchers design better-performing VANET, especially for preexisting problems. Being the open medium of communication between vehicles, VANET and VSN are crucially exposed to both inside and outside threats which tend to affect the reliability of those communications. Nonetheless, we categorized the most common threats amidst all the violations of network security in the vehicular networks and discussed the solutions to those issues as well. This section outlines the problems focused on by the researchers, the techniques they followed, and the experiment environment they used while solving those problems.

3.1 Problem Addressed

Considering the enormous possibilities AI is bringing to develop a sustainable future, it has new opportunities for Intelligent Transportation Systems (ITS). Also, blockchain technology-based transactions in a distributive way have the prospects for a better vehicular system. Both of these technologies reportedly have promising potential to resolve vehicular network security attacks and to establish the better performance of the system as well. Nowadays, researchers are therefore trying to explore AI and Blockchain technology for the vehicular network paradigm. As a result, several networks and application-based problems of VANET have been resolved by dint of AI and blockchain technologies.

Existing problems of VANET and VSN on the mentioned papers have been categorized in Fig. 1. As an illustration of that, we could consider two terms: Application-based issues and Network-based issues of vehicular networks. Application-based attacks can be defined as the attacks occurring upon the features of vehicular systems whereas Network-based attacks can be referred to as the attacks occurring upon the transmission networks of vehicular communication. Here, Network-based attacks include Sybil attacks, DoS attacks, MITM attacks, malicious security attacks,

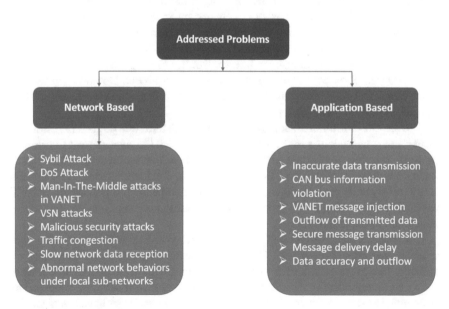

Fig. 1 Categorization of the addressed problems

abnormal network behaviors. Besides, Application-based attacks include inaccurate data transmission, message injection, violation of message transmission, message delivery delay, data inaccuracy, and outflow.

By adopting the machine learning and AI techniques, researchers solved different problems related to application and network-based issues of vehicular networks such as Sybil attacks [15–18], MITM attacks [19], traffic congestion, and network security issues [20–22], location privacy violation [16], CAN bus ID security [23], abnormal network behaviors under local sub-networks [24], message injection [25], etc. Throughout the last decade, researchers are utilizing AI techniques to resolve those cybersecurity issues of vehicular networks.

Furthermore, some scientists also solved VANET problems including message delay issues [26], maintenance of data accuracy [27], attacks in Vehicular Social Networks (VSN) [28], secure message transmission [29], security mechanism for vehicular communication [30], outflow of transmitted data of vehicular network [31], malicious security attacks [26] using blockchain technology, and cloud computing as well. Common issues related to vehicular networks on those mentioned papers have been described below:

Sybil attack allows multiple false identities to slip into an established network and thus manipulate data confidentiality, integrity, and anonymity. This attack, oftentimes plays the role of the main constraint [32] behind other network attacks and dilutes the quality of the network by creating illusional heavy traffic in the network bandwidth. By forging various identities this attack lets the drivers of the vehicle assume there are multiple vehicles on the nearby roads. It is also able to a create black hole attack which causes leakage of messages via multipath routing.

Denial of Service (DoS) attack can be referred to as a cyber-attack which causes failure of functionalities of a network by rendering malicious attacks to the network. However, distributed denial of service attacks has distributed sources for DoS attacking. DoS attacks can be categorized [33] based on network layers such as volume-based DoS attacks: Internet Control Message Protocol (ICMP) and User Datagram Protocol (UDP) flood attacks; protocol-based DoS attacks: Ping of Death attack and SYN Flood attack; application-based DoS attacks: zero-day DDos, Apache, HTTP Flood attacks, etc.

Man-in-the-middle-attack (MITM) attack - the term was derived [19] from an incident of a basketball game. In VANET context, MITM attacks tend to create a catastrophe in communication which eventually causes alteration of data. This attack leads both active and passive attackers to emanate confidential data. Through this attack, third parties (outsiders) can intrude [34] on authentic information of VANET passively and can also delay, drop, or tamper message transmission actively in the middle of the communication.

Controller Area Network (CAN) ID security attack refers to the message spoofing and tampering attacks CAN ID unintentionally allows in their network systems. Due to some historical reasons [35], controller area networks do not have secure communication mechanisms within their systems, and they are supposed to be assumed as secure communication media. Such vulnerabilities have leveraged the communication methods of CAN systems and thus let the security be compromised.

Other addressed in the mentioned research works include Traffic congestion, Message manipulation, Malicious security attacks, Abnormality of the sub-network system, Message injection and spoofing, Outflow of data transmission, etc. Such issues are initiated by several DoS, malicious attacks in the VANET system.

Tables 1 and 2, respectively, show network-based and application-based vehicular network's recent works; addressed existing problems, proposed solutions, and the main contribution of their approach. We have already discussed the addressed problems in this section and we will continue to illustrate the proposed methods in the following section. In terms of the contribution, in most of the cases, researchers offered models or methods to solve specific problems of vehicular networks using AI and Blockchain based frameworks. The proposed methods or models worked properly on predefined environments in some cases with high accuracy.

3.2 Methods

To date, researchers explored several AI and blockchain-based algorithms to resolve the problems mentioned in the tables. The most popular were ML-based intrusion detection approaches. They adopted some popular types of supervised and unsupervised learning algorithms such as K-nearest neighbor (kNN) [16, 20] and Support vector machine (SVM) classifier [15]. They also implied deep learning [24] based approaches to build classifiers for VANET security. In contrast with AI techniques, many researchers also deployed blockchain technologies [26, 27, 29, 31] intending to solve some major security issues of vehicular social networks. Other methods such as received signal strength indicator (RSSI) [17], software-defined vehicular networks [28], cloud computing, fuzzy i.e., clustering [20] have also been used in those papers. We categorized the used methods of those researches as mentioned below in Fig. 2:

Learning-based algorithms are methods that are generally used in machine learning to emulate humans' learning and decision-making process. Usually, the models use a large amount of data to learn different attributes and patterns between them and apply the gained knowledge in unknown scenarios to make decisions [36]. There are many learning algorithms available nowadays designed for different tasks. For example, regression, decision tree, Support Vector Machine (SVM), k Nearest Neighbor (kNN), Random Forest, etc. can be applied to almost all data-related problems. On the other hand, for a huge dataset, different deep learning methods can be used as they use multiple artificial neural network layers in the structure and can make intelligent decisions.

Table 1 Recent work on network-based problems

Author	Category	Year	Addressed Problem	Proposed Solution	Contribution
Gu, P. [11]	Conference paper	2017	Sybil attack	Proposed a Support Vector Machine (SVM) based Sybil attack detection method	A sybil attack detection method with low error rate has been presented here
Gu, P. [12]	Conference paper	2017	Sybil attack and location privacy violation	Proposed a K- Nearest Neighbor classification-based Sybil attack detection method	A sybil attack detection method with good error rate control has been enlightened here
Kim, M. [14]	Conference paper	2017	DoS attack	Proposed a collaborative security attack detection mechanism using multi-class SVM in a software-defined vehicular networks	Multi-class SVM based detection mechanism has been implemented to identify the several kinds of vehicular DoS attacks
Yao, Y. [13]	Journal paper	2018	Sybil attack	Proposed a Received Signal Strength Indicator (RSSI) based Sybil attack detection method for VANETs	A sybil attack detection method using Received Signal Strength Indicator (RSSI) which does not rely on radio propagation and work as independent detection method
Lyamin, N. [17]	Conference paper	2018	Traffic congestion	Proposed an AI based method for real-time jamming detection	By combining statistical network traffic analysis using data mining a hybrid jamming detection method has been evaluated here
Yahiatene, Y. [24]	Conference paper	2018	Vehicular Social Networks (VSN) attacks	Proposed a framework to secure VSN using Software-Defined Vehicular Networks (SDVN) and Blockchain technology	A Distributed Miners Connected Dominating Set algorithm (DM-CDS) based on blockchain technology has been presented here for maintaining security in VSN
Ahmad, F. [15]	Journal paper	2018	Man-In-The-Middle (MITM) attacks in Vehicular Ad Hoc Network (VANET)	Studied the effects of MITM attacks on VANET through simulations: message delayed, message dropped and message tampered	The proposed simulation results concluded that MITM attacks have crucial impacts on VANET nodes which causes high number of compromised messages, message delays and packet losses
Mourad, A. [18]	Journal paper	2020	Malicious security attacks	Proposed a solution including Vehicular Edge Computing (VEC) fog-enabled scheme which allows offloading intrusion detection tasks to be executed with minimal latency with the help of cloud computing	Vehicular Edge Computing (VEC) fog-enabled scheme allowing offloading intrusion detection tasks to federated vehicle nodes located within nearby formed ad hoc vehicular fog to be cooperatively executed with minimal latency
Shu, J. [20]	Journal paper	2020	Abnormal network behaviors under local sub-networks	Proposed a collaborative intrusion detection system (CIDS) using deep learning techniques to detect anomaly in sub-network flows	The correctness of the designed CIDS in both IID (Independent Identically Distribution) and non-IID situations on a real-world dataset has been evaluated here

(continued)

Table 1 (continued)

Author	Category	Year	Addressed Problem	Proposed Solution	Contribution
Lv, Z. [16]	Journal paper	2020	Traffic congestion and slow network data reception	Proposed a function-based fuzzy mean clustering algorithm theory (FCM) for improving the performance of the electrical vehicle network	An algorithm using K-means and fuzzy theory in big data analysis technology has been indicated here

Table 2 Recent work on application-based problems

Author	Category	Year	Addressed problem	Proposed solution	Contribution
Sharma, P. [21]	Conference paper	2017	Vehicular Ad Hoc Network (VANET) message injection	Proposed a method for message authentication using context-adaptive signature verification including AI filters	An AI algorithm for signature verification has been implemented here to detect spoofed DoS messages
Singh, M. [26]	Conference paper	2017	Inaccurate data transmission	Proposed a blockchain technology based secure peer-to-peer communication environment between intelligent vehicles	An Intelligent Vehicle Trust Point (IV-TP) method for secure and reliable communication among IVs using Blockchain technology
Han, M. L. [19]	Journal paper	2018	Controller Area Network (CAN) bus information violation	Proposed a method of identifying malicious CAN messages to detect the normality and abnormality of a vehicle network	Based on the survival analysis model an anomaly intrusion detection method has been developed and validated here
Yahiatene, Y. [27]	Special issue article	2018	Outflow of transmitted data of vehicular network	Proposed a framework using blockchain paradigm to not only enable the certification of transactions but also to ensure full data anonymity	An IoT device using blockchain technology has been presented here and its performance has been analyzed using simulation with different parameters such as nodes density, radio range, node density, and trust metric
Shrestha, R. [22]	Journal paper	2019	Message delivery delay	Proposed a regional blockchain based intelligent solution to maintain V2V, V2I networks	The implementation of this regional blockchain based method showed a consistency of 51% attack detection accuracy
Singh, M. [23]	Conference paper	2020	Data accuracy and outflow	Proposed a blockchain technology-based trust environment for intelligent vehicle (IV) information sharing	A trust environment based Intelligent Vehicle framework using blockchain technology has been outlined here
Malik, N. [25]	Journal paper	2020	Secure message transmission	Proposed an innovate hybrid algorithm termed Sea Lion Explored-Whale Optimization Algorithm using blockchain technology for securing message transmission nodes	The trustability of the node of the vehicular network was computed based on the "two-level evaluation process" such as rule based and machine learning-based evaluation process

Methods

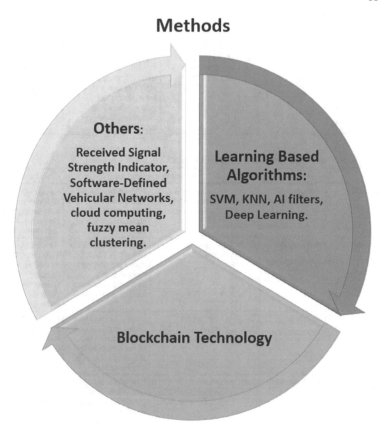

Fig. 2 Methods used by the researchers while solving the problems

Blockchain Technology is a system based on distributed databases with some ledge characteristics and it is reportedly created by some anonymous developers for secure cryptocurrency. In blockchain, different data can be stored on multiple systems around the world having the same blockchain network. Besides, the data in the blockchain network deliberately remains [37] immutable, transparent, and irreversible. Hence, the data in the block cannot be changed due to the univocal connection of the cryptocurrency algorithms of the network. It has been anticipated [38] that blockchain preferably can bring better results in VANET security, healthcare technology, and supply chain management system because of its being distributed and decentralized. Applications of blockchain technology could be categorized as shown in Fig. 3:

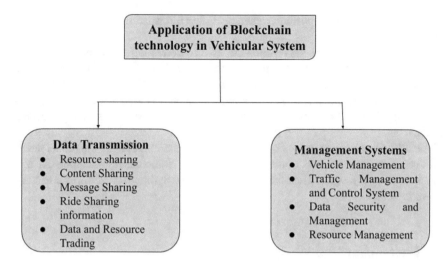

Fig. 3 Applications of blockchain technology in vehicular systems

Other technologies that have been used in the mentioned research projects are fuzzy mean clustering, software-defined vehicular networks, received signal strength indicator systems, and cloud computing. Solutions based on cloud computing have the potential to work on a real-time basis in a cost-efficient way. However, software-defined techniques supposedly need testing to work on real-life scenarios.

4 Open Challenges

Although researchers are doing outstanding work on different vehicular network-related problems using learning-based algorithms, blockchain technology, and other well-known algorithms, there are still multiple open challenges that need to be overcome. First of all, most of the proposed methods were tested using different simulated environments [20, 25, 28, 29, 31]. Simulated environments let those projects work in a limited environment whereas real-time simulation is needed to solve actual scenarios on multi-platform technologies. Despite the fact that the outcomes of those studies were considerable, there are still some chances that the proposed methods might face some challenges in real-life traffic data. Secondly, the experiments were performed using a limited number of scenarios. Therefore, in many proposed methods, robustness and mobility are not settled yet. Thus, the performance of those methods lacks trust metrics due to connectivity and fitness. Thirdly, a unified method to block an attacker from the vehicular network is still an open challenge [7]. Researchers have been able to mitigate the effects of intrusion on networks but an approach to both detect and block the attacker still remains a concern. Fourthly, when the density of the traffic increases, few algorithms lose their peak performance, which might cause a huge problem for those algorithms when obtaining information and detecting attacks [17].

Although different ML algorithms are able to detect intrusions in VANET and VSN, defined instructions and appropriate knowledge should be interpreted for enhancing the accuracy of those algorithms. There can be several prospects of expanding intelligent systems on autonomous vehicles in terms of sensory data analysis, recognition of accurate navigation paths, and thus, ensuring safety-oriented tasks.

Finally, since vehicle networks deal with a huge amount of data, the model needs a huge computational cost and storage facility in some cases, including AI and blockchain technologies. Therefore, additional research should be performed to scale down the cost.

5   Conclusion

Over the last few decades, human civilization has encountered tremendous advancements in almost every sector of living, especially in transportation systems. From man-operated vehicles to intelligent vehicles, there have always been revolutionary changes due to technological developments in the systems. It has been projected that the major accomplishment of intelligent vehicles is its ability to connect with other networks (other vehicles, devices, infrastructures) in a fully automatic way. There are several scopes to improve vehicular communications in terms of computations cost and resource management as well. It is also supposed to assure safety measures and privacy of vehicular communications between the networks. But the medium of such communications can easily be violated and thus can cause several network breaches. Sybil attacks, man-in-the-middle attacks, and malicious network attacks are some of them which cause both security and privacy contraventions. In recent times, there are many technical solutions that have been implemented to address those problems. However, some issues require more detailed solution providers and therefore researchers are now working on AI and blockchain-based solutions to vehicular network issues. Lately, it has been envisioned that providing solutions combining both AI and blockchain technology can ensure better performance of vehicular communications in terms of privacy and computational cost.

In this chapter, we have analyzed a total of seventeen papers that are related to AI and blockchain-based solutions of vehicular network security issues. Since we have gathered recent research works upon vehicular network's existing solution, it should have a greater impact on future reference. Although most of the papers possessed legitimate solutions using AI and blockchain, some of them lacked in performance on a real-time basis. It is also noticeable that most of the fraud detection methods of those papers have been prototyped using learning algorithms. Besides, blockchain-based solutions ensured the security of VANET communication as the experiment of those research. To summarize, both AI and blockchain technology-based solutions for VANET and VSN related issues are increasing in popularity due to their consistency of better performance. Despite all the improvements in vehicular communication, there are still some issues to address in terms of stimulation and backgrounds. It regardless needs deeper understanding because there are a many more papers that we have not considered here. Also, further research is needed to solve such matters so that intelligent vehicular communication can be performed on real-time applications.

References

1. Singh, P.K., Nandi, S.K., Nandi, S.: A tutorial survey on vehicular communication state of the art, and future research directions. Veh. Commun. **18**, (2019)
2. Hammoud, A., Sami, H., Mourad, A., Otrok, H., Mizouni, R., Bentahar, J.: AI, blockchain and vehicular edge computing for smart and secure IoV: challenges and directions. IEEE Internet Things Mag. **3**, 68–73 (2020)
3. Lu, Z., Qu, G., Liu, Z.: A survey on recent advances in vehicular network security, trust, and privacy. IEEE Trans. Intell. Transp. Syst. **20**(2), 760–776 (2018)
4. Thing, V.L., Wu, J.: Autonomous vehicle security: a taxonomy of attacks and defences. In 2016 IEEE International Conference on Internet of Things (iThings) and IEEE Green Computing and Communications (GreenCom) and IEEE Cyber, Physical and Social Computing (CPSCom) and IEEE Smart Data (SmartData), pp. 164–170. IEEE, December 2016
5. Faezipour, M., Nourani, M., Saeed, A., Addepalli, S.: Progress and challenges in intelligent vehicle area networks. Commun. ACM **55**(2), 90–100 (2012)
6. Oh, S.I., Kang, H.B.: Object detection and classification by decision-level fusion for intelligent vehicle systems. Sensors **17**(1), 207 (2017)
7. Tong, W., Hussain, A., Bo, W.X., Maharjan, S.: Artificial intelligence for vehicle-to-everything: a survey. IEEE Access **7**, 10823–10843 (2019)
8. Ye, H., Liang, L., Li, G.Y., Kim, J., Lu, L., Wu, M.: Machine learning for vehicular networks: recent advances and application examples. IEEE Veh. Technol. Mag. **13**(2), 94–101 (2018)
9. Pilkington, M.: Blockchain technology: principles and applications. In: Research Handbook on Digital Transformations. Edward Elgar Publishing (2016)
10. Singh, M., Kim, S.: Branch based blockchain technology in intelligent vehicles. Comput. Netw. **145**, 219–231 (2018)
11. Elagin, V., Spirkina, A., Buinevich, M., Vladyko, A.: Technological aspects of blockchain application for vehicle-to-network. Information **11**(10), 465 (2020)
12. Mollah, M.B., Zhao, J., Niyato, D., Guan, Y.L., Yuen, C., Sun, S., Koh, L.H.: Blockchain for the Internet of vehicles towards intelligent transportation systems: a survey. IEEE Internet Things J. **8**(6), 4157–4185 (2020)
13. Singh, S., Sharma, P.K., Yoon, B., Shojafar, M., Cho, G.H., Ra, I.H.: Convergence of blockchain and artificial intelligence in the IoT network for the sustainable smart city. Sustain. Cities Soc. **63**, (2020)
14. Google Scholar (n.d). https://scholar.google.com/. Accessed 05 Jan 2021
15. Gu, P., Khatoun, R., Begriche, Y., Serhrouchni, A.: Support vector machine (SVM) based sybil attack detection in vehicular networks. In: 2017 IEEE Wireless Communications and Networking Conference (WCNC), pp. 1–6. IEEE, March 2017
16. Gu, P., Khatoun, R., Begriche, Y., Serhrouchni, A.: k-Nearest neighbours classification based sybil attack detection in vehicular networks. In: 2017 Third International Conference on Mobile and Secure Services (MobiSecServ), pp. 1–6. IEEE, February 2017
17. Yao, Y., Xiao, B., Wu, G., Liu, X., Yu, Z., Zhang, K., Zhou, X.: Multi-channel based sybil attack detection in vehicular ad hoc networks using RSSI. IEEE Trans. Mob. Comput. **18**(2), 362–375 (2018)

18. Kim, M., Jang, I., Choo, S., Koo, J., Pack, S.: Collaborative security attack detection in software-defined vehicular networks. In: 2017 19th Asia-Pacific Network Operations and Management Symposium (APNOMS), pp. 19–24. IEEE, September 2017

19. Ahmad, F., Adnane, A., Franqueira, V.N., Kurugollu, F., Liu, L.: Man-in-the-middle attacks in vehicular ad-hoc networks: evaluating the impact of attackers' strategies. Sensors **18**(11), 4040 (2018)

20. Lv, Z., Qiao, L., Cai, K., Wang, Q.: Big data analysis technology for electric vehicle networks in smart cities. IEEE Trans. Intell. Transp. Syst. **22**(3), 1807–1816 (2020)

21. Lyamin, N., Kleyko, D., Delooz, Q., Vinel, A.: AI-based malicious network traffic detection in VANETs. IEEE Network **32**(6), 15–21 (2018)

22. Mourad, A., Tout, H., Wahab, O.A., Otrok, H., Dbouk, T.: Ad-hoc vehicular fog enabling cooperative low-latency intrusion detection. IEEE Internet Things J. (2020)

23. Han, M.L., Kwak, B.I., Kim, H.K.: Anomaly intrusion detection method for vehicular networks based on survival analysis. Veh. Commun. **14**, 52–63 (2018)

24. Shu, J., Zhou, L., Zhang, W., Du, X., Guizani, M.: Collaborative intrusion detection for VANETs: a deep learning-based distributed SDN approach. IEEE Trans. Intell. Transp. Syst. (2020)

25. Sharma, P., Liu, H., Wang, H., Zhang, S.: Securing wireless communications of connected vehicles with artificial intelligence. In: 2017 IEEE International Symposium on Technologies for Homeland Security (HST), pp. 1–7. IEEE, April 2017

26. Shrestha, R., Nam, S.Y.: Regional blockchain for vehicular networks to prevent 51% attacks. IEEE Access **7**, 95021–95033 (2019)

27. Singh, M., Kim, S.: Blockchain based intelligent vehicle data sharing framework. arXiv preprint arXiv:1708.09721 (2017)

28. Yahiatene, Y., Rachedi, A.: Towards a blockchain and software-defined vehicular networks approaches to secure vehicular social network. In: 2018 IEEE Conference on Standards for Communications and Networking (CSCN), pp. 1–7. IEEE, October 2018

29. Malik, N., Nanda, P., He, X., Liu, R.P.: Vehicular networks with security and trust management solutions: proposed secured message exchange via blockchain technology. Wirel. Netw. **26**, 4207–4226 (2020)

30. Singh, M., Kim, S.: Introduce reward-based intelligent vehicles communication using blockchain. In: 2017 International SoC Design Conference (ISOCC), pp. 15–16. IEEE, November 2017

31. Yahiatene, Y., Rachedi, A., Riahla, M.A., Menacer, D.E., Nait-Abdesselam, F.: A blockchain-based framework to secure vehicular social networks. Trans. Emerg. Telecommun. Technol. **30**(8), (2019)

32. Preetha, M.: A survey of sybil attack detection in vanets (2020)

33. Bouzoubaa, K., Taher, Y., Nsiri, B.: Dos attack forecasting: a comparative study on wrapper feature selection. In: The 2020 International Conference on Intelligent Systems and Computer Vision (ISCV), pp. 1–7. IEEE, June 2020

34. Al-shareeda, M.A., Anbar, M., Manickam, S., Hasbullah, I.H.: Review of prevention schemes for man-in-the-middle (MITM) attack in vehicular ad hoc networks. Int. J. Eng. Manage. Res. **10** (2020)

35. Karray, K., Danger, J.L., Guilley, S., Elaabid, M.A.: Identifier randomization: an efficient protection against can-bus attacks. In: Koç, Ç.K. (ed.) Cyber-Physical Systems Security, pp. 219–254. Springer, Cham (2018)

36. Lahmiri, S.: On simulation performance of feedforward and NARX networks under different numerical training algorithms. In: Handbook of Research on Computational Simulation and Modeling in Engineering, pp. 171–183. IGI Global, 2016

37. Astarita, V., Giofrè, V.P., Mirabelli, G., Solina, V.: A review of blockchain-based systems in transportation. Information **11**(1), 21 (2020)

38. Patil, P., Sangeetha, M., Bhaskar, V.: Blockchain for IoT access control, security and privacy a review. Wirel. Pers. Commun. **117**, 1–20 (2020)

Privacy-Preserving Multivariant Regression Analysis over Blockchain-Based Encrypted IoMT Data

Rakib Ul Haque and **A. S. M. Touhidul Hasan**

Abstract Most of the studies related to privacy-preserving linear regression training with the Internet of Medical Things (IoMT) data from various entities do not satisfy all the privacy issues of the data owner. This article proposes a secure design in order to protect privacy issues of IoMT data at the time of training a linear regression model. Blockchain is employed with a partially homomorphic cryptosystem known as Paillier to protect all participant's data privacy. To eliminate the territory on a third-party, the proposed study unites secure building blocks in secure linear regression. Firstly, a guarded data-sharing platform is developed among various data providers, where encrypted IoMT data is registered on a shared ledger. Secondly, secure polynomial operation (SPO), and secure comparison (SC) are outlined using the homomorphic property of Paillier. Secure linear regression does not need any trusted third-party. It requires only three interplays in each iteration. Severe security inquiry proves that secure linear regression preserves sensitive data privacy for each data provider and analyst. The secure linear regression achieved 0.78, 0.066, and 0.196 adjusted R^2 on BCWD, HDD, and DD datasets respectively. The performance of secure linear regression is nearly similar to the general linear regression.

R. U. Haque
School of Computer Science and Technology, University of Chinese Academy of Sciences, Shijingshan District, Beijing 100049, China
e-mail: rakibulhaqueraj@mails.ucas.ac.cn

R. U. Haque · A. S. M. T. Hasan
Institute of Automation Research and Engineering, Dhaka 1205, Bangladesh

A. S. M. T. Hasan (✉)
Department of Computer Science and Engineering, University of Asia Pacific, Dhaka 1205, Bangladesh
e-mail: touhid@uap-bd.edu

© The Author(s), under exclusive license to Springer Nature Switzerland AG 2021   45
Y. Maleh et al. (eds.), *Artificial Intelligence and Blockchain for Future Cybersecurity Applications*, Studies in Big Data 90,
https://doi.org/10.1007/978-3-030-74575-2_3

1   Introduction

An enormous volume of data has been gathered by individuals and many medical institutions due to the widespread utilization of the Internet of Medical Things (IoMT) [1]. This results in an expansion of collaborative extraction of information among various data owners. Most of the data owners do not have the resources and professional skills required for data mining. Generally, data owners outsource those tasks to a service provider. However, in many cases, data owners are reluctant to share their data because of privacy issues related to data ownership, integrity, and privacy, such as, any pharmaceutical company's desire to estimate the impact of various prescription strategies for inmates with a critical illness. Individual prescription strategies could be a mixture of diverse medicines in particular symmetries. Pharmaceutical companies are ready to share their data for analysis on the consolidated dataset following the assumption that all participant's secrecy is well preserved.

Differential privacy [2], cryptographic [3], and privacy-preserving data publishing [4–7] etc., studies consider data secrecy concerns, and each of them has limitations in efficiency, time complexity, and data analysis, respectively. On the other hand, records of participant's information are not stored in these methods. SecureSVM [8], and secure $k-$nn [9] etc. are some recent resolutions, where blockchain is consolidated to retain participant's information at the time of model training and establish the nearest possible accuracy like standard methods. Again, some works directly focused on privacy-preserving linear regression [10–14] but none of them cover all the privacy requirements for real life, and none of them utilized blockchain technology in their work.

In this work, we introduce cryptosystem-based privacy-preserving secure linear regression in order to alleviate the concerns mentioned earlier. A partial homomorphic public-key cryptosystem [15] known as Paillier integrates with blockchain and encrypted IoMT data to employ secure linear regression in order to protect the data owner's privacy. The proposed method applies secure building blocks using the homomorphic features of Paillier as linear Regression has basic arithmetic operations and comparisons as discussed below:

- Secure Polynomial Operation (SPO): For calculating addition and subtraction on Paillier.
- Secure Comparison (SC): For comparing any numbers in Paillier.

Secure linear regression requires only three interactions for each iteration, and there is no need for a trusted third party. The main contributions are as follows.

- Blockchain technology establishes secure and reliable data sharing between the data owner and data analyst. A unique transaction is developed for recording the encrypted data on a blockchain.
- Secure building blocks (i.e., SPO, and SC) are employed using Paillier for assembling secure linear regression algorithm with three interactions in each iteration and trusted third-party is not required.

Table 1 Notations

Symbol	Expositions	Symbol	Expositions
P	Data owner	A	Data analyst
D	Data-set	x	Independent variable in D
y	Dependent variable in D	m	No. of records in D
n	No. of independent variable in D	α	Learning rate
$\sum_{i=1}^{m} \theta_i$	Regrassion co-efficient	$\sum_{i=1}^{m} h_\theta^i$	Hypothesis function
J_θ	Cost function	$\sum_{i=0}^{n} \theta_i$	Gradient descent function
PK	Public key	SK	Privet key
π	Protocol	δ	Bias
x_i, y_i	i^{th} record in data-set	$\phi(N)$	Euler phi-function
l_i	Class label	$[[message]]$	Encrypted message with Paillier

- Meticulous investigation shows that the secure linear regression can shield data privacy and achieve similar performance just like standard linear regression

The rest of the study determines as follows. Sections 2 and 3 illustrate preliminaries and system overview, respectively. Sections 4 and 5 present model construction and performance evaluation respectively. This article concludes in Sect. 6.

2 Preliminaries

All symbols, background technologies are discussed in this section.

2.1 Notations

x_i and y_i are the ith attributes in the data-set D with m records. After classification attributes x_i and y_i get a label l_i. P and A indicate the data owner and data analyst respectively. $[[m]]$ represents the encryption of message under Paillier. All notations are described in Table 1.

2.2 Homomorphic Cryptosystem

A pair of keys $(PK; SK)$ is used in public-key cryptosystems. For example, a Private key SK and Public key PK are used for encryption and decryption. Without knowing the decryption key, if the feature of a cryptosystem can map the computation over ciphertext to the respective plaintext, it is known as Homomorphic. In the proposed schema, a partial homomorphic cryptosystem (Paillier [15]) is used which allows polynomial operations like, secure addition, and subtraction. Let, n-bit primes are p and q and $N = pq$. N is the public key and $(N, \phi(N))$ is the private key. Paillier's encryption function is $c := [[(1 + N)^m r^N \ mod \ N^2]]$, where decryption function is $m \in \mathbb{Z}_N$ and $m := [[\frac{[c^{\phi(N)} \ mod \ N^2 - 1]}{N} \times \phi(N)^{-1} mod \ N]]$.

2.3 Blockchain

A connected increasing record of transactions, where blocks are joined and shielded using cryptosystem is known as the blockchain [16]. The Peer-to-Peer (P2P) protocol is adopted in blockchain to handle the single point of failure. The consensus mechanism ensures general, unambiguous regulation of transactions and blocks.

2.4 Linear Regression

Linear regression [17] is a simple estimation analysis. These approximations are generally utilized in order to describe the association between one dependent variable and one or more independent variables. Three main techniques for regression analysis are:

- Measuring the accuracy of predictors
- Effect forecasting
- Forecasting trend

Formulas required for regression analysis are listed below.
Hypothesis function's Eq. 1:

$$h_\theta = \theta_0 + (\theta_1 * x_1) + ... + (\theta_n + x_n) \tag{1}$$

Cost function's Eq. 2:

$$J_\theta = \frac{1}{2m} * \sum_{i=1}^{m}(h_\theta^i - y^i)^2 \tag{2}$$

Gradient Descent function, when $j = 0$ follows Eq. 3 and when $j > 0$ follows Eq. 4:

$$\theta_{j=0} = \theta_{j=0} - \alpha * \frac{1}{m}\sum_{i=1}^{m}(h_\theta^i - y^i) \tag{3}$$

$$\sum_{j=1}^{n}\theta_j = \sum_{j=1}^{n}\theta_j - \alpha * \frac{1}{m}\sum_{i=1}^{m}(h_\theta^i - y^i)x_j^i \tag{4}$$

3 System Overview

This section demonstrates the used system model, thread model, and security definitions.

3.1 System Model

The main focus of the model is to ensure secure data sharing between P and A. Individual P share their encrypted data to A and registered them in the blockchain-based distributed record by forming transactions. A can train it's linear regression algorithm by accumulating encrypted data recorded in the public ledger. A can assemble a secure method based on secure building blocks, such as SPO, and SC. It is essential to have interplay between the A and P for sharing intermediate results while training. P will add a small amount of bias (δ) while sharing the intermediate data. This bias will shield the data privacy and lessen the time and space complexity of the algorithm. There is no adverse effect on the classification results for this bias. This bias is mainly used in Secure comparison for comparing encrypted numbers. Figure 1 illustrates the entire process.

- **IoMT Devices:** A device that transfer IoMT data wireless networks.
- **Data Owners P:** An entity or individual, who gather all data from IoMT devices
- **Data Analyst A:** An entity or individual, who can perform data analysis.

Generally, the proposed system is consist of one untrusted data analyst A and n data owners $P_i (i \in 1, ..., n)$. A data set D_i containing vital information is held by

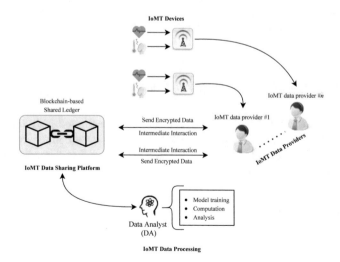

Fig. 1 Data-driven IoMT ecosystem

individual P_i. Horizontal data sharing [18] is considered, where n data-sets $\{D_i\}_{i=i}^n$ share alike feature spaces but distinct in samples. n encrypted data are sequentially gathered by A and linear regression model is trained upon the sample set $D :=$ $(D_1 \cup \text{ ... } \cup D_n)$. A can achieve the desired model after running privacy-preserving training protocols π.

Security Goals: The privacy-preserving training protocols π satisfies the following requirements.

- The A unable to learn any vital information in the data-set D.
- Every P cannot distinguish the model parameters.
- Each P unable to learn about other P's sensitive information.

3.2   Threat Model

All participants in the proposed models do not trust each other. This study considers all participants as honest but curious adversaries. This section will discuss the thread model.

The A is honest in following the pre-designed ML training protocol π. A is also interested in the details of the data and attempts to obtain additional knowledge by analyzing the intermediate data of computation. Again, P might attempt to recognize the design parameter of the A from the intermediate data.

- Recognized Ciphertext Model. A can merely get the encrypted information recorded in the blockchain platform. The A can record intermediate outputs while training secure methods, such as iteration steps.
- Recognized Backend Model. A is expected to be knowledgeable of details than whichever can be distinguished in the disclosed ciphertext model. A can scheme with different P to gather the vital data of another P.

3.3   Encrypted Data Sharing via Blockchain

This study considers that all similar data instances are assigned with the respective feature vectors and are locally preprocessed. A transaction is defined to accumulate the encrypted D in the blockchain. The intended transaction formation is mainly based on two domains: input and output. The input field consists of:

- The address of the sender
- The encrypted version of data
- Source IoMT device name

The corresponding output terminal holds:

- The address of the receiver

- The encrypted version of data
- Source IoMT device name

Addresses of the sender and receiver will be hash value. The encrypted data is determined from the Partial Homomorphic Cryptosystem (Paillier). 128 bytes is the length of the private key and the length of each encrypted data instance and they are recorded in the blockchain. The length of segments for the IoMT device type is 4 bytes. After assembling a new transaction, it is broadcasted in a P2P system of the blockchain network by the sender node. The correctness of the operation is being validated by the miner nodes. The transaction is packaged in a new block by a specific miner node. The block is added to the current chain using traditional consensus protocols, i.e. the PoW mechanism. A single block may register with multiple transactions.

3.4 Security Definitions

This study employs the Secure Two-party Computation framework [19]. Modular Sequential Composition [20] was applied to compose the secure building blocks into a PPML training protocol in a modular way.

Secure Two-Party Computation. For two-party protocols, to ensure security, we have to show that whatever A (B) can compute from its interactions with B (A) can be computed from its input and output, which leads to a commonly used definition, i.e., secure two-party computation [19]. Let $F = (f_A, f_B)$ be a (probabilistic) polynomial function. π is a protocol computing F. A and B want to compute $F(a, b)$ where a is A's input and b is B's input. The view of party A during the execution of π is the tuple $view_A^\pi(a, b) = (a, r, m_1, m_2, ..., m_n)$ where $m_1, m_2, ..., m_n$ are the messages received from B, r is A's random tape. The view of B is defined similarly. Secure Two-party Computation is stated formally as follows:

Definition 1 (Secure Two-Party Computation [19]). A two-part protocol π privately computes f if for all possible inputs (a, b) and simulators S_A and S_B hold the following properties:

$$S_A(a, f_A(a, b)) \equiv_c view_A^\pi(a, b)$$
$$S_B(b, f_B(a, b)) \equiv_c view_B^\pi(a, b)$$

where $\equiv_c$ denotes computational indistinguishability against Probabilistic Polynomial Time (PPT) adversaries with negligible advantage in the security parameter λ [15].

Modular Sequential Composition. Since all our protocols are designed and constructed in a modular way, we employ Modular Sequential Composition [20] for justifying the security proofs of our protocols.

Definition 2 (Modular Sequential Composition [20]). Let $f_1, ..., f_n$ be two-party probabilistic polynomial time functionalities and $\rho_1, ..., \rho_n$ protocols that securely compute respectively $f_1, ..., f_n$ in the presence of semi-honest adversaries. Let F be a probabilistic polynomial time functionality and π a protocol that securely computes F with $f_1, ..., f_n$ in the presence of semi-honest adversaries. Then $\pi^{\rho_1,...,\rho_n}$ securely computes F in the presence of semi-honest adversaries.

4 Model Construction

This section presents the construction details of the proposed model. The goal is to protect the privacy of distinct P and A while training a linear regression model over multiple private datasets from various P.

4.1 Secure Polynomial Operations (SPO)

Secure polynomial addition and secure polynomial subtraction is developed to train the proposed secure linear regression method using Paillier. Reliablity can be achieved at the time of additions, subtractions, and multiplication. Paillier's homomorphic property of addition is derived as: $[[m_1 + m_2]] = [[m_1]] \times [[m_2]] \ (mod \ N^2)$, and subtraction is derived as: $[[m_1 - m_2]] = [[m_1]] \times [[m_2]]^{-1}(mod N^2)[[[m]]^{-1}$ represents the modular multiplicative inverse. It can calculate $[[m]] \times [[m]]^{-1}(mod N^2) = 1$ in Paillier. $[[m]]^{-1}$ can be calculated by $\phi(N)$ function, $[[m]]^{-1} = [[m]]^{\phi(N)-1}$. Again, ciphertext manipulation can achieve secure polynomial multiplication represented in Eq. (5).

$$[[am_1 + bm_2]] = [[m_1^a]] \times [[m_2^b]](mod \quad N^2) \tag{5}$$

Yet, secure polynomial subtraction and addition are needed for this research. Paillier is statistically indistinguishable, so the secure polynomial subtraction and addition are also alike [15].

4.2 Secure Comparison (SC)

It refers to the secure comparison among encrypted numbers. Suppose, A and B involve in the secure comparison algorithm in order to compare $[[m_1]]$ and $[[m_2]]$ following protocol π and neither individual can get original m_1 and m_2. Algorithm 1 represents the secure comparison algorithm.

Algorithm 1: Secure comparison

---

1 *P:* **Input**: $D = \{m_1, m_2\}$
2 *A:* **Input**: PK, SK
3 *P:* **Output**: $flag$
4 P computes $([[m_1 + \delta]], [[m_2 + \delta]])$ by SPO as $([[m_1']], [[m_2']])$;
5 P send $([[m_1']], [[m_2']])$ to A;
6 A decrypts and compares $([[m_1']], [[m_2']])$;
7 **if** $[[m_1']] \geq [[m_2']]$ **then**
8 $\quad | \quad$ A send flag 0 to P;
9 **end**
10 **else**
11 $\quad | \quad$ A send flag 1 to P;
12 **end**

Proposition 1 *(Security of Secure Comparison Algorithm). Algorithm 1 is secure in the curious-but-honest model.*

***Proof (Proof of Proposition** 1).* In Algorithm 1 two entities $(P$ and $A)$ involved. The function F:

$$F([[m_1']]_A, [[m_2']]_A, PK_A, SK_A) = (\phi, (m_1 \geq m_2))$$

The view of P is

$$view_P^\pi = ([[m_1']]_A, [[m_2']]_A, PK_A)$$

Hence, the simulator: $S_P^\pi((m_1, m_2))$;

$$F(m_1, m_2)) = view_P^\pi([[m_1]]_A, [[m_2]]_A, [[\delta]]_A, PK_A)$$

where $[[m_1]]_A$ and $[[m_2]]_A$ are encrypted by PK_A and the confidentiality of $[[m_1]]_A$ and $[[m_2]]_A$ are alike to Paillier. So, P has no opportunity to infer the original value.
The view of A is

$$view_A^\pi = (([[m_1']]), ([[m_2']]), PK_A, SK_A)$$

Then, S_A^π runs as follows:

$$F(m_1', m_2') = view_A^\pi(m_1', m_2', PK_A, SK_A)$$

A will never retrieve original m_1 and m_2 from (m_1') and (m_2') because A is unaware of bias δ. A will compare (m_1') and (m_2') and return 0 or 1 based on the condition $(m_1') \geq (m_2')$ or $(m_1') < (m_2')$ as A is honest in obeying the protocols. $\quad\square$

Algorithm 2: Proposed $Protocol_\pi$

---

1 **P Input**: $D = \sum_{i=1}^{m} \{y^i \times \sum_{j=0}^{n} x_j^i\}$

2 **A Input**: $(PK, SK)_A, \alpha, \sum_{i=0}^{n} \theta_i$

3 **A Output**: updated $\sum_{i=0}^{n} \theta_i'$

4 A computes $([[\alpha]]_{PK_A}, [[\sum_{i=0}^{n} \theta_i]]_{PK_A})$ and send to P ;

5 P computes $[[\sum_{i=1}^{m} h_\theta^i]]_{PK_A}$ using SPO, SC;

6 P computes $[[\sum_{i=1}^{m} (h_\theta^i - y^i)^2]]_{PK_A}$ using SPO, SC and sends to A;

7 A decrypt $[[\sum_{i=1}^{m} (h_\theta^i - y^i)]]_{PK_A}$ using SK_A;

8 A computes J_θ by $\frac{(\sum_{i=1}^{m} (h_\theta^i - y^i))}{2 \times m}$;

9 P computes $[[\{\alpha \times (\sum_{i=1}^{m} (h_\theta^i - y^i) \times x_j^i)\}]]_{PK_A}$ as $[[draft1]]_{PK_A}$ using SPO, SC;

10 P sends $[[draft1]]_{PK_A}$ to A ;

11 A decrypt $[[draft1]]_{PK_A}$ using SK_A;

12 A computes $\frac{draft1}{m}$ as $draft2$;

13 A sends $[[draft2]]_{PK_A}$ to P;

14 P computes $[[\sum_{j=0}^{n} \{\theta_j - draft2\}]]_{PK_A}$ as $[[\sum_{j=0}^{n} \theta_j']]_{PK_A}$ using SPO, SC ;

15 P send $[[\sum_{j=0}^{n} \theta_j']]_{PK_A}$ to A;

16 A decrypt $[[\sum_{j=0}^{n} \theta_j']]_{PK_A}$ using SK_A;

4.3 Training Algorithm of Secure Linear Regression

This study employs lightweight protected linear regression training protocols for protecting the model parameters of all parties. Assume there is a single A and n number of P. Algorithm 2 specifies the proposed training protocols. In Algorithm 2, model parameters of A and sensitive data of P are secret. Each participant will never be able to infer any sensitive data of other participants from intermediate results of the algorithm while confronting any curious-but-honest adversaries or collusions.

Proposition 2 *(Security of $Protocol_\pi$). $Protocol_\pi$ in Algorithm 2 is secure in the curious-but-honest model.*

***Proof (Proof of Proposition** 2). P and A, are the roles involved in $Protocol_\pi$ of Algorithm 2. P function for everyone is similar. If one of the P satisfies the security specifications then all P will satisfy the security requirements. The function F:

$$F(D, (PK, SK)_A, \alpha, \sum_{i=0}^{n} \theta_i) = (\phi, (\sum_{i=0}^{n} \theta_i'))$$

Individual IoMT data owner's P view is

$$view_P^{Protocol_\pi} = (D, [[\alpha]]_{PK_A}, [[\sum_{i=0}^{n} \theta_i]]_{PK_A}, PK_A)$$

where $[[\alpha]]_{PK_A}$ and $[[\sum_{i=0}^{n} \theta_i]]_{PK_A}$ are encrypted by PK_A, the confidentiality of $[[\alpha]]_{PK_A}$ and $[[\sum_{i=0}^{n} \theta_i]]_{PK_A}$ are equivalent to the cryptosystem Paillier. So none of the P can infer the value directly.

The view of A is

$$view_A^{Protocol_\pi} = (PK_A, SK_A, \sum_{i=1}^{m}(h_\theta^i - y^i), (\alpha \times (\sum_{i=1}^{m}(h_\theta^i - y^i) \times x_j^i)))$$

Now, the confidentiality of $\sum_{i=1}^{m}(h_\theta^i - y^i)$ and $(\alpha \times (\sum_{i=1}^{m}(h_\theta^i - y^i) \times x_j^i))$ needs to be discussed, i.e., whether the A can predict the private D of individuals P from the values. Clearly, $\sum_{i=1}^{m}(h_\theta^i - y^i)$ and $(\alpha \times (\sum_{i=1}^{m}(h_\theta^i - y^i) \times x_j^i))$ are no-solution for the unknown D. The A may try to calculate unknown D using the known values α and $\sum_{i=0}^{n} \theta_i$. At the time of division, A has some intermediate values and also knows m. Still, A, will never be able to guess the exact D of P. There is no more reliable method to perceive the genuine value of D except for brute force cracking. Assume, each P has 2-dimensional limited dataset consists of 100 instances. Each dimension is 32 bits [Typically, 4 bytes (32-bit) memory space is occupied by single-precision floating-point]. Based on this condition, A's successful guessing probability is $\frac{1}{2^{(n \times 6400)}}$. It is a minute achieving possibility [15]. So, π is secure in the honest but curious model. □

Security of $Protocol_\pi$ can be achieved by modular sequential composition, like SC, and SPO, which are used in Algorithm 2, so in the honest-but-curious scenario, it is secure.

5 Performance Evaluation

This segment represents the performance analysis of the proposed system.

5.1 Testbed

Each P collects all data from the IoMT devices and encryption them in the suggested system. All operations are being executed on MacBook Pro implemented with memory (4 GB 1600 MHz DDR3), Intel Core i5 processor (2.5 GHz), laboring as A and P concurrently. The SC, SPO, and the secure linear regression are implemented in the Browser: Google Chrome; Language: Python 3; Platform: Google's Collaboratory.

Table 2 Statistics of datasets

Datasets	Instances number	Attributes number	Discrete attributes	Numerical attributes
DD	768	9	0	9
HDD	303	13	13	0
BCWD	699	9	0	9

5.2 Dataset

Three real-world datasets, namely Diabetes Data Set (DD), Heart Disease Data Set (HDD), and Breast Cancer Wisconsin Data Set (BCWD) [21, 22]. Description of the cell nuclei and breast mass from the image has represented by the features of BCWD. Individual data instances are indicated as malignant or benign. 13 discrete and 9 numeric attributes are contained by HDD and DD, respectively. Heart diseases and diabetes symptoms are the types based on which instances are classified. Table 2 represents the statistics of Datasets. 80% and 20% from the dataset are selected for training and testing of the model, respectively.

5.3 Float Format Conversion

Standard linear regression can be trained on floating-point and integers numbers, but cryptosystems perform their operations on whole numbers. So, format conversion must be performed and convert all numbers into an integer. According to the global standard IEEE 754 representation of a floating-point binary number is D is $D = (-1)^s \times M \times 2^E$ [where the sign bit is s, a significant number is M and exponential bit E].

5.4 Key Length setting

The public key cryptosystem's security is strictly correlated with the length. Some issues are:

- Vulnerable encryption may cause by a condensed key
- The efficiency of the homomorphic operation may be reduced by a long key.
- The plaintext space's overflow may cause by a too-short key.

Hence, it is essential to estimate the size of the key to avoid the probability of congestion. In secure linear regression, the key size of Paillier cryptosystem N is fixed to 1024-bit.

5.5 Evaluation parameters

Four most popular method for the evaluation regression algorithms are as follows:

- R-squared (R^2)is the relationship of variation in the outcome that is explained by the predictor variables. The Higher the R-squared, the better the model. Adjusted R-squared is a version of R^2, which regulates the R^2 for begetting multiple variables in the design.
- Root Mean Squared Error (RMSE) measures the average error performed by the model in predicting the outcome for an observation. $MSE = mean((observeds - predicteds)^2)$ and $RMSE = \sqrt{MSE}$. The lower the RMSE, the better the model.
- AIC stands for (Akaike's Information Criteria). AIC penalizes the incorporation of extra variables into a model. It combines a discipline that enhances the failure when adding extra terms. The lower the AIC, the better the model. AICc is a version of AIC adjusted for little unit sizes.
- BIC (or Bayesian information criteria) is an alternative to AIC with a greater penalty for holding extra variables in the model.

The outcomes are shown in Table 3. Here, on BCWD dataset, secure linear regression achieved 0.79, 0.78, 0.031, −424.03 and −2169.72 scores on R^2, Adjusted R^2, RMSE, AIC and BIC respectively, which is almost similar to standard linear regression.

Table 3 Performance analysis

Data-set	Measures	Standard linear regression	Secure linear regression
BCWD	R^2	0.83	0.79
	Adjusted R^2	0.814	0.78
	RMSE	0.038	0.031
	AIC	−438.66	−424.03
	BIC	−2224.51	−2169.72
HDD	R^2	0.54	0.43
	Adjusted R^2	0.08	0.066
	RMSE	0.115	0.136
	AIC	−69.84	−123.83
	BIC	−477.75	−655.13
DD	R^2	0.26	0.37
	Adjusted R^2	0.214	0.196
	RMSE	0.171	0.193
	AIC	−253.94	−537
	BIC	−1296.36	−1468.43

Table 4 Time consumption's

Data-set	SC	SPO	Secure linear regression
BCWD	1245 s	3842 s	4131 s
DD	985 s	2988 s	3766 s
HDD	542 s	1679 s	2133 s

5.6 Efficiency

Table 4 shows the execution time of the SPO with encrypted datasets on $Protocol_\pi$. It also illustrates the total time consumption of P and A.

Table 4 shows the outcomes of secure linear regression. It consumes less than an hour on DD, HDD, and BCWD datasets in encrypted form for training. It is an adequate performance in terms of time consumption. In Python, multi-threading is utilized at the time of implementation to constrain the execution time of a larger dataset. We simulated various P linearly. So, Table 4 shows the collective time consumed of distinct P. P can run their methods parallelly so that the execution time of SPO and SC is reduced.

6 Conclusion

This study introduces a novel privacy-preserving framework for training linear regression over encrypted IoMT data. Secure data sharing between the data provider and data analyst to train linear regression algorithm is the primary focus of this study. The proposed secure linear regression method make sure the privacy and integrity of the IoMT data. Numerous IoMT data provider sends their data and blockchain technology is applied in order to train the ML algorithm in a scenario of multiparty. In order to erect a better model, a partially homomorphic cryptosystem known is utilized. Blockchain is used to records all transactions. This study demonstrates the performance and safety of secure linear regression. This recommended approach succeeds approximately comparable accuracy in comparison to standard linear regression.

Acknowledgements Authors thanks the school of computer science and technology of the University of Chinese Academy of Science, Beijing, China, and the Department of Computer Science and Engineering of University of Asia Pacific, Dhaka, Bangladesh for their support towards this study.

References

1. Joyia, G.J., et al.: Internet of Medical Things (IOMT): applications, benefits and future challenges in healthcare domain. J. Commun. **12**(4), 240–247 (2017)
2. Abadi, M., Chu, A., Goodfellow, I., McMahan, H.B., Mironov, I., Talwar, K., Zhang, L.: Deep learning with differential privacy. In: Proceedings of the 2016 ACM SIGSAC Conference on Computer and Communications Security, pp. 308–318. ACM, New York (2016)

3. Bost, R., Popa, R.A., Tu, S., Goldwasser, S.: Machine learning classification over encrypted data. In: Proceedings of Network and Distributed System Security Symposium, San Diego, California, 23–26 February 2014 (2014)
4. Hasan, A.S.M.T., Qu, Q., Li, C., Chen, L., Jiang, Q.: An effective privacy architecture to preserve user trajectories in reward-based LBS applications. ISPRS Int. J. Geo-Inf. **7**, 53 (2018)
5. Hasan, A.S.M.T., Jiang, Q., Chen, H., Wang, S.: A new approach to privacy-preserving multiple independent data publishing. Appl. Sci. **8**, 783 (2018)
6. Hasan, A.S.M.T., Jiang, Q., Li, C.: An effective grouping method for privacy-preserving bike-sharing data publishing. Future Internet **9**, 65 (2017)
7. Hasan, A.S.M.T., Jiang, Q., Luo, J., Li, C., Chen, L.: An effective value swapping method for privacy-preserving data publishing. Secur. Commun. Netw. **9**, 3219–3228 (2016). https://doi.org/10.1002/sec.1527
8. Shen, M., Tang, X., Zhu, L., Du, X., Guizani, M.: Privacy-preserving support vector machine training over blockchain-based encrypted IoT data in smart cities. IEEE Internet Things J. **6**, 7702–7712 (2019). https://doi.org/10.1109/JIOT.2019.2901840
9. Haque, R.U., Hasan, A.S.M.T., Jiang, Q., Qu, Q.: Privacy-preserving K-nearest neighbors training over blockchain-based encrypted health data. Electronics **9**, 2096 (2020). https://doi.org/10.3390/electronics9122096
10. Senavirathne, N., Torra, V.: Approximating robust linear regression with an integral privacy guarantee. In: 2018 16th Annual Conference on Privacy, Security and Trust (PST), Belfast, pp. 1–10 (2018). https://doi.org/10.1109/PST.2018.8514161
11. Qiu, G., Gui, X., Zhao, Y.: Privacy-preserving linear regression on distributed data by homomorphic encryption and data masking. IEEE Access **8**, 107601–107613 (2020). https://doi.org/10.1109/ACCESS.2020.3000764
12. Giacomelli, I., et al.: Privacy-preserving ridge regression with only linearly-homomorphic encryption. In: International Conference on Applied Cryptography and Network Security. Springer, Cham (2018)
13. Dong, X., et al.: Privacy-preserving locally weighted linear regression over encrypted millions of data. IEEE Access **8**, 2247–2257 (2019)
14. Gascón, A., et al.: Privacy-preserving distributed linear regression on high-dimensional data. Proc. Priv. Enhanc. Technol. **2017**(4), 345–364 (2017)
15. Katz, J., Lindell, Y.: Introduction to modern cryptography. In: CRC Cryptography and Network Security Series. CRC Press, Boca Raton (2014)
16. Nakamoto, S.: Bitcoin: a peer-to-peer electronic cash system (2008). https://bitcoin.org/bitcoin.pdf. Accessed 19 Dec 2020
17. Montgomery, D.C., Peck, E.A., Vining, G.G.: Introduction to Linear Regression Analysis, vol. 821. Wiley, Hoboken (2012)
18. Shokri, R., Shmatikov, V.: Privacy-preserving deep learning. In: Proceedings of the 22nd ACM SIGSAC Conference on Computer and Communications Security, CCS 2015, pp. 1310–1321. ACM, New York (2015). https://doi.org/10.1145/2810103.2813687
19. Goldreich, O.: Foundations of Cryptography: Volume 2, Basic Applications. Cambridge University Press, Cambridge (2009)
20. Canetti, R.: Security and composition of multiparty cryptographic protocols. J. Cryptol. **13**(1), 143–202 (2000). https://doi.org/10.1007/s001459910006
21. Dheeru, D., Karra, T.E.: UCI Machine Learning Repository. University of California, Irvine, CA, School of Information and Computer Science (2017)
22. Detrano, R., Janosi, A., Steinbrunn, W., Pfisterer, M., Schmid, J., Sandhu, S., Guppy, K.H., Lee, S., Froelicher, V.: International application of a new probability algorithm for the diagnosis of coronary artery disease. Am. J. Cardiol. **64**, 304–310 (1989)

Blockchain for Cybersecurity in IoT

Fatima Zahrae Chentouf and Said Bouchkaren

Abstract In the space of a few years, the concept of the Internet of Things (IoT) has become increasingly large and diversified. In addition, the use of new technologies is paramount and especially for businesses and institutions, also IoT considered as the engine of smart cities because it can be applied in different smart city services to facilitate the human daily lifestyle. However, this evolution brings cybersecurity challenges. Thus, security and privacy must be the main subject in this scenery. Also making the protection and sustainability of The Internet of Things issue essential by putting in place effective security, so that it can guarantee the availability of the services offered by a smart city ecosystem. In this chapter, we present a review of Blockchain technology and how it can help in shaping a safer IoT system, and we explore the major challenges and security issues stunting the growth of this concept.

Keywords Internet of Things · Smart city · Cybersecurity · Blockchain

1 Introduction

As we know, people nowadays are experiencing a huge change in communication technologies, indeed they use a lot of things like smartphones, smartwatches, cars, and so many other devices. So how can people connect all these devices together? Here comes the concept of the Internet of Things (IoT).

Recently, this concept had received substantial attention in the research and academic community, especially after the industrial development in the manufacture of "things" that have the ability to identify themselves within the network, communicate over the Internet and interact with other things connected to the Internet, without human intervention. In fact, the Internet of Things (IoT) considered an emerging technology that plays an important role in our daily life. James (Jim) E. Heppelmann

F. Z. Chentouf (✉) · S. Bouchkaren
ERMIA Laboratory, Department of Computer Science, ENSAT, Abdelmalek Essaadi University, Tangier, Morocco

S. Bouchkaren
e-mail: sbouchkaren@uae.ac.ma

© The Author(s), under exclusive license to Springer Nature Switzerland AG 2021
Y. Maleh et al. (eds.), *Artificial Intelligence and Blockchain for Future Cybersecurity Applications*, Studies in Big Data 90,
https://doi.org/10.1007/978-3-030-74575-2_4

PTC CEO answering to the question of why should businesses investing in the IoT he said there is about 7 billion people on earth, one-third of them are connected to the internet with their smartphones, tablets, computers, and so on, but in 2010 there were more things connected to the internet than people on earth and these things are not only smartphones and computers, they could be cars, thermostats, Fitbit, building, hospitals, etc. which means that the trend of connected things is accelerating and with the growth of the number of connected devices by 2020 it predicting to have 50 billion connected devices [1].

In 1982 the expression of the Internet of Things has been proposed by Kevin Ashton. His point was to offer the facilities of humans to communicate with the virtual or fanciful condition [2]. A thing or object in Internet of things (IoT) can be any computing device to whom we can appoint an IP address and that device has an ability to transfer the information over the network. Sensors are embedded in these devices so that these devices can detect the environment and gather the data, then transfer this collected data over the internet. Internet of Things allows being detected and controlled remotely. So we can say that IoT is a global network of things that have the ability to connect, collect, exchange data and Communicate through a Standard protocol. To be clear, it means the communication that allows the exchange of data between machines without human intervention. In other words, the Internet of Things (IoT) is used now in various topics, it is a system physical and virtual that contains a lot of sensors and other devices collecting, communicating, and exchanging information over the Internet.

The goal of the Internet of things is to connect everything to the Internet in order to make Smart Things. Like Smart Cities, Smart Schools, Smart Hospitals, and Smart Homes [3]. So the issue here is the Authentication of each new device, which is trying to connect with other Smart Things. So IoT applications covered different fields in a smart city, and also wearable devices that humans wear such as glasses and watches, connected cars, smart agriculture, healthcare. This technological evolution, especially information and communications technology (ICT), has changed our society. As a result, the emergence of the concept of the "smart city" has come to be seen as a more efficient and sustainable way to fight against the challenges of the modern city and the vital needs. So it is necessary to have a complete overview of the available opportunities and to link them to the specific challenges of the city, hence the goal of the Smart City is to make better use of public resources by improving the quality of service to citizens while reducing costs; and operational aspects of the public administration.

In order to improve a city's smart services, it is necessary to detect and collect data on its environments, infrastructure, events, and people. So, the intervention of IoT is not surprising as it is the key driver of smart cities. In fact, a smart city requires the integration of ICT and IoT in order to improve citizens' quality of life. This means that a smart city will lead to the improvement of many fields such as healthcare, economy, business, agriculture, transportation, and education [4]. These technologies depend on the interconnectivity of devices. Nevertheless, those connected devices might face several security challenges, if this interconnectivity is vulnerable people will lose trust in the system. So a security plan should be made at the planning phase if later it will cause cascading effects that could have a bad effect.

Blockchain technology becomes important in the past 10 years, this concept emerged with the foundation of Bitcoin, a digital cryptocurrency, in 2008 by Satoshi Nakamoto [5]. So, Blockchain can help in reducing costs and barriers, removing a single point of failure and preventing censorship, and ensuring transparency and trust between all parties that are involved in an interaction. Blockchain technology comes as a solution for that by providing a shared ledger technology that allows any participant in the network to see the one system ledger. In other words, Blockchain consists of sharing a digital ledger that records transactions in a public or private peer-to-peer network. Distributed to all members' nodes in the network, the ledger permanently records, in a sequential chain of cryptographic hash-linked blocks, so the blocks of data are organized and chained together. The integration of Blockchain technology in IoT can enforce security by using a trusted public ledger without the need for a third party. Because of that, a lot of companies invested in deploying Blockchain, and also much research has been carried out in the field of Blockchain in the past few years [6].

The rest of the chapter is organized as follows. Section 2 contains a background of IoT and Blockchain. In Sect. 3 some IoT challenges are presented. Section 4 gives a picture of the integration of Blockchain technology in IoT systems to provide security and some related works. And finally, we conclude this chapter with a conclusion.

2 Background

2.1 Internet of Things

2.1.1 History

The history of connected objects began in 1999 when Kevin Ashton [7], a pioneer of Radio Frequency Identification (RFID) technology, coined the term "Internet of Things". In the same year, the concept was born in the United States, notably at MIT (Massachusetts Institute of Technology). This laboratory is dedicated to the creation of connected objects using radio frequency identification and wireless sensor networks. The Internet of Things (IoT) was born of mechanization and standardization, applied to the automation of document and information processing on hardware and then on digital media. It has spread rapidly with globalization. Little by little, objects were modified (with RFID chips for example) and becoming "connected objects" related to centralized servers capable of communicating with each other or with networks of servers and various actors, in a less and less centralized manner.

In 2003, Rafi Haladjian, inventor of the first Internet operator in France (Francenet), created the DAL lamp. A mood lamp equipped with 9 LEDs, offering different colors and sold for 790 euros. Two years later, the creator's company launched the Nabaztag, a rabbit connected by Wi-Fi that reads e-mails aloud, emits visual signals, and broadcasts music. However, it was in 2007 that the IoT phenomenon took off, with the democratization of smartphones and the release of the first iPhone by Apple. Dematerialization is on the way. This revolution was finally the first step towards the exodus from the Internet. This technology was no longer accessible only from home or the office, but anywhere in the world and at any time. The world's population has gradually taken up the challenge, whether with the ease of use of laptops, or tablets that have been accessible to the general public since 2010. The first smartphones or tablets date back much earlier (1989 for Samsung's GridPad tablet) but the lack of dematerialization of data and technological advances were not yet ready for the boom it is currently experiencing. Other connected objects appeared during the same period. For example, the Nabastag (or Karotz) rabbit, launched in 2005, was the first connected entertainment object. But the desire to connect everything to the internet and to connect all objects to each other has existed for many years. According to Gartner, in 2009 there were 2.5 billion connected objects and by 2020 there will be almost 30 billion objects connected to the internet. So we are experiencing exponential growth in the Internet of Things sector.

The IoT can measure everything remotely, instantaneously, and automatically. This is why it is a revolutionary technology.Dates to know:

- 1999: Kevin Ashton coined the term Internet of Things;
- 2003: Creation of the first connected object, the DIAL lamp;
- 2007: Appearance of smartphones;
- 2008: Creation of IPSO addresses, IP addresses of connected objects that allow them to interact with each other;

• Concrete applications [8]:

- Home automation: This technology allows the remote management of many functions in the home: opening and closing shutters, starting the heating system, electrical appliances, light, etc.
- Autonomous car: The autonomous car is a car that can move without the intervention of a driver. The first to have launched it are Google, in 2010. Many obstacles still prevent its democratization, such as cost, the current slowness of vehicles and safety issues.
- Connected dumpsters: Today, we operate with a regular garbage collection system, regardless of how full the dumpsters are. The fact of having connected bins makes it possible to optimize the removal of garbage which, rather than being done at regular intervals, is done as soon as there is a need and only if there is a need.
- The connected toothbrush: It collects data on the way you brush your teeth and analyzes it. This helps to instill good brushing habits in children and why not in their parents;
- The connected keys: Admit it, you've had to look for your keys everywhere when you leave the house. With the keys connected, this will never happen again because your phone will allow you to locate them wherever they are.

2.1.2 Definition

According to the International Telecommunication Union [9], the Internet of Things (IoT) is "a global infrastructure for the information society, providing advanced services by interconnecting objects (physical or virtual) using existing or evolving inter-operable information and communication technologies" [10]. In reality, the definition of the Internet of Things is not fixed. It cuts across conceptual and technical dimensions [11]. Conceptually, the Internet of Things characterizes connected physical objects that have their own digital identity and are able to communicate with each other. This network creates a kind of gateway between the physical and the virtual world.

On a functional level, the Internet of Things refers to a computer system that blends into our daily lives to simplify our lives, save us time, and relieve our brains of the task of memorizing logistical data (itineraries, diaries, etc.). It allows us to create new uses, such as, for example, real-time information on the location of our friends. It also makes it possible to make exhaustive measurements, where in the past I was content with a simple panel, such as measuring car traffic in the streets of the capital. The new applications of the Internet of Things (IoT) are enabling Smart City initiatives around the world.

From a marketing point of view, IoT has an impact on product policy, consumer relations and research, and can lead to the emergence of new distribution methods based, for example, on supply devices, or rather automatic replenishment [12]. IoT also contributes to the development of the large data phenomenon and its challenges through the volumes of collected data it generates.

2.1.3 Features

Connected objects are objects with virtual identities and personalities, operating in real life spaces but also using intelligent interfaces to connect to the Internet and communicate with other objects. There are 5 characteristics associated with the object.

2.1.3.1 Identification
Identification by type or entity is a fundamental concept of the IoT. In general, identifiers are numerical. For example, consumer products usually have a barcode, ISBN books, etc. Isolated objects may also have Assigned numbers: RFID chips store electronic product codes using 96-bit suites.

2.1.3.2 Sensitivity to One's Environment
While it can report its condition, an object can also communicate information about its environment: temperature, humidity, vibration level, noise level or geolocation. If bandwidth permits, an object can also record or play an audio and video stream.

2.1.3.3 Interactivity
The latest technological advances have made it possible to interconnect a wide variety of objects and equipment. Most of the time, it is not necessary for objects to be

permanently connected to the network(s) to which they are attached. Many so-called "passive" objects such as RFID chips need only be activated when they need to exchange information. "Active" objects can be connected continually or when a connection is available.

2.1.3.4 Virtual Representation

It characterizes the possibility for a program present on the cloud to act on behalf of a physical object to which it is attached and of which it is perfectly aware. Thus, even an object carrying no physical intelligence can theoretically have a complex virtual representation. This virtual representation is sometimes referred to as a cyber-object or virtual agent.

2.1.3.5 Autonomy

Objects are processed individually, usually from a single point, and operated independently of remote control. The notion of statelessness is extremely important here: there must be no central intelligence controlling all individual objects in a totalitarian manner. On the contrary, each object is somehow autonomous and independent, with the ability to be interrogated and to interact with other objects in the network when necessary.

2.1.4   Life Cycle

A complete system of a connected object integrates four distinct components, which perform in a life cycle as shown in Fig. 2.1.

2.1.4.1 Data Collection (The Sensors)

First, sensors collect data from their environment. This can be a simple noise level reading as well as a complete reading of a photo or a video stream. Sensors are used because they can often be grouped together or be part of a device that does much more

Fig. 2.1 IoT life cycle

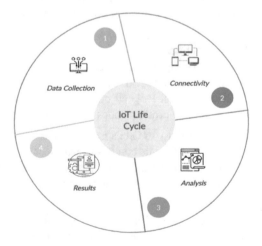

than just sense its environment. For example (your phone is a device that has several sensors: camera, accelerometer, GPS, etc.), but your phone is not just a sensor. In any case, whether it's a stand-alone sensor or a device, the first step is always the same: something is capturing information from its environment.

2.1.4.2 Communication (Connectivity)

Next, the information is sent to the cloud, but it needs a way to get there. Sensors/devices can be connected to the cloud through a variety of methods including: cellular network, satellites, WIFI, Bluetooth, LPWAN (LowPower Wide-Area Network) or direct connectivity to the internet via Ethernet. Each option has its own set of constraints in terms of energy consumption and bandwidth. Choosing the best connectivity depends on the features and application of the connected object. In the end, all options offer the same result: sending information to the cloud.

2.1.4.3 Analysis (Information Processing)

Once the information is in the cloud, software performs processing on it. The processing can be very simple, such as checking if the temperature read is within an acceptable range defined previously. The processing may be also very complex, such as identifying a particular object through a video read by a computer. This application is used to identify intruders in a house.

2.1.4.4 Results (User Interface)

The information is first made useful and readable to the end-user. It can be in the form of an alert to the user (email, SMS, notification, etc.). An SMS is sent to the user, when an intruder is detected in the house, for example. Also, users often have an interface that allows them to proactively check into the system. For example, our user will be able to see a video stream from inside his house on a mobile application or a web page. Depending on the scope of the connected object, the information flow is not always a one-way flow. Also, the user can be able to prompt action on the system. For example, the user can adjust the temperature of the cold room when he detects a heatwave thanks to the sensors, directly from his phone. Other actions are triggered automatically. For example, on some systems, the user can create scripts that will trigger by predefined rules. Thus, instead of alerting you directly, the system of the connected object will be able to alert the competent authorities directly.

2.1.5   IoT Applications

2.1.5.1 Smart Home

Home automation is the automation of the buildings in a house, called "smart home". A home automation system controls lighting, air conditioning, entertainment systems and appliances. It can also include home security, such as access control and alarm systems. When connected to the Internet, household appliances are an important part of the Internet of Things. In 2017, Terence K.L. Huia et al. [13] proposed in their paper the use of IoT technology that interconnects objects to build Smart Homes according to major requirements such as heterogeneity, self-configuration, scalability, context awareness, user-friendliness, security, privacy and intelligence.

2.1.5.2 Structural Health of Buildings

The proper maintenance of a city's historic buildings requires the continuous monitoring of the actual conditions of each building and the identification of areas most susceptible to the impact of external agents. The Urban IoT can provide a distributed database of structural integrity measurements, collected by appropriate sensors located in the buildings, such as vibration and deformation sensors to monitor building stresses, atmospheric sensors in the surrounding areas to monitor pollution levels, and moisture sensors for a complete characterization of environmental conditions. This database should reduce the need for costly periodic structural testing by human operators and allow targeted and proactive maintenance and restoration actions. Finally, it will be possible to combine vibration and seismic measurements to better study and understand the impact of light earthquakes on the city's buildings. This database can be made accessible to the public in order to make citizens aware of the care taken to preserve the city's historical heritage. The practical implementation of this service, however, requires the installation of sensors in buildings and surrounding areas and their interconnection with a control system, which may require an initial investment to create the necessary infrastructure.

2.1.5.3 Health

Smart cities have many innovations to improve the lives of their inhabitants but also their health, either directly or indirectly. The doctor/patient relationship is already facilitated thanks to video communication via smartphones and the latest generation internet networks. In [14], the authors proposed a solution to the problem of monitoring the status of health and safety products within companies. To solve this problem they used a central cloud registration system, a central hub, an analysis data system and detection systems for devices. That's why they installed sensors to collect the data for each product and they used an analysis software in the cloud, it can manage the huge amount of data produced. This allows the service manager to understand what health and safety issues need to be maintained, what their maintenance priorities are, and how they need to plan their maintenance staff to keep them in good working order.

2.1.5.4 Waste Management

Waste management is a major problem in many modern cities because of the cost of the service and the problem of storing waste in landfills. However, deeper penetration of ICT solutions in this area can lead to significant cost savings and economic and environmental benefits. For example, the use of intelligent waste containers, which detect the load level and optimize the routing of collection trucks, can reduce the cost of waste collection and improve the quality of recycling. To provide such a waste management service, the IoT must connect the terminal devices, i.e. The smart waste containers has a control center where the data get treated by an optimization software with the determination of the optimal management of the collection truck fleet.

2.1.5.5 Air Quality

In order to fix some reduction changes in climate in the next 10 years, the European Union has adopted a 20-20-20 Renewable Energy Directive. The goal is attain a 20%

reduction in greenhouse effect gas emissions by 2020, with a reduction in energy consumption because of the improvement of energy efficiency and a 20% increase in the use of renewable energy.

To such an extent, an urban IoT can provide a means of monitoring air quality in overcrowded areas; health pathways. In addition, communication facilities can be provided to allow medical applications running on jogging devices to be connected to the infrastructure. In this way, people can always find the healthiest route to go outside and can be permanently connected to their preferred personal training application. The realization of such a service requires that air quality and pollution sensors be deployed throughout the city and that the data from the sensors be made public to citizens.

2.1.5.6 Traffic Congestion

A possible smart city service that the Urban IoT can provide is to monitor traffic congestion in the city. Although camera-based traffic monitoring systems are already available and deployed in many cities, large-scale, low-power communications can provide a denser source of information. Traffic monitoring can be achieved by using the detection capabilities and GPS installed in modern vehicles, and also by adopting a combination of air quality and acoustic sensors along a given route. Cars can communicate with other systems like weather systems and traffic management systems, which can help in counting traffic in real-time 24/7, this data we collect not only can be used but also put it on the cloud and shared with citizens who would like to consume it.

2.1.5.7 Urban Logistics

In [15], the problem presented was to calculate driving routes in North Jutland, Denmark covering 100% of the roads in the region. This problem was part of a project focused on measuring the coverage and quality of mobile networks, driving with special equipment mounted on the roof of two cars. This paper presents the methodology used to solve the road planning problem, but more importantly, it illustrates an example of how to move from theory to a real-world practical application of graph theory and combinatorial optimization.

2.1.5.8 City Energy Consumption

With the air quality monitoring service, an urban IoT can provide a service for monitoring the city's energy consumption, allowing authorities and citizens to own a transparent and detailed view of the number of energy required by different services, transport, traffic lights, control cameras, heating/cooling of public buildings, etc.). This may successively enable the most sources of energy consumption to be identified and priorities to be set so as to optimize their behavior. This is often in line with the European directive for improving energy efficiency within the coming years. So as to get such a service, electricity consumption monitoring devices must be integrated into the city's electricity grid. Additionally, it'll even be possible to boost these services with active functionalities to manage local energy production structures (e.g. photovoltaic panels).

2.1.5.9 Smart Grids

Smart grids are systems capable of intelligently recording the actions taken by consumers/users and producers of resources (electricity, gas, water) in order to maintain an efficient, economical, sustainable and safe energy distribution. In [2], Authors aim to select the location and design of a sustainable city in Jordan. This selection was based on important indicators which are energy and water. In order to meet these criteria, they chose a location with exceptional renewable energy resources (Wind energy was used to produce electricity) and is close to the sea (Drinking water was covered by desalination).

2.1.5.10 Smart Parking

By using parking sensors, instead of driving in circles looking for a spot to park, drivers can get real-time information on an application which locates free parking spots. In fact, Intelligent Parking Service is based on road sensors and intelligent displays that direct motorists to the best route to park in the city. The benefits of this service are manifold: a shorter time to find a parking space means less CO emissions from the car, less traffic jams and happier citizens. Moreover, by using short-range communication technologies such as RFID (Radio Frequency Identifier) or NFC (Near Field Communication), it is possible to implement an electronic verification system for parking permits in parking spaces reserved for residents or persons with disabilities at the service of citizens who can legitimately use these locations and an effective tool for the early detection of offences.

2.1.5.11 Intelligent Lighting

Optimizing the efficiency of street lighting is an important feature. In particular, this service makes it possible to optimize the intensity of street lamps according to the time of day, weather conditions, and the presence of people. In order to function properly, such a service must include the street lamps in the Smart City infrastructure. It is also possible to exploit the increased number of connected points to provide Wi-Fi connectivity to citizens. In addition, a fault detection system will be easily implemented above the street lighting controllers. Let's an example of public lighting that adapts and dims when there are no activities but brightens up when sensors detect motion.

Automation and Healthiness of Public Buildings: Another important application of IoT technologies is the monitoring of energy consumption and environmental health in public buildings (schools, administrative offices, and museums) by means of different types of sensors and actuators controlling lights, temperature, and humidity. By controlling these parameters, it is indeed possible to improve the comfort level of people living in these environments, which can also have a positive return in terms of productivity, while reducing heating/cooling costs.

2.2 Blockchain

2.2.1 History

BitTorrent is a file-sharing network had developed in the early 2000s, it is arguably the first decentralized application who have been created. BitTorrent allows anyone to share any kind of file with anyone else in the world, allowing people to distribute the content quickly and easily even if they do not have the resources to pay for their own website or server.

Five years later, Satoshi Nakamoto came up with the idea of Blockchain, which is a sort of distributed database and used it to build Bitcoin, the first decentralized currency [16]. So decentralized currency like Bitcoin allows people to send money instantly anywhere around the world with the regard for national borders was negligible fees, Bitcoin is increasingly being used there for international remittances, micropayments and commerce online.

The objective of Blockchain is to deploy a Peer-2-Peer network that keeps tracking the occurrence of events. For example, if Alice wants to send any amount of money to Bob, the Bank guarantees the transaction so Alice and Bob must trust the third party which is the Bank to ensure the transfer. The problem here is that the financial institution may be malicious. Blockchain can resolve this problem by implementing a P2P network that does not rely on a third party to validate the transactions, which means that Blockchain is a separate and distributed network of nodes that all communicate with each other directly.

Blockchain can store any kind of data not only money transaction. So, decentralized applications can be created for finance, cloud computing, messaging and distributed governance. Ethereum is a platform that is specifically designed for people to build these kinds of Decentralized applications. In 2013, Ethereum was proposed by Vitalik Buterin, and the network went live on 30 July 2015, with 72 million coins premised [17]. The Ethereum Virtual Machine (EVM) can execute Turing-complete scripts and run decentralized applications. The Ethereum clients which we are calling the Ether-browser when court ability and P2P network for sending messages and a generalized Blockchain with a built-in programming language allowing people to use the Blockchain for any kind of decentralized application that they want to create.

2.2.2 Blockchain Categories

There are two main categories concerning Blockchain, with permission (private) and without permission (public). The first category imposes restrictions on consensus contributors. Only those from trusted and selected have the right to validate transactions. It does not require a lot of computation to reach a consensus, so it is economical in terms of execution time and energy. Usually transactions are private and are only accessible by authorized objects. The second category (public Blockchain) uses an unlimited number of anonymous objects. Based on cryptography, each actor can communicate in a secure way. Each object is represented by a key pair (public/private) and has the right to read, write and validate transactions in the Blockchain. The Blockchain is secure if 51% of the objects (or more) are honest and when the network consensus is reached. Usually Blockchains without permission consume a lot of power and time, as they require an amount of computation to enhance system security (e.g. using PoW).

2.2.3 Cryptography

2.2.3.1 Hash Function

A fingerprint (or hash) is a value obtained after applying a checksum (a mathematical calculation called a hash function) to a starting datum. The hash function is such that a small change in the input data produces a completely different hash. For example, the hash function is such that a tiny change in the starting data produces a completely different fingerprint than the first one.

Due to the nature of this hash function, it is impossible to guess the origin of a hash: all current cryptography relies on this inability to reproduce the original data from a hash. Having said that, let's return to the 3rd principle of the Blockchain, which protects the immutability of the latter: Modifying an element within any block would change its hash considerably. Since the next block would have to reference this fingerprint, it too would have to be modified to hide the attack; otherwise the whole network would realize that the forged block was fake and reject it. But if the attacker actually decides to modify the next block, he would have to continue until he has modified all the blocks.

Since the attacker has less strength than the rest of the network, he will always generate fewer blocks than the others and his chain would be shorter. From there, it is then simple to decide which chain is "right": The longest chain. The first principle here ensures that the attacker would take time to generate a new block, and the second principle ensures that the rest of the network will be faster than him.

2.2.3.2 Hash Pointer

The hash pointer is a pointer to where the data is stored and a summary of the data, it is just a hash that is used to reference other known information that can be used to verify the summary of the data (whether the data has changed or not). The hash pointer can be used to build data structures such as blockchain which is a linked list of hash pointers and Merkle tree which is a binary tree of hash pointers.

2.2.3.3 Digital Signature

The digital signature is another component of the blockchain. It uses public-key cryptography to ensure the integrity, non-repudiation, and authenticity of a message and its source [18]. A message signed with a digital signature can be verified by other users, but the message can only be signed by the owner of the signature. Digital signatures can be created by a public-key, this public-key uses a key combination of public and private keys. The private key is saved only by the owner while the public key is distributed to other users. Other users can encrypt the message with the owner's public key, and the message can only be decrypted by the owner with his or her private key.

Blockchain uses a digital signature algorithm such as the Elliptic Curve Digital Signature Algorithm (ECDSA) to generate the digital signature [19]. It involves three steps to create, sign and verify the message with the digital signature [18]. The secret key (SK) and the public key (PK) are generated by the "generate Keys" method. The SK is kept only by the owner and the PK is distributed over nodes of the block chain. The message is signed using the SK. The signature method takes the SK and the message as input and generates the signature of the message. This signature can be verified with nodes using the verification method which takes PK, message and signature as input. If it is true, the message is verified, otherwise it is invalidated. Thus, the public key guarantees that the message was created by the owner of the signature and with the message verification, the identity of the user is verified. The public key is therefore used as the user identity in the blocking chain. By using the distributed blocking chain, users do not need to provide their social security number, phone number or email address to a central server or authority. They can create their digital identity themselves and distribute their public key over the distributed network. This allows users to benefit from anonymous, decentralized and distributed identity management.

2.2.3.4 Merkle Trees
Merkle trees are a data structure based on binary trees. Unlike a binary tree, only leaves are used to store data and the leaf nodes are the only children of their parents. The parent leaf nodes contain a hash of their leaf nodes. The nodes of the next generation, however, each have two child nodes and contain a hash of the hash their child nodes contain. If the tree is unbalanced because there is an uneven amount of leaves, the solitary leaf hash will be used twice. The advantage of this data structure is that the integrity of the entire tree can still be verified even if redundant leaves have been removed. An example of a Merkle tree can be seen in Fig. 2.2.

Fig. 2.2 Merkle trees

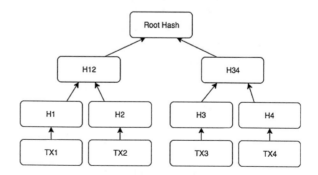

2.2.4   Distributed Consensus

In Blockchain, each node has a copy of the code and can have a different fork. So it is necessary that all nodes in the distributed network agree, at some point, on the correct version.

2.2.4.1 Proof of Work

Bitcoin uses an consensus algorithm called proof of work [16]. The general idea is to prove the validity of your ledger by performing calculations to obtain a result that fulfills certain requirements. In the case of Bitcoin, the calculation targets find a nonce which, when hashed, returns a hash value less than a global target value defined. Due to the unidirectional nature of hash algorithms, where it is impossible to compute a value that would result in an acceptable hash, finding's optimal method of calculating such values is simply to iterate through the numbers and hash them. This means that the only way to increase the chances that finding has the right properties is to increase the hash power of the computer. Since each node in the network that is being operated works on its own proof of work, they must agree on the choice of computer. Therefore, nodes that have found an acceptable hash, broadcast it and the other nodes in the network can easily verify that it is correct by performing the calculation themselves. If a longer valid chain of blocks is propagated to the nodes, the nodes always choose this one instead of their current chain. The longest block chain, which is also the one that requires the most computational work, will eventually be accepted as the main block chain by each node in the network.

2.2.4.2 Proof of Stake

In 2012, another method proposed by S.King et al. called Proof of-stake [20]. It is defined because the number of Ncoin coins multiplied by the number of time units a user has had Tcoin in his possession. To generate a new block, the user must perform a transaction called a "coinstake". The sender and receiver of this transaction are the same, i.e. the user himself. Instead of working on a target value using computing power, the target is reached using the age of the coin. The higher the age of the part, the faster the block is likely to be generated. In addition, the chain of blocks with the highest number of years of life of the invested parts is considered valid. Thus, we end up with a system entirely regulated by stakes and the authors claim that this could replace proof of work.

2.2.5   Smart Contract

The concept of smart contract was first theorized by Nick Szabo in 1994, in a scientific paper soberly entitled "Smart contracts" [21]. So we can say that a smart contract is an agreement between several parties in the form of computer code. They are distributed and therefore stored in a public ledger and cannot be modified after we deployed it. This smart contract can allow transactions to be carried out automatically without having recourse to a third party, thus not depending on anyone. The Ethereum blockchain is the most widely used to deploy smart contract and they are based on a programming language called Solidity.

2.2.6 Bitcoin

Bitcoin is a digital payments system based on a public blockchain. It allows you to create a cryptocurrency called Bitcoin. Each block of the Bitcoin blockchain contains a hash of its transactions called the root Merkle stored in its header [22]. The latter also contains the hash of the header of the previous block. Each participant in the Bitcoin network can be a minor or not, and stores a copy of the current blockchain. In the mining operation, transactions are ordered and time-stamped, and then stored in blocks. Then a consensus mechanism is executed. Indeed, in order to validate the transactions, Bitcoin uses its own rules. Specifically, transactions have version numbers that tell Bitcoin objects the appropriate set of rules that must be used to validate them. To share the same blockchain and avoid conflicts between minors, Bitcoin uses the longest chain rule. A conflict occurs when multiple (competing) miners generate blocks at the same time, and each of those miners considers his or her block to be the legitimate block that should be added to the blockchain.

2.2.7 Ethereum

Ethereum is an open-source protocol for a distributed network based on block chain technology. Ethereum supports both Smart Contracts and simple monetary transactions of the Ether protocol, a cryptographic currency. There are accounts belonging to third parties, which are controlled by private keys, and contract accounts, which are controlled by the code of the Smart Contracts mentioned earlier in Ethereum.

2.2.7.1 Solidity
Solidity is a high level language [23], static, complete and contract-oriented, created and developed by the Ethereum project. The Solidity code runs in an isolated environment known as the Ethereum Virtual Machine (EVM) [24]. The executed code has no access to any other processes on the machine and has limited access to other intelligent contracts. The contracts written in Solidity have some similarities with objects in object-oriented languages such as Java [25]. Contracts support state variables as well as functions and events, etc. To deploy contracts written in Solidity in the block chain, they first must be compiled in bytecode, which is done with the solc compiler. The contracts can be deployed in bytecode, which is done with the solc compiler.

2.2.7.2 Gas
Each externally held account has a balance that is measured in ether. Each transaction, including contract functions or deployments, costs a certain amount of gas to perform. Users sending the transaction can specify how much Ether they are willing to pay per gas but if no price is specified then the current average price of gas will be used.

This decouples the cost of the transaction from the volatile cryptographic currency and also creates an auction-like system where users willing to pay more will have their transactions processed faster [26].

2.2.8 Vulnerabilities

The Blockchain is faster and cheaper than a centralized system because of its decentralized and distributed design. Although it is reliable and secure due to its consensus protocol, cryptography and anonymity, but it still has some potential vulnerabilities. Sybil attack is one of the most famous attacks that threaten the Blockchain. In fact, Blockchain has no central authority to manage the identity of participants [27]. As a result, the attacker can create multiple copies of himself, which may look like separate participants although they are all controlled by the same node. So, other nodes are likely to connect only to the attacker's nodes. The attacker can then refuse to relay the blocks and transactions of others, disconnect the connecting node from the network or relay only the blocks created by himself. This attack can be avoided by trusting only the blockchain with the best proof of work, as it cannot be easily falsified due to the mining power requirement of significant.

Another known attack is Identity Theft, even if blockchain ensures user owner ship, the private key that supports this digital identity must be secure and kept private. If this private key is stolen or the device that stores it is hacked, the victim will lose all their digital assets as well as their digital identity. Furthermore, this digital identity will not be recoverable and it will be almost impossible to find the perpetrator. There are various applications that encrypt and synchronize private keys on different devices in order to recover the private key. But if these applications contain malicious code or are hacked, the user is again faced with identity theft. In addition, synchronizing keys between multiple devices increases the risk of hacking. In addition, with the development of quantum computing, it may become possible to crack the cryptographic keys used by block string technology [28].

In addition, recording the data on the Blockchain can be safer because immutability is one of the Blockchain features that can make it powerful, in other words data on the Blockchain is unchangeable whenever we deployed it we cannot modify or change the code. However, the code base and system that implements the Blockchain can be modified because, depending on the company or organization, the code may be open source [29] For example, the most popular Blockchain platforms like Bitcoin and Ethereum, are open source software. Therefore, any user can contribute to the development of these applications and if these contributors provide vulnerable code or if there is human error in the code base because of the contributor, it could eventually end up in the production system which, in turn, could cause the system to be hacked.

The immutability and pseudonymity of Blockchain transactions can make it difficult to track and monitor. Hence, the system can be misused for money laundering, illegal movement of funds. For example, Silk Road, a website to buy and sell illegal drugs used bitcoin for its payments [30].

3   IoT Challenges

3.1   Privacy

When billions of sensors around the world constantly collect data about their environment, which includes human beings, privacy concerns in the IoT world take center stage. Most of the developed world has attempted to protect consumers from the illicit use of confidential information, but in many cases legislation is not adequate to address the multitude of new ways in which information is acquired and used. The recent attempt by the EU to update the law on the protection of intellectual property rights is a symptom of the obsolescence of many laws in the developed world. At an earlier stage in the development of the Internet, consumers became familiar or even completely accustomed to tracking files, also known as cookies. Knowing that there was no law restricting the use of cookies by websites to track users' browsing behavior, many companies simply adopted this practice without really taking into account users' concerns. In fact, browsers have responded to these concerns with tools that restrict the use of cookies and delete them at the end of a browsing session.

European laws now govern how cookies are used and what types of data they are allowed to collect from users, however with the advent of mobile technology, plenty of these laws are becoming inadequate in the IoT field. Similarly, the United States is relying on older model legislation for new IoT devices and systems. Despite that, there is no federal law governing the use of personal data. Instead, the U.S. relies on a patchwork of federal and state laws to protect consumer privacy. Public outrage at the federal government, particularly the National Security Agency, for its "data-mining" activities related to law enforcement and counter-terrorism, bodes well for future public policy debates.

3.2   Cyber Security

Cybercrime is a major danger for today's businesses. According to one estimate, cybercrime costs businesses 400 billion USD annually. From IoT point of view, the most worrying aspect is the fact that cybercriminals break into systems that are apparently secured by several layers of protection. The complexity of securing IoT devices remains an area for improvement for businesses, especially in preparation for the day when the "IoT ecosystem" will emerge in which billions of objects will be connected to the Internet and to each other. We must keep in mind that any device with an Internet connection is a potential pathway for a hacker. For example, in 2014 a hacker managed to break into a baby monitor to harass a two-year-old child. Follow-up research on the product, manufactured by the Chinese-based company Focsam, revealed that of the 46,000 systems on the market, 40,000 had not received the security update that would have prevented the intrusion. We must also bear in mind that the more we automate and connect certain systems, especially industrial ones, the more vulnerable we make them to hacking. A city building a smart power grid can save a lot of money by streamlining problem solving. At the same time, this same system allows a potential hacker to easily interrupt the power supply to an entire city from his computer.

Concerning security requirements for IoT there is Data privacy, confidentiality and integrity along with Authentication, authorization and accounting also Availability of services and energy efficiency.

In [31] they categorized security levels in 3 main levels: Low-level security issues, Intermediate-level security issues and High-level security issues. The first level is concerned physical and data link layers of communication and hardware level (Jamming adversaries, insecure initialization, Low-level Sybil and spoofing attacks, insecure physical interface, Sleep deprivation attack). The second level is related to communication, routing and session management at transport layers of IoT (Replay or duplication attacks due to fragmentation, Insecure neighbor discovery, Buffer reservation attack, RPL routing attack, Sinkhole and wormhole attacks, Sybil attacks on intermediate layers, Authentication and secure communication, Transport level end-to-end security, Session establishment and resumption, Privacy violation on cloud-based IoT). The third level which is the high-level security issues it related to applications that are executed on IoT such as (CoAP security with internet, insecure interfaces, insecure software/firmware, and Middleware security).

3.3   Responsibility

When it comes to autonomous vehicles such as unmanned cars, we are faced with an ethical dilemma: seconds before an accident, should an autonomous vehicle do everything it can to protect its passengers, even if this means harming other motorists or pedestrians?

A human being in danger cannot be held responsible when his or her survival instinct causes him or her to swerve toward a pedestrian. But when machines make the decisions, is a pedestrian injured in an accident entitled to blame the car manufacturer? Does a driver have the right to sue a car manufacturer after an accident in which the driver is injured? As pointed out in a European Commission report on the ethical dilemmas inherent in IoT technology, "People are not used to objects with their own identity or acting on their own, especially if they act unpredictably". With billions of devices collecting data, it becomes more difficult to know who is responsible for what data. IoT objects operate autonomously and in association with many other objects. Data is quickly shared, processed, shared again and processed before it can be seen by human eyes. In other words, it is too simple to associate a device with a unit of data, because too much of the potential of the IoT lies in the smooth transfer of this data between objects. For example, an IoT cardiac monitoring device will not simply monitor a patient's heart for signs of an impending heart attack. It is likely to access data from another object monitoring the patient's sports activities, which in turn uses data from a device monitoring his or her diet.

3.4 Energy Consumption in WSNs

Sensor networks can be considered as a remarkable technology to activate the IoT. They can shape the world by providing capabilities to measure, infer and understand environmental indices. The current development and improvement of technologies has provided effective and inexpensive devices for application to large-scale remote sensing uses. In addition, smartphones contain different types of sensors and therefore enable different types of mobile uses in different areas of the IoT. To this end, the main challenge is how to analyze the power consumption characteristics of a wireless sensor node. This systematic analysis of the energy of a sensor node is extremely important to identify problems in the energy system to enable effective optimization. The energy consumption of a sensor is as a result of the following operations: detection, processing and communication.

- Capture Energy: The sources of energy consumption of nodes for detection or capture operations are: sampling, analog-to-digital conversion, signal processing and activation of the capture probe.
- Processing energy: Process energy is composed of two kinds of energy: switching energy and leakage energy. The switching energy is determined by the supply voltage and the total capacity switched at the software level (when running software). The leakage energy, on the other hand, is the energy consumed when the computing unit does not perform any processing. In general, the processing energy is low compared to the energy required for communication.
- Communication energy: Communication energy is divided into three parts: reception energy, transmission energy, and standby energy. This energy is determined by the amount of data to be communicated and the transmission distance, as well as by the physical properties of the radio module. The transmission of a signal is characterized by its power; when the transmission power is high, the signal will have a long range and the energy consumed will be higher. Actually, communication energy considered as the largest portion of the energy consumption of a sensor node.

4 Blockchain with IoT

As we know Blockchain technology has an important role in IoT security solutions, so Blockchain uses elliptic curve cryptography (ECC) and SHA-256 hash function in order to provide data confidentiality and integrity. In fact, Blockchain is a solution that provides a shared ledger technology that allows any participant in the network to see the one system of ledger and the block data contains a list of all transactions and a hash to the previous block. And each transaction in the public ledger is verified as a majority consensus of miner nodes and the blocks of data are immutable which means it could not be altered or erased. So a blockchain design contains a header block (Version, timestamp, block size, and the number of transactions, Merkle root, the nonce, the difficulty target), and the block body contains the list of transactions [31]. Moreover Bitcoin considered as an application that runs on the top of blockchain infrastructure. Also, Ethereum blockchain implements smart contracts which means it store, record and run smart contract, and recently other smart contract blockchain platforms have been emerged (Hyperledger, Eris, Stellar, Ripple, and Tendermint). In addition, there are various applications of smart contract blockchain like trading to autonomous machine-to-machine transactions, asset tracking to automated access control and sharing, digital identity and voting to certification, management, and governance of records and data.

IBM launched its blockchain framework and it is used in banks, supply chain systems, and cargo shipping companies. Here are some useful features of blockchain that can be beneficial for IoT security devices: First, Address space: 160 bit (20 bytes) address space so 160 bit hash of public key, this public key generated by using ECDSA (Elliptic Curve Digital Signature Algorithm). Second, Identity of Things (IDoT) and Governance: IoT device owner could be changed in addition to other attributes of the device such as type, serial number, deployment GPS coordinates, and location and so on, also the device could have a relationship with human or other devices or services. So blockchain comes up with a solution to those challenges of identity and access management by providing trustworthy and authorized identity registration, ownership tracking and monitoring of products, goods, and assets. The approaches like TrustChain in order to have secure transactions and giving identity to the connected IoT device with the management and governance of the device life cycle. In addition, Data authentication and integrity: all the data transmitted in the blockchain network are e cryptographically proofed and signed by the true sender that has a unique public key and GUID. Also, Authentication, Authorization, and Privacy: with smart contacts single and multiparty authentication to an IoT Device is provided along with authorization access rules and data privacy, Finally, Secure Communications: the protocols used in communication and routing between IoT devices are not secure enough (MQTT, CoAP, XMPP, RPL, 6LoWPAN) that's why we used them with other security protocols like DTLS and TLS. But with blockchain, there is no need for those protocols since every IoT device has its own unique GUID and asymmetric key pair [31].

Concerning IoT security requirements there is Data privacy, confidentiality and integrity along with Authentication, authorization and accounting also Availability of services and energy efficiency. Bahga et al. [32] In their survey they propose a blockchain-based framework for industrial IoT (or IIoT). That is used to communicate with the cloud and the blockchain network. Christidis et al. [33] they also discussed the benefits of blockchain for IoT and propose a scenario where blockchain can facilitate the buying and selling of energy automatically among IoT device like smart meters.

Minhaj et al. [31] in their survey they discuss some security issues and security requirements, they also mention the use of Blockchain to solve security problems. So they categorized security levels into 3 main levels: Low-level security issues, Intermediate-level security issues and High-level security issues. The first level is concerned physical and data link layers of communication and hardware level (Jamming adversaries, insecure initialization, Low-level Sybil and spoofing attacks, insecure physical interface and Sleep deprivation attack). The second level is related to communication, routing and session management at transport layers of IoT (Replay or duplication attacks due to fragmentation, Insecure neighbor discovery, Buffer reservation attack, RPL routing attack, Sinkhole and wormhole attacks, Sybil attacks on intermediate layers, Authentication and secure communication, Transport level end-to-end security, Session establishment and resumption and Privacy violation on cloud-based IoT). The third level which is the high-level security issues is related to applications that are executed on IoT (CoAP security with the internet, insecure interfaces, insecure software/firmware, and Middleware security).

Francesco Buccafurri et al. [34] In their survey they focus on authentication problems and integrate Blockchain technology to enhance authentication for MQTT protocol. Their solution adopts the use of Blockchain as another channel with the use of Ethereum to enforce security without using TLS by using hashes as user pseudonyms in smart contracts. Also using Blockchain here is about to have a trusted public ledger without the need of a third party, the client sends a CONNECT message to the broker and the authentication process offers two fields to transmit username and password to the broker but the fields are not encrypted which make it exposed to attacks. To solve this problem they use One Time Password (OTP) which means a password valid for only a session but it cannot be sent in the same channel that's why they use Blockchain as another channel to implement two-factor authentication.

IoT technology relies on the communication of devices, So IoT devices need to interact with others, and we are talking about thousands of devices. The use of the basic model of a server-client may have some limitations, and it's not safe because this model is based on a centralized system where the server and a local database contain all the code, so anyone has access to it can manipulate or change the code.

Since the basic model faces some limitations because of the growing number of IoT devices, Seyoung Huh et al. [35] proposed a model that is based on Blockchain technology to monitor and control IoT devices. They used RSA algorithm to manage keys, Ethereum platform is used here to store public keys while private keys are protected in the IoT devices. Ethereum is one of the biggest cryptocurrency that supports smart contracts, so the smart contract contains all the code of the Turing-complete that runs on Ethereum so we can control IoT devices.

Agrawal et al. [36] in their survey they present a solution for providing continuous security in IoT based on the distributed nature of Blockchain through IoT-Zone identification. So each user in an IoT system is stored as a node on a Blockchain network, and each interaction of this node considered as a transaction. In order to make this transaction legitimate, it requires a unique crypto-token to avoid unauthorized access which makes the system more secure and safe. Three main phases are needed in this process, IoT-Zone identification by the activation of user IoT trails, IoT-token generation where the permission is checked with the Enrollment Certificate Authority (ECA), then the IoT-token is validated.

5 Conclusion

In this chapter, we presented a study on the importance of blockchain technology in providing a secure environment for IoT users. So, we began our chapter with an overview of IoT and blockchain. And then, we discussed some challenges that might IoT systems face. Finally, we discussed some security issues, security requirements also the use of blockchain to solve those security problems.

To sum up, some challenges are facing the implementation of security in IoT devices such as resource limitations, Heterogeneous devices, Interoperability of security protocols, single points of failure, Hardware/firmware vulnerabilities, Trusted updates and management, and also some blockchain vulnerabilities because the mechanism that depends on miner's hashing can be compromised which allows attackers to host the blockchain. Research on the application of blockchain technology in IoT and smart environments is quite extensive, and there are many challenges awaited them. This chapter shortly introduces how blockchain technology can be used to solve security problems in IoT systems. We hope that this discussion and exploration can pave a new path for the development and implementation of IoT.

References

1. Nordrum, A.: The internet of fewer things [news]. IEEE Spectr. **53**(10), 12–13 (2016). https://doi.org/10.1109/MSPEC.2016.7572524
2. Alkhalidi, A., Qoaider, L., Khashman, A., Al-Alami, A.R., Jiryes, S.: Energy and water as indicators for sustainable city site selection and design in Jordan using smart grid. Sustain. Cities Soc. **37**, 125–132 (2018). https://doi.org/10.1016/j.scs.2017.10.037
3. Zanella, A., Bui, N., Castellani, A., Vangelista, L., Zorzi, M.: Internet of Things for smart cities. IEEE Internet Things J. **1**(1), 22–32 (2014). https://doi.org/10.1109/JIOT.2014.2306328
4. Farahat, I.S., Tolba, A.S., Elhoseny, M., Eladrosy, W.: Data security and challenges in smart cities. In: Hassanien, A.E., Elhoseny, M., Ahmed, S.H., Singh, A.K. (eds.) Security in Smart Cities: Models, Applications, and Challenges, pp. 117–142. Springer, Cham (2019)
5. Dabbagh, M., Sookhak, M., Safa, N.S.: The evolution of blockchain: a bibliometric study. IEEE Access **7**, 19212–19221 (2019). https://doi.org/10.1109/ACCESS.2019.2895646
6. Sharma, P.K., Chen, M., Park, J.H.: A software defined fog node based distributed blockchain cloud architecture for IoT. IEEE Access **6**, 115–124 (2018). https://doi.org/10.1109/ACCESS.2017.2757955
7. Ashton, K.: That 'Internet of Things' thing. RFID J. **22**(7), 97–114 (2009)
8. Focus sur l'Internet of Things (IoT)... l'essentiel à savoir. https://www.welcometothejungle.com/fr/articles/focus-sur-l-internet-of-things-iot-l-essentiel-a-savoir. Accessed 08 July 2020
9. L'Internet des Objets. Effleurant la surface | Guido Noto La Diega, PhD - Academia.edu. https://www.academia.edu/10928378/LInternet_des_Objets._Effleurant_la_surface. Accessed 08 July 2020
10. Bitaillou, A., Parrein, B., Andrieux, G.: Synthèse sur les protocoles de communication pour l'Internet des objets de l'industrie 4.0, LS2N, Université de Nantes; IETR, Université de Nantes, Technical Report, January 2019. https://hal.archives-ouvertes.fr/hal-02365063. Accessed 08 July 2020
11. Challal, Y.: Sécurité de l'Internet des Objets: vers une approche cognitive et systémique. Thesis, Université de Technologie de Compiègne (2012)

12. Pallec, S.L.: La convergence des identifiants numériques, p. 12 (2005)
13. Hui, T.K.L., Sherratt, R.S., Sánchez, D.D.: Major requirements for building Smart Homes in Smart Cities based on Internet of Things technologies. Future Gener. Comput. Syst. **76**, 358–369 (2017). https://doi.org/10.1016/j.future.2016.10.026
14. Al-Dulaimi, J., Cosmas, J.: Smart safety & health care in cities. Procedia Comput. Sci. **98**, 259–266 (2016). https://doi.org/10.1016/j.procs.2016.09.041
15. Gutierrez, J.M., Jensen, M., Riaz, T.: Applied graph theory to real smart city logistic problems. Procedia Comput. Sci. **95**, 40–47 (2016). https://doi.org/10.1016/j.procs.2016.09.291
16. Nakamoto, S.: Bitcoin: a peer-to-peer electronic cash system, Manubot, November 2019. https://git.dhimmel.com/bitcoin-whitepaper/. Accessed 20 August 2020
17. Buterin, V.: Ethereum: Platform Review, p. 45 (2016)
18. Wang, L., Ohta, K., Kunihiro, N.: Near-collision attacks on MD4: applied to MD4-based protocols. IEICE Trans. Fundam. Electron. Commun. Comput. Sci. **E92-A**(1), 76–86 (2009)
19. Information Technology Laboratory: Digital Signature Standard (DSS), National Institute of Standards and Technology, NIST FIPS 186-4, July 2013. https://doi.org/10.6028/NIST.FIPS. 186-4
20. King, S., Nadal, S.: PPCoin: peer-to-peer crypto-currency with proof-of-stake, p. 6 (2012)
21. Mohanta, B.K., Panda, S.S., Jena, D.: An overview of smart contract and use cases in blockchain technology. In: 2018 9th International Conference on Computing, Communication and Networking Technologies (ICCCNT), Bangalore, pp. 1–4, July 2018. https://doi.org/ 10.1109/ICCCNT.2018.8494045
22. A Digital Signature Based on a Conventional Encryption Function. https://link.springer.com/ chapter/10.1007/3-540-48184-2_32. Accessed 26 Aug 2020
23. Solidity — Solidity 0.7.1 documentation. https://solidity.readthedocs.io/en/latest/. Accessed 15 Aug 2020
24. Introduction to Smart Contracts — Solidity 0.4.22 documentation. https://solidity.readthedocs. io/en/v0.4.22/introduction-to-smart-contracts.html. Accessed 15 Aug 2020
25. Structure of a Contract — Solidity 0.4.21 documentation. https://solidity.readthedocs.io/en/v0. 4.21/structure-of-a-contract.html. Accessed 15 Aug 2020
26. Ether — Ethereum Homestead 0.1 documentation. https://ethdocs.org/en/latest/ether.html. Accessed 15 Aug 2020
27. Bitcoin and Cryptocurrency Technologies, PDFDirectory.com. https://pdfdirectory.com/765-tutorial-bitcoin-and-cryptocurrency-technologies.pdf. Accessed 15 Aug 2020
28. Underwood, S.: Blockchain beyond bitcoin. Commun. ACM **59**(11), 15–17 (2016). https://doi. org/10.1145/2994581
29. Xu, J.J.: Are blockchains immune to all malicious attacks? Financ. Innov. **2**(1), 25 (2016). https://doi.org/10.1186/s40854-016-0046-5
30. Hong, N.: Silk Road creator found guilty of cybercrimes, MarketWatch. https://www.mar ketwatch.com/story/silk-road-creator-found-guilty-of-cybercrimes-2015-02-04-151035739. Accessed 15 Aug 2020
31. Khan, M.A., Salah, K.: IoT security: review, blockchain solutions, and open challenges. Future Gener. Comput. Syst. **82**, 395–411 (2018). https://doi.org/10.1016/j.future.2017.11.022
32. Bahga, A., Madisetti, V.K.: Blockchain platform for industrial Internet of Things. J. Softw. Eng. Appl. **9**(10), Art. no. 10 (2016). https://doi.org/10.4236/jsea.2016.910036
33. Christidis, K., Devetsikiotis, M.: Blockchains and smart contracts for the Internet of Things. IEEE Access **4**, 2292–2303 (2016). https://doi.org/10.1109/ACCESS.2016.2566339
34. Buccafurri, F., De Angelis, V., Nardone, R.: Securing MQTT by blockchain-based OTP authentication. Sensors **20**(7), 2002 (2020). https://doi.org/10.3390/s20072002
35. Huh, S., Cho, S., Kim, S.: Managing IoT devices using blockchain platform. In: 2017 19th International Conference on Advanced Communication Technology (ICACT), Pyeongchang, Kwangwoon Do, South Korea, pp. 464–467 (2017). https://doi.org/10.23919/ICACT.2017.789 0132
36. Agrawal, R., et al.: Continuous security in IoT using blockchain. In: 2018 IEEE International Conference on Acoustics, Speech and Signal Processing (ICASSP), Calgary, AB, pp. 6423–6427, April 2018. https://doi.org/10.1109/ICASSP.2018.8462513

Blockchain and the Future of Securities Exchanges

Zachary A. Smith, Mazin A. M. Al Janabi, Muhammad Z. Mumtaz, and Yuriy Zabolotnyuk

Abstract In this paper, we analyze blockchain technology as an alternative to facilitate securities market transactions. We argue that the ability to lower transaction costs combined with the reduction of intermediaries and improvements in transaction efficiency will promote the implementation of blockchain technology in these markets. We use platform economics to illustrate how blockchain-based securities exchanges can reduce unfair rents that platform operators extract from the investing public under the current regulatory regime, and indicate how the economics of platforms might be used to implement a blockchain solution to enhance the exchange process. We argue that in order to reduce transaction costs and increase the potential for innovation, blockchain platforms should be moved from the supply-side to an intermediary position and operate as a neutral party that straddles both sides of the market. Finally, we find that open routing procedures are likely to positively impact the market and enhance the efficiency of securities market transactions.

JEL Classification G20 · G23 · G29

Keywords Blockchain · Securities exchanges · Securities markets · Platform economics

Z. A. Smith
Saint Leo University, St. Leo, Florida, USA
e-mail: Zachary.Smith@saintleo.edu

M. A. M. Al Janabi (✉)
Tecnologico de Monterrey, EGADE Business School, Mexico City, Mexico
e-mail: mazin.aljanabi@tec.mx

M. Z. Mumtaz
National University of Sciences and Technology (NUST), Islamabad, Pakistan
e-mail: zubair@s3h.nust.edu.pk

Y. Zabolotnyuk
Carleton University, Ottawa, Ontario, Canada
e-mail: yuriy.zabolotnyuk@carleton.ca

Y. Maleh et al. (eds.), *Artificial Intelligence and Blockchain for Future Cybersecurity Applications*, Studies in Big Data 90,
https://doi.org/10.1007/978-3-030-74575-2_5

1 Introduction

Security exchanges based on blockchain technology can offer a viable alternative to the currently organized securities markets. In the recent years, blockchain technology garnered a lot of attention after it was successfully implemented in cryptocurrency markets, most notably with Bitcoin. Only recently, Bitcoin and the idea that a cryptocurrency, built on a distributed ledger network, could offer the world an alternative to traditional forms of currencies that are controlled by central banks was once reserved for "nerds, libertarians and drug dealers" [31]. However, more and more financial markets are adopting the blockchain technology and it is only a matter of time when securities exchanges will follow the suit. For example, in September 2016, the Australian Securities Exchange (ASX) released a request for consultation paper that communicated their desire to receive assistance in analyzing and evaluating their current business requirements and potentially transition from their current order processing, clearing, and settlement system to a new system that is built on the blockchain technology [3]. ASX had begun discussing the implementation of an alternative to the CHESS system, which they currently use to process, clear, and settle trades on their exchange, at the end of 2017. ASX plans to have the blockchain system operational by 2023.

To illustrate how blockchain technology can help to facilitate securities transactions, one can think about what happens when a transaction is made over a traditional securities market. Over traditional financial exchanges, the investor calls a broker or inputs an order, the broker routes that order to a specialist or a market maker, and the market maker matches a buy and a sell order to complete the transaction. After this occurs, the shares and the purchase price associated with those shares are exchanged through an intermediary. This process may require the presence of multiple intermediaries and may, potentially, suffer from cost and time inefficiencies. Fortunately, Bitcoin highlighted how a distributed network ledger system, commonly referred to as blockchain technology, could be used to help facilitate market transactions without the presence of an intermediary and execute these transactions in a more efficient manner (i.e., direct transactions, which are unencumbered by an intermediary).

Most innovative uses of technology introduce new questions and challenges in terms of adoption and we would be remiss if we did not shed light on some of these potential issues before highlighting how advances in blockchain-based technologies could revolutionize trading in securities markets. According to Chang et al. [10], the primary issues that delay the implementation of a blockchain solution across securities markets on a broader scale are the following: (a) Scalability, (b) Security, (c) Privacy, and (d) Energy Consumption. Further, Guo and Liang [22] question whether disintermediation is even possible. They contend that some level of centralization is likely required to ensure that certain information is safeguarded. However, a multi-centered weakly intermediated solution is most likely to prevail. So, there are real questions regarding the scalability, security, privacy, and the energy consumption associated with the implementation of a blockchain solution, but we believe that in time markets will provide novel solutions to these problems and it is unlikely that an entirely disintermediated trading solution will arise, but movement towards this disintermediated solution will likely reduce costs and make markets more efficient.

In this paper, we discuss how the blockchain technology could be applied to securities markets to improve the user experience, potentially decrease transaction costs, and create new ways to transact across disparate markets. In our analysis, we attempt to provide, first, a reason for a change in the current structure of securities markets across the globe. Second, we provide some indication of how a securities trading market built on the blockchain infrastructure may work. Third, we parse through the literature on the economics of platforms to build a case for an additional trading platform to exist. Fourth, we illustrate how the routing of securities market transactions across the different securities exchanges is likely to affect the markets. Fifth, we examine regulatory hurdles associated with the implementation of a blockchain-based solution to securities market transaction and, sixth, we highlight some potential challenges that will need to be addressed before a blockchain solution could be implemented at scale and across the globe.

The paper is structured as follows: we review the current literature on use of blockchain in Sect. 2. Section 3 reviews the mechanics and potential cost benefits of the blockchain technology. Section 4 uses platform economics to outline rationales for blockchain adoption. Section 5 provides our vision on how the blockchain-based securities markets could be implemented. We conclude the paper in Sect. 6.

2   Literature Review

Blockchain technology has gathered a lot of public attention recently [36] because it replaces a need to hire a 'trust intermediary' to facilitate some types of exchange [21]. Researchers have explored how blockchain technology could be used to change: (a) how we transfer equity in the crowdfunding industry [51], (b) how we perceive money or currencies such as the Bitcoin and other cryptocurrencies [9, 32], (c) how we access traditional banking and financial services [1, 14, 22, 42, 50], (d) how the insurance industry operates [14, 42], (e) how notary services are provided [14], (f) how the music industry operates [14], (g) how we store data in the cloud [14], (h) how we access public records [32], (j) the future of online gambling [18], and (k) voting in proxy or political contexts [50] among other uses [49]. In the literature, few papers have taken the time to illustrate why and how the use of the blockchain technology may provide a real alternative [34] to the current structures that institutions across the globe choose to use to facilitate exchange and monitor transactions.

According to Geranio [19], secondary market transactions are likely to undergo the most significant transformation due to the introduction of blockchain technology that "will allow for a true redesign of current procedures for clearing, settlement, and custody, no more anchored to the presence of a central counterparty." The author contends that the secondary market transactions benefit from "a sort of natural monopoly granted by available technology and regulation. Blockchain could disrupt such monopoly, promoting higher efficiency, shorter duration and cost reduction in post-trading processes." The study claims that the adoption of the blockchain technology to process and settle transactions in the secondary market could reduce the cost associated with equity analysts by 25%, which is estimated to be a 7% decrease in the aggregate transaction costs on the European exchanges and bring a 15% reduction of the exchange-related costs in Australia. We believe that this significant cost reduction potential of blockchain-based exchanges will speed the transition from the traditionally organized exchanges to blockchain-based exchanges.

The potential move of securities exchanges to blockchain technology may be affected by a country's legal system characteristics. As pointed by Block et al. [5], countries with stronger regulation can lower the cost of entry and ensure contractual certainty thus encouraging development of financial technology firms. On the other hand, Hornuf and Schwienbacher [24] argue that very strong investor protection may harm financial innovations.

Lee [33] provides us with a useful analogy to consider by discussing massive changes that we have seen in the U.S. Postal Service as a result of the introduction of email as the preferred medium of communication. The most appropriate and likely route to adoption of blockchain technologies is a slower, more thoughtful and patient adoption of a blockchain solution. According to the author, a complete replacement of the institutions and systems that are currently available to broker exchange related transactions in the securities markets would engender significant institutional inertia. First, many jobs are dependent on the current structure and an ouright replacement would cause massive job market dislocations in these industries. Second, laws that regulate the securities markets would have to receive a massive overhaul, which provides a framework to transact over these new exchanges.

Andolfatto [2] raises some questions about the immutability of the blockchain network. The author also looks at whether the blockchain or distributed ledger system is worth it and if the traditional trust-based methods of transacting are superior to these more consensus-driven solutions. Berentsen and Schar [4] illustrate that through a consensus-based framework where the miners in a system (i.e., auditors) are incentivized to reach agreements about additions to the blockchain because that is how they generate compensation and that this leads to efficiency. However, according to the authors "mining is expensive, as the computations use large amounts of electricity and are increasingly dependent on highly specialized hardware." Andolfatto [2] and Berentsen and Schar [4] also look at the potential benefits associated with a blockchain solution to securities market transactions, and question whether the benefits that society potentially may receive from a system built on the blockchain network would outweigh the costs from an economic and social perspective. Further, Cai [8] suggests that advances in blockchain technologies will likely be applied by financial intermediaries to obtain further cost advantages, which they may share with their clients, but their applications of these advances in technology will likely help them to retain their monopolistic position.

Set against this background, one of the goals of this paper is to examine the current state of securities exchanges with a specific emphasis on the dynamics of supply and demand in terms of processing, settling, and clearing trades and apply these ideas globally. In addition, we would like to highlight the weaknesses of the current regulatory environment and identify potential changes that would have to be initiated to facilitate a movement towards a more efficient and effective means to facilitate exchange.

3 Blockchain and Distributed Ledger Technology

3.1 Structure of Blockchain-Based Transactions

Before we start to imagine the future of blockchain-based securities exchanges, we should focus briefly on the underlying structure. Three major features of a blockchain ledger are openness, decentralization, and continuous competitive record validation. In an open public blockchain ledger, past transactions are saved on multiple computers in a peer network. To prevent tampering with records, transactions are periodically bundled into blocks. A cryptographic hash is calculated from the contents of each block and a unique hash identifier is assigned to that block. Each block contains the most recent transactions as well as a link, in terms of the previous block's hash, to information from all past blocks. Bundling of the blocks creates a chain of blocks, or blockchain. Modifying any part of the past transaction record changes the hash related to the block where the transaction was originally recorded as well as hashes related to all subsequent blocks. Therefore, each node verifies hashes locally, but the validation of transactions is performed by comparing hash identifiers stored on multiple computers in the peer network (i.e., based on consensus throughout the network). If anyone alters any block in the chain, the hashes associated with the altered blockchain will not match the peers' records of the blockchain and, therefore, the altered record will be discovered.

In the case of the Bitcoin blockchain network, the hash identifiers must start with a certain number of zeros. To create such a hash identifier for each respective block of data, a certain number, called "nonce", is added to the block of data. Peers on the blockchain network (called miners) try to compute (or mine) the "nonce" as quickly as possible. The peer that computes the "nonce" first closes the block of data and receives some compensation in form of Bitcoins. The closed block is then added to the blockchain and distributed to network peers. If anyone wanted to alter any existing block of data, they would have to re-compute "nonces" for the respective and all subsequent blocks more efficiently than other peers which is very costly (or almost impossible) by design. The miners are, therefore, not really mining anything—they exist to validate that the blocks are accurate and that they have not been tampered with.

In Fig. 1, we illustrate how a transaction on a blockchain-based securities exchange would occur using the structure provided by Brownworth [6].

Figure 1 has two parties that are interested in exchanging money for security ownership, the supply-side and the demand-side. The supply of a security on the supply-side is verified by a distributed group of auditors or miners just as the amount

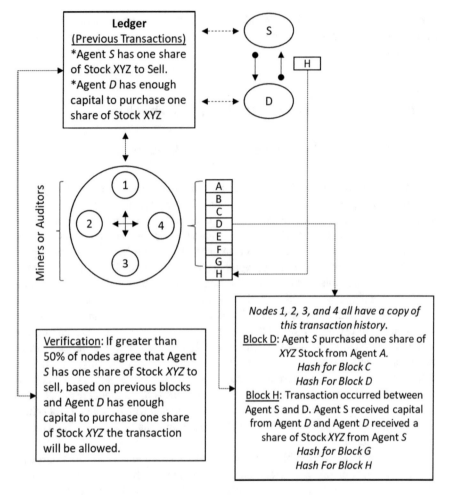

Fig. 1 Illustration of the distributed ledger technology applied to securities markets. *Source* Fig. 1 is designed by authors

of capital available to purchase the security from the demand side is verified based on previous records contained within the existing blockchain. Once agents *D* and *S* agree to transact, a record of their transaction is amended to the previous block and becomes a portion of the blockchain. From a recordkeeping and auditing perspective, the utilization of the blockchain solution seems to be a more efficient way to execute securities market transactions and a natural evolution of the centralized system would be to move to a decentralized network, which eliminates the participation of an unnecessary intermediary to complete a transaction.

It is important, at this point, to step back and address an issue of security and transparency associated with using a distributed ledger to store transaction information in recent applications of the blockchain technology to cryptocurrency markets.

Nakamoto [38] highlighted how the transactions on the Bitcoin cryptocurrency network would be secure and private by indicating that even though all transactions are part of the public record, like trades that are currently displayed in order books, "privacy can be maintained by breaking the flow of information in another place: by keeping public keys anonymous." According to Nakamoto [38], by creating and retaining a new cryptographic key pair for each transaction, an additional firewall could be added so that "the public can see that someone is sending an amount to someone else, but without information linking the transaction to anyone".

As with any new technology, issues arise that need to be rectified in the implementation process. According to Irrera [26], digital wallets that are used to store the one-time cryptographic transaction keys can be hacked, which is an issue that prevents large financial firms from using the current blockchain infrastructure. However, digital security firms are working to integrate highly secure processors that are designed to secure passwords and digital keys so that larger firms are more comfortable participating in a more distributed networked environment [26]. As the blockchain technology progresses, we believe that the security associated with transactions over a blockchain network will be similar or even stronger to security that exists in securities markets today.

Disintermediation and the innovation achieved based on advances in blockchain technology and cryptography have the power to redirect efforts in the field of finance and economics and focus energy on solving more novel problems. What was once considered critically important value-added tasks like record-keeping, auditing, brokering relationships, and other tasks that require humans to intervene on behalf of both parties that seek to enter into a contract to exchange goods or services will no longer be necessary if our societies and institutions embrace the potential power of these recent innovations [37, 41]. Thus, a more sustainable global financial regime should encourage the adoption and exploration of cases of disintermediation and implementation of blockchain technology to allow the markets to operate more efficiently. Iansity et al. [25] argue that implementation of blockchain will help to identify, validate, store, and share processes and tasks, and would enable individuals, organizations, machines, and algorithms to freely transact with one another with little friction.

In order to illustrate how blockchain-based transactions are more efficient than the traditional brokered transactions, we present the diagrams that explain the workings of each type of transaction. We present a typical brokered transaction in Fig. 2, where the securities broker's primary role in facilitating transactions is to record, route, verify, and validate the transaction. Iansiti et al. [25] argue that blockchain technology serves as an important breakthrough that can provide us with an opportunity to reduce the number of intermediaries involved in these transactions by one, which is the reduction of the securities' broker.

An example of a blockchain-based transaction is provided in Fig. 3.

The initial structures associated with these blockchain-based networks are likely to be run on private blockchains, which restrict the access to network only to authorized users but, eventually, to benefit consumers, these private blockchain networks would benefit from evolving into public blockchain networks. According to the Australian

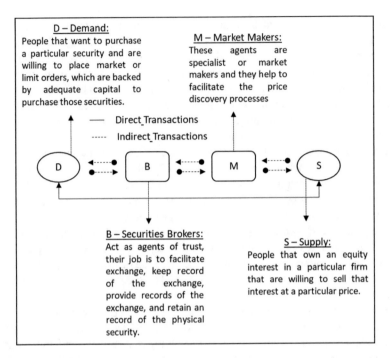

Fig. 2 Illustration of the typical brokered transaction. *Source* Fig. 2 is designed by the authors

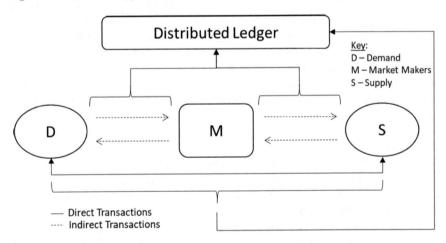

Fig. 3 Illustration of the exchange relationship using blockchain or distributed ledger technology. *Source* Fig. 3 is designed by the authors

Securities Exchange [3], the ASX plans to replace their current trading platform with a private-permissioned, secure blockchain network where only known, licensed participants would be authorized to access the system. Similarly, according to Tepper [47], NASDAQ's blockchain-based system Linq will proceed through a number of steps, and not necessarily go straight from one writer to a totally permission-less environment. In this paper, we will highlight the benefits of a permission-less environment that maintains market-makers to ensure order flow and platforms such as NASDAQ and the NYSE to allow people to access the securities markets but remove the broker who acts as an agent of trust and as a record keeper.

3.2 Blockchain and Reduction of Transaction Fees

One of the benefits of moving to blockchain-based securities markets is reduction in transaction costs. According to Tapscott and Tapscott [46], our financial "system is rife with problems, adding costs through fees and delays, creating friction through redundant and onerous paperwork, and opening up opportunities for fraud and crime." Authors argue that the system inefficiency caused by its antiquity, centralization, and exclusivity creates an opportunity to significantly reduce transaction costs.

The need to reduce cost of transacting by market participants should serve as a significant motivator for the traditional exchanges to move to the blockchain-based systems. We can illustrate how this may happen by looking at the evolution of portfolio management that moved from active management to passive strategies. The value provided by active portfolio management was early questioned by Cowles [13] who cast doubt on the assumption that beating the market was probable for people that were hard working and well-informed and found that the 'experts,' on average, could not beat a randomly drawn sample of security selections. Ellis [16] outlines the history of ever-increasing fees that advisors have historically charged to manage assets and adds that while the decreased value added combined with increased costs should move markets towards structures with lower costs, such as passive portfolio management, nevertheless, the acceptance of social innovation, in general, can be problematic since leading members of the establishment are often dismissive of *all* new ideas as they have much to lose in institutional stature, reputations as experts, and earning power.

Fanning and Centers [17] claim that the savings from transferring to blockchain-based solutions in global securities markets could exceed 20 billion dollars. Other studies, such as Khapko and Zoican [30] claim that the savings potential, based on estimates of implementing blockchain solutions into the trade settlement process for secondary market transactions globally could exceed $100 billion.

4 Platform Economics and Blockchain-Based Securities Markets

Many securities exchanges are organized as intermediated markets. Intermediated markets are different from search markets as in search markets the buyer and the seller search to transact and in intermediary markets products are bought and sold by an intermediary. Intermediaries charge a spread and are costly to use, which leads us to question whether they are needed. Gerhig [20] argues that the intermediaries offer immediacy and they can often obtain access to goods (in our case, shares) at the bid price, if purchasing, or the ask price, if selling, and generate a profit from the spread. If the spread is too high, a prospective purchaser is likely to seek an alternative to going through the intermediary and attempt to go directly to the seller and trade in the frictional market.

As the frictions inherent in the marketplace vanish so does the need for an intermediary and their market power. The central discussion in Gerhig [20] hinges on the contention of how much and to what extent organized markets reduce trading costs. On the one hand, the author finds that competition amongst intermediaries and the frictional market leads to a Walrasian outcome and pushes transaction costs to zero. On the other hand, if the intermediation business has fixed entry cost that provides a prohibitive barrier to entry, then the industrial structure is more properly defined as a natural monopoly. The intermediary acts akin to a monopolist to maximize profits but is constrained to some extent by the competition from the frictional market. Intermediation in the equity markets, in which market makers and specialists trade their own inventories to assist in price discovery, adds value. However, if intermediaries extract unfair 'profits' from engaging in this activity, it would be beneficial if the costs associated with accessing this platform, were reduced.

Blockchain will be able to change the nature of intermediation in industries where "platform operators…[are]…enjoying uncompetitive rents from their position as trusted nodes above and beyond their added value to transactions, or the privacy risk and censorship risk must be substantial" [9]. The way that the current exchange market is structured (and older markets were set up) is that there are significant barriers to entry created by regulatory bodies that purport to protect consumers. These barriers were important as markets needed an intermediary to retain a record of their transactions, to hold the security certificates that they had in 'street form' so that they could make transactions instantaneously, to provide them with quotes on the securities that they were interested in purchasing or liquidating, and to guarantee that the security that they were purchasing was what they believed it was. Under the current regime, these services would not be needed in most markets because all of the above can be incorporated into a blockchain distributed ledger system.

To further discuss how disintermediation should occur, we will focus on exploring economics associated with platforms. First, when we think about platforms, we are often interested in exploring a two-sided market in which "(i) two sets of agents interact through an intermediary or platform, and (ii) the decision set of agents affects the outcomes of the other set of agents, typically through an intermediary"

[45]. Second, when we refer to the platform economics, we are often interested in, according to Rysman [45], exploring the choices made "by market intermediaries, particularly pricing, when there is some kind of interdependence or externality between groups of agents that the intermediary serves." So, the focus of this section is to explore the intermediary's choice of pricing and how that choice affects the agent's decision of whether to transact over that platform.

To highlight how two-sided networks work, Parker and Van Alstyne [40] in their comparison of QWERTY and VHS, and the network effects generated on the demand-side, indicated that the incumbent supplier of a particular good, be it the QWERTY or the VHS, does not welcome competition because there would be no way to benefit initially from that additional supplier entering the market. Nevertheless, consumers prefer competition because it gives them more choices about who they purchase their goods from. It also puts downward price pressure on the goods and offers an alternative if the incumbent firm goes out of business. In this case, the externality runs from producers to consumers. According to Parker et al. [40], initially the consumer is not really concerned about how many other consumers adopt a particular platform because additional consumers might price them out of the market. However, producers are really concerned about how many consumers use their platform because the number of consumers that adopt their chosen platform affects their production opportunities. In this case, the externality runs from consumers to producers. Similarly, the network externalities, according to Parker et al. [40] can run either way and both producers and consumers prefer growth in their respective markets; however, "this may be mediated by the indirect effect of the internetwork externality. At issue is whether own-market entry expands participation on the other side of each transaction." When thinking about platforms and platform economics, Parker et al. [40] introduce the 'platform' as a third participant in this marketplace that straddles "both markets, can set prices more efficiently by internalizing these two-sided externalities. Independent firms serving either market separately lose this advantage."

Rochet and Tirole [44] state that "platforms must choose a price structure and not only a price level for their service." Furthermore, according to the authors, "the choice of a business model seems to be a key to the success of a platform and receives much corporate attention." The study also explores the idea of platform competition and how 'multihoming,' which occurs when consumers and producers use multiple platforms, "intensifies price competition on the other side as platforms use low prices in an attempt to 'steer' end users on the latter side toward an exclusive relationship."

How could this be applied to our base case, which is the introduction of a new exchange run on the blockchain technology? The introduction of a blockchain-based securities exchange could supplement the use of the traditional brokered exchange markets and offer capabilities to bring additional sources of supply and demand into the marketplace by lowering transaction costs and increasing efficiency through competition. Also, if the platform was built so that it encourages the development of functional applications in a decentralized manner, developers could be allowed to experiment with adding additional functionality to this blockchain exchange. Presumably, by allowing experimentation, breakthroughs in the application and the capabilities of the blockchain exchange will likely spur competition in terms of exploration of how to create more effective and efficient mechanism to transact across the traditional networks.

4.1 Network Externalities

An important element to consider when exploring the potential idea of establishing a blockchain solution to security market transactions is that there are potential network externalities that could arise because of the implementation of such a solution. In this section, we will consider how positive or negative externalities may affect a platform operator's profit opportunities. Consider the model of externalities developed in Katz and Shapiro [29]. The authors develop a general model of network competition and devise a partial equilibrium oligopoly model in which there are no income effects and the consumers' objective is to maximize their individual surplus. However, the consumers' ability to maximize their surplus is dependent on their understanding and proper forecasting of the future size of the network associated with whatever brand (or in our case, platform) they choose to purchase.

According to Katz et al. [29], the process for developing expectations about the future size of the platforms and which platform to choose develops as follows: (i) consumers form their expectations about the future size of the platform, (ii) platforms play an output or volume game and propose a series of prices, and (iii) consumers choose a platform by comparing their reservation prices and the prices set by the 'n' firms. The authors propose the following model using firms instead of platforms:

$$\gamma_i^e = x_i^e \text{ (Without the network effect)} \tag{1}$$

$$\gamma_i^e = \sum_{j=1}^{m} x_i^e \text{ (With the network effect)} \tag{2}$$

where γ_i^e is the consumer's prediction about the size of the network that the platform will have, x_i^e is the number of customers that it expects platform i to have, and m is the number of compatible platforms.

Of further interest to our present analysis, Katz et al. [29] indicated that there are three consumption externalities that one should consider when analyzing the network effects associated with the choice of a platform or exchange: (i) a direct network effect that is associated with an investor's willingness (reluctance) to trade on a particular platform because a friend or someone that she knows had a positive (negative) experience trading on that platform, (ii) an indirect externality resulting from the demand and supply-side belief that the exchange will be adopted by other agents, and (iii) an externality resulting from the bolt-on or add-on services associated with the marketplace. The first two network effects are critical to establishing the blockchain based solution in any specific geographic location. The third effect seems to be the most critical when contemplating broader scale adoption and integration on a global scale. Katz et al. [29] argue that the output of an industry, or a number of transactions in our case, increases by ensuring that platforms are compatible.

Adding more securities, countries, and exchanges to a blockchain platform would likely create positive network effects. The introduction of an innovation in the exchange market could affect the current markets in a number of ways by creating mechanisms that aid in the transfer of shares across nations, simultaneously converting the ask prices for securities traded in foreign markets into the domestic currency, and allowing access to securities that were more costly to purchase under the current system (e.g., allow investors to purchase securities traded on foreign exchanges instead of ADRs—American Depository Receipts).

4.2 Routing Rules

Hermalin and Katz [23] took a somewhat different approach in their analysis of two-sided markets. Instead of analyzing intermediaries, they focused their attention on the routing implications associated with platform choice. The main conclusions reached by the authors is that "membership decisions can override formal rules and lead to counterintuitive equilibrium outcomes." They continue by explaining that, in equilibrium, if party A is considering a transaction with party B and there are multiple networks to transact over, and party A is also given formal authority to route the order, the transaction tends to be routed through party B's preferred network. In addition, all else equal, if one network imposes routing restrictions on its participants, that network receives a lower share of transactions than its competitors. Thus, allowing prospective parties to choose the platform on which they will transact will increase transaction volume throughout the system and minimize the potential problem of trade inefficiency [23]. To relate this to our primary consideration, which is whether a blockchain alternative to the current securities transaction markets should exist, we need to look at whether it will positively impact the current securities market and whether such a market is likely to increase efficiency. Based on the above, we argue that routing by choice is preferred to mandated routing procedures both in terms of welfare maximization and in terms of the total volume of trade that would occur over both networks.

5 Implementation of Blockchain-Based Securities Markets

While claiming that we are still a long way from implementation and mass adoption, Geranio [19] highlights the steps for adoption of the blockchain technology which are: (a) experimentation, (b) implementation on a smaller scale, and (c) involvement of the regulatory bodies to establish a set of governing rules. Lipton [35] offers a warning against the implementation of a blockchain-based solution to trading and settlement and highlights the advantages with the current system, which include:

counterparty credit risk management, netting, credit risk and margin considerations, and the ability to borrow stocks, which are important considerations, but seem outside the purview of the process of executing transactions and exchanging money or capital for ownership in a firm. As a counterpoint, Egelund-Muller et al. [15] contend that there are additional opportunities associated with putting the entire financial system on a ledger and these opportunities are related to assessing systemic risks, identifying cases of tax evasion and fraud, and affording economists and analysts access to real-time minute-by-minute transaction data, which would presumably aid them in their forecasts of key economic indicators.

5.1   Legal Regimes and Blockchain-Based Securities Markets

Brummer [7] contends that the U.S. federal government "has enjoyed a virtual monopoly over the provision of securities laws" ever since they passed the Securities Exchange Acts of 1933 and 1934. Further, the author explains that because U.S. securities exchanges were the most liquid in the world, there was little danger of multinational corporations going elsewhere to raise capital and, therefore, they needed to follow the rules associated with raising debt and equity in U.S. capital markets. However, over time the public market for the provisioning of securities laws is evolving because the services offered by exchanges are becoming commoditized as floor-based exchanges are being replaced by electronic trading systems. To compete globally, national regulators (and legislatures) are motivated to provide attractive, cost-effective rules for foreign and domestic companies in order to protect or grow their domestic exchanges and financial centers. The study concludes that the consolidation of stock exchanges across the globe creates a market for the provisioning of securities laws.

According to Brummer [7], the three different paths that securities' market regulation could take are to: (i) allow companies to select a regulatory regime and list on another countries exchange but abide by the securities laws of the chosen regulatory regime, (ii) allow the exchanges to choose the regulatory regimes under which issuers on their exchange abide by, or (iii) under the 'substituted compliance' regime foreign companies would be able to list in another country but operate under their home country's legal regime if the home country's regime is deemed to be compliant with the listing country's legal regime. The adoption of these proposals should encourage regulators to implement regulations that protect investors and boost innovation and efficiency. On the other hand, this would discourage burdensome regulations because these regulatory regimes will discourage listing and the revenues that are associated with listing and trading on an exchange.

As access to capital markets becomes more commoditized, nations will be forced to adopt regulatory regimes that facilitate global transactions. Governments will be slow to adopt rules and regulations primarily because, according to Karajovic, Kim, and Laskowski [28], there is a lack of interoperability across global security markets and regulators do not seem to be sure about whether, in its current state, the technology is secure enough to use throughout the economy. Although a discussion of what this regulatory regime change should look like is beyond the scope of this paper, any attempts to change the regulatory regime should consider how those changes will help connect international securities markets, promote innovation within this industry, and provide a path towards integration of a global network of financial exchanges.

5.2   A Path Forward

Using the four quadrants associated with the adoption of foundational technologies, as illustrated in Iansiti et al. [25], we think that the most direct path towards the adoption of blockchain technology to help facilitate securities transactions will likely start through single use and localization applications, which are not complex, but move from a low to high degree of novelty. According to Iansiti et al. [25], the single use applications create better, cheaper, and highly focused solutions while the localization applications contain innovations that are relatively high in novelty but need only a limited number of users to create immediate value which promotes their prompt implementation. Similar to how the knowledge associated with the average portfolio manager's inability to generate abnormal returns has led investors to move capital into more passive index or exchange-traded fund solutions, we believe that ETFs could lead the transition towards a blockchain-based securities network by creating private blockchain solutions to facilitate local exchange in their products. ETFs are currently able to externalize the costs associated with their transactions and with the distribution of their investments, a next logical step is for the low-cost investment instruments to minimize the costs associated with processing and settling transactions in these securities, and to potentially externalize these costs as well. This could be done using the blockchain and the fact that the ETF managers match buyers and sellers of their ETFs makes a private blockchain network a meaningful and logical step towards a more efficient and cost-effective method to settle and process these transactions, especially with ETFs that trade frequently.

Once investors experience the reductions in costs associated with trading and purchasing securities over these private blockchain networks, we may be able to think about scale, network effects, and externalities that could result from public blockchain networks. Iansiti et al. [25] highlighted how this single use and localized blockchain networks evolve into networks that act as potential substitutes for the incumbent methods that people rely on to transact and move towards transforming the industry. According to the authors, 'substitution' occurs when there is a high level of coordination and complexity associated with transactions, but the degree of novelty is low. Eventually, transformation occurs when the amount of complexity and coordination involved with transactions increases, but the degree of novelty also increases. When we think about substitution, we realize that there will be substantial barriers to changing how investors think about transacting over securities markets. As mentioned earlier in this paper, there are institutions that have a substantial vested interest in ensuring that things stay the same, that investors continue to transact over their platforms, and that the regulators that have the authority to devise rules and regulations that monitor transactions have no incentive to change how we transact, even if a new transaction mechanism could enhance the investor's experience and increase efficiency. Further, Cichonczyk [11] claims that disintermediation will probably never occur because the current market intermediaries are incentivized to overcome any potential disintermediation in the financial markets. Cohen [12] points out what happened to the taxi industry as they failed to innovate and respond to advances in technology and watched their market power disappear as Uber disrupted their control over the 'taxi-cab' market, meaning that there is an opportunity to disrupt this market, but the question of who will disrupt it is very important. Public blockchain-based exchanges could be used as a substitute to the traditional brokered exchanges and this is the point of technological innovation at which this would begin to occur. During this step, we envision smaller private blockchain networks integrating into a more global public blockchain network but maintaining the traditional brokered exchanges so that investors are able to choose which medium they use to complete their transactions. As earlier examples of Australian Stock Exchange's and Nasdaq's move towards blockchain-based systems have illustrated, the first step towards innovation has occurred, but the transition from the private to public blockchain solution to the typical brokered exchange will likely take time.

Finally, the network effects and the externalities associated with the adaption of a new technology are likely to make the adoption of a blockchain solution fundamental to how investors participate in securities market transactions. As mentioned previously, to really transform securities markets and securities market transactions, investors, regulators, and investment professionals are going to need to embrace and adopt the use of the blockchain technology. More local introductions of blockchain-based systems similar to the ones illustrated in Kabi and Franqueira [27] and Olsen et al. [39] are needed to determine how efficient and cost-effective these alternative solutions could be.

6   Challenges

As mentioned earlier, there is a number of challenges or obstacles that need surmounting to disintermediate securities markets and offer a global blockchain solution that would reduce cost and improve efficiency across the global financial markets. First, as securities transactions become more commoditized, regulators should look to install and adopt governance mechanisms that better facilitate global transactions as the current governance mechanisms create unnecessary costs and execution delays. Scalability seems to be the next challenge, as Chang et al. [10] suggested, as even more advanced networks, such as Ethereum and Bitcoin, can only handle 20 transactions per second. However, according to Ventura [48], some blockchain solutions can exceed 1 million transactions per second, so a solution could be built to solve some of the scalability issues. Security associated with the consensus mechanism is likely another challenge that a blockchain solution would have to solve. According to Reyna et al. [43], there is a variety of security attacks (i.e., majority, double-spend, race, denial of service, and man in the middle attacks) that the builders of a system to facilitate trades across a global blockchain network should be aware of and protect against. Anonymity and privacy are also a concern; however, encrypting transactions could ensure anonymity [10, 43]. Finally, according to Chang et al. [10] energy consumption associated with maintain a blockchain solution is a concern. For example, 3 Bitcoin transactions use as much energy as 300.000 Visa transactions [10]. Therefore, when developing a potential blockchain solution that would provide the infrastructure to facilitate securities market transactions over a global exchange the amount of energy and cost associated with consuming that energy should be considered. In summary, there is a number of challenges that developers and planners need to contend with before a global blockchain solution to facilitate securities market transactions is developed and deployed. However, these challenges do not seem insurmountable.

7 Conclusion

This paper highlights the need to explore a blockchain-based alternative to the current process used to facilitate securities market transactions. We discuss how the blockchain technology could be applied to securities markets to improve the user experience, decrease transaction costs, and create new ways to transact across disparate markets. We argue that cost reduction could serve as one of the main reasons for a change in the current structure of securities markets across the globe. We provide some indication of how a securities trading market built on the blockchain infrastructure may work.

Our study illustrates how platform operators extract unfair rents on the investing public under the current regulatory regime and system and indicates how the economics of platforms might be used to think about implementing blockchain technologies more broadly to enhance the traditional process of exchange in securities markets. We argue that blockchain-based securities trading platforms should operate as a neutral party that straddles both sides of the market and act as a bridge between the supply-side and the demand-side and should not be positioned on one side of the transaction. Moving the blockchain-based platform from the supply-side to an intermediary position decreases the costs of the transaction for the customers and increases the potential for innovation in the process of securities market transactions. Implementation of open routing procedures should also positively impact the market in terms of welfare maximization and enhance efficiency.

Based on the literature, it seems likely that competition brought by implementation of blockchain-based securities exchanges alongside traditional exchanges will spur innovation, increase transparency, and increase efficiency, which all would likely benefit the participants in the marketplace in the long term. However, even though there seem to be significant benefits that could be realized by implementing a blockchain solution, we believe that, as is the case with the implementation of any new technology, further research must be completed to determine if the benefits associated with this type of solution outweigh the cost of maintaining the blockchain network from both a social and an economic standpoint.

Compliance with Ethical Standards:

Funding: This study did not receive any funding from any entity or organization.

Conflict of Interest: The authors declare that they have no conflict of interest.

References

1. Ammous, S.: Can cryptocurrencies fulfill the functions of money? Q. Rev. Econ. Financ. **70**, 38–51 (2018)
2. Andolfatto, D.: Blockchain: what it is, what it does, and why you probably don't need one. Federal Reserve Bank St. Louis Rev. **100**(2), 78–96 (2018)
3. Australian Securities Exchange: ASX's replacement of CHESS for equity post-trade services: Business requirements. Consultation Paper (2016). http://www.asx.com.au/documents/public-consultations/ASX-Consultation-Paper-CHESS-Replacement-19-September-2016.pdf
4. Berentsen, A., Schar, F.: A short introduction to the world of cryptocurrencies. Federal Reserve Bank St. Louis **100**(1), 1–16 (2018)
5. Block, J.H., Colombo, M.G., Cumming, D.J., Vismara, S.: New players in entrepreneurial finance and why they are there. Small Bus. Econ. **50**(2), 239–250 (2018)
6. Brownworth, A.: Blockchain 101 – A visual demo. YouTube.com (2016). https://www.youtube.com/watch?time_continue=690&v=_160oMzblY8. Accessed 17 Dec 2017
7. Brummer, C.: Stock exchanges and the new markets for securities laws. Univ. Chicago Law Rev. **75**(4), 1435–1491 (2008)
8. Cai, C.: Disruption of financial intermediation by FinTech: a review on crowdfunding and blockchain. Account. Financ. **58**, 965–992 (2018)
9. Catalini, C., Gans, J.: Some simple economics of blockchain. Rotman School of Management Working Paper No. 2874598; MIT Sloan Research Paper No. 5191-16 (2017)
10. Chang, V., Baudier, P., Zhang, H., Xu, Q., Zhang, J., Arami, M.: How blockchain can impact financial services - the overview, challenges and recommendations from expert interviewees. Technol. Forecast. Soc. Chang. **158**, 120–166 (2020). https://doi.org/10.1016/j.techfore.2020.120166
11. Cichonczyk, M.: On the future of markets driven by blockchain. In: Abramowicz, W., Paschke, A. (eds.) Business Information Systems Workshops. BIS 2018. LNBIP, vol. 339. Springer, Cham (2019)
12. Cohen, B.: The rise of alternative currencies in post-capitalization. J. Manage. Stud. **54**(5), 739–746 (2017)
13. Cowles, A.: Can stock market forecasters forecast? Econometrica **1**(3), 309–324 (1933)
14. Crosby, M., Nachippan, P.P., Verma, S., Kalyanaranman, C.: Blockchain technology: beyond bitcoin. Appl. Innov. Rev. **2**, 1–19 (2016)
15. Egelund-Muller, B., Elsman, M., Henglein, F., Ross, O.: Automated execution of financial contracts on blockchains. Bus. Inf. Syst. Eng. **59**(6), 457–467 (2017)
16. Ellis, C.D.: The rise and fall of performance investing. Financ. Anal. J. **70**(4), 14–23 (2014)
17. Fanning, K., Centers, D.: Blockchain and its coming impact on financial services. J. Corp. Account. Financ. **27**(5), 53–57 (2016)

18. Gainsbury, S., Blaszczynski, A.: How blockchain and cryptocurrency technology could revolutionize online gambling. Gaming Law Rev. **21**(7), 482–492 (2017)
19. Geranio, M.: Fintech in the exchange industry: potential for disruption? Masaryk Univ. J. Law Technol. **11**(2), 245–266 (2017)
20. Gehrig, T.: Intermediation in search markets. J. Econ. Manage. Strat. **2**(1), 97–120 (1993)
21. Glaser, F.: Pervasive decentralisation of digital infrastructures: a framework for blockchain enabled system and use case analysis. In: 50th Hawaii International Conference on System Sciences (HICSS 2017), Waikoloa (2017)
22. Guo, Y., Liang, C.: Blockchain application and outlook in the banking industry. Financ. Innov. **2**(24), 1–12 (2016)
23. Hermalin, B.E., Katz, M.L.: Your network or mine? The economics of routing rules. RAND J. Econ. **37**(3), 692–719 (2006)
24. Hornuf, L., Schwienbacher, A.: Market mechanisms and funding dynamics in equity crowd-funding. J. Corp. Financ. **50**, 556–574 (2018)
25. Iansiti, M., Lakhani, K.: The truth about blockchain. Harvard Bus. Rev. **95**(1), 118–127 (2017)
26. Irrera, A.: Blockchain startup Chain teams with Thales to bolster security. Reuters (2017). https://www.reuters.com/article/us-blockchain-security-idUSKBN1711KA
27. Kabi, O.R., Franqueira, V.N.L.: Blockchain-based distributed marketplace. In: Abramowicz, W., Paschke, A. (eds.) Business Information Systems Workshops. BIS 2018. LNBIP, vol. 339. Springer, Cham (2019)
28. Karajovic, M., Kim, H., Laskowski, M.: Thinking outside the block: projected phases of blockchain integration in the accounting industry. Aust. Account. Rev. **29**(2), 319–330 (2019)
29. Katz, M.L., Shapiro, C.: Network externalities, competition, and compatibility. Am. Econ. Rev. **75**(3), 424–440 (1985)
30. Khapko, M., Zoican, M.: Smart settlement. Rotman School of Management Working Paper No. 2881331; EFA 2017 Mannheim Meetings Paper; Society for Financial Studies (SFS) Cavalcade, 2017; Swedish House of Financial Studies Research Paper No. 17–4 (2017)
31. Kharif, O.: All you need to know about Bitcoin's rise from $0.01 to $15,000. Bloomberg Businessweek, 1 December 2017. https://www.bloomberg.com/news/articles/2017-12-01/und erstanding-bitcoin-s-rise-0-01-to-11-000-quicktake-q-a
32. Kiviat, T.: Beyond Bitcoin: issues regulating blockchain transactions. Duke Law J. **65**(3), 569–608 (2015)
33. Lee, L.: New kids on the blockchain: how bitcoin's technology could reinvent the stock market. Hastings Bus. Law J. **12**(2), 81–132 (2016)
34. Lewis, R., McPartland, J., Ranjan, R.: Blockchain and financial market innovation. Econ. Perspect. **41**(7), 1–13 (2017)
35. Lipton, A.: Blockchains and distributed ledgers in retrospective and perspective. J. Risk Financ. **19**(1), 4–25 (2018)
36. Miau, S., Yang, J.: Bibliometrics-based evaluation of the blockchain research trend: 2008 – March 2017. Technol. Anal. Strateg. Manag. **30**(9), 1029–1045 (2017)
37. Nair, M., Sutter, D.: The blockchain and increasing cooperative efficacy. Independent Rev. **22**(4), 529–550 (2018)
38. Nakamoto, S.: Bitcoin: A peer-to-peer electronic cash system. Unpublished Manuscript (2008). https://bitcoin.org/bitcoin.pdf
39. Olsen, R., Battiston, S., Caldarelli, G., Golub, A., Nikulin, M., Ivliev, S.: Case study of Lykke exchange: architecture and outlook. J. Risk Financ. **19**(1), 26–38 (2018)
40. Parker, G., Van Alstyne, M.: Two-sided network effects: A theory of information product design. Manage. Sci. **51**(10), 1494–1504 (2005)
41. Peterson, M.: Blockchain and the future of financial services. J. Wealth Manage. **21**(1), 124–131 (2018)
42. Puschmann, T.: Fintech. Bus. Inf. Syst. Eng. **59**(1), 69–76 (2017)
43. Reyna, A., Martin, C., Chen, J., Soler, E., Diaz, M.: On blockchain and its integration with IoT. Challenges and opportunities. Future Gener. Comput. Syst. **88**, 173–190 (2018)

44. Rochet, J., Tirole, J.: Platform competition in two-sided markets. J. Eur. Econ. Assoc. **1**(4), 990–1029 (2003)
45. Rysman, M.: The economics of two-sided markets. J. Econ. Perspect. **23**(3), 125–143 (2009)
46. Tapscott, A., Tapscott, D.: How blockchain is changing finance. Harvard Bus. Rev. **1**, 2–5 (2017)
47. Tepper, B.: Building on the blockchain: Nasdaq's vision of innovation. NASDAQ (2016)
48. Ventura, T.: The World's Fastest Blockchain Exceeds 1 Million Transactions Per Second. Medium, 2 June 2020. https://medium.com/predict/the-worlds-fastest-blockchain-exceeds-1-million-transactions-per-second-8931df09320d
49. Williams, P.: Does competency-based education with blockchain signal a new mission for universities? J. High. Educ. Policy Manage. **41**(1), 104–117 (2018)
50. Yermack, D.: Corporate governance and blockchains. Rev. Financ. **21**(1), 7–31 (2017)
51. Zhu, H., Zhou, Z.: Analysis and outlook of applications of blockchain technology to equity crowdfunding in China. Financ. Innov. **2**(29), 1–11 (2016)

Artificial Intelligence and Blockchain for Cybersecurity: Applications and Case Studies

Classification of Cyber Security Threats on Mobile Devices and Applications

Mohammed Amin Almaiah, Ali Al-Zahrani, Omar Almomani, and Ahmad K. Alhwaitat

Abstract Mobile devices and applications are prone to different kinds of cyber threats and attacks that affect their users' privacy. Therefore, there is critical need to understand all cyber threats characteristics in order to prevent their risks. However, most of cyber threats classifications are usually limited and based on one or two criteria in the classification process of threats. In addition, the current frameworks did not present an exhaustive list of cyber threats on mobile devices and applications. According to above reasons, this study proposes an exhaustive framework for mobile devices and applications-cyber security threat classifications, which includes most cyber threats classification and principles. The main purpose of our framework is to systematically identify cyber security threats, show their potential impacts, draw the mobile users' attention to those threats, and enable them to take protective actions as appropriate.

Keywords Mobile devices · Mobile applications · Cyber threats · Malicious attacks

1 Introduction

Mobile applications have become a well-known and popular tool of doing human life activities such as online shopping, bank transactions, etc. This huge increase in the

M. A. Almaiah (✉)
Department of Computer Networks and Communications, King Faisal University, Al-Ahsa 31982, Saudi Arabia
e-mail: malmaiah@kfu.edu.sa

A. Al-Zahrani
Department of Computer Engineering, King Faisal University, Al-Ahsa 31982, Saudi Arabia

O. Almomani
Computer Network and Information Systems Department, The World Islamic Sciences and Education University, Amman 11947, Jordan

A. K. Alhwaitat
Department of Computer Science, University of Jordan, Amman, Jordan

© The Author(s), under exclusive license to Springer Nature Switzerland AG 2021
Y. Maleh et al. (eds.), *Artificial Intelligence and Blockchain for Future Cybersecurity Applications*, Studies in Big Data 90,
https://doi.org/10.1007/978-3-030-74575-2_6

use of mobile applications has resulted in a large rising in cyber security attacks [7, 8, 10]. Security problems in mobile applications are still a serious concern for many researchers due to the lack security of mobile devices [12, 34, 45]. This makes cyber attackers exploit these vulnerabilities to access the systems illegally [26]. Mobile devices have several limitations in terms of lower power, computational processing and limited resources [36]. Mobile devices have many security vulnerabilities that can put users and organizations at high risk [59]. Mobile users face several types of security risks such as loss of data, invasion of privacy and financial losses. These risks are happened when malicious attacks exploit vulnerabilities in the operating systems of mobile devices. Mobile devices include many applications, which collect data according to their assigned task and share it with other applications. This interconnection of these devices in a heterogeneous environment makes them more vulnerable to the cyber security issues and threats. Therefore, cyber-attacks have become a serious concern, and this led to offer many security solutions from research community [60].

Cyber security is defined as a combination of security procedures, techniques, tools and guidelines that employed to protect the applications and devices over the internet [18]. Cyber security now is one of the most important issues for users and organizations, through protecting their assets and securing their information through detecting and mitigating the various cyber threats and attacks [19]. Despite many researchers have presented many countermeasures to address various types of security issues and problems in mobile application platforms, but are still not sufficient to protect the mobile applications from the ever-increasing security vulnerabilities and attacks. Thus, protecting mobile applications from cyber-attacks and threats has become one of the important role that has been motivating many researchers to conduct more research in the recent years [3, 4, 6, 9, 20].

In the literature, despite several studies have been conducted for identifying and classifying of cyber security attacks and threats for several technologies such as Internet of Things (IOT) [24], Cloud computing [45] and wireless networks [3, 4, 6, 9]. Each one of these studies has different tools and countermeasures to tackle various security attacks and breaches. In addition, there is limited research conducted into identification and classification of cyber security threats for mobile application platforms. Where, threats classification helps to understand the risk and nature of the cyber attacks, which is the major step in effective threat mitigation [11]. Therefore, this study aimed:

(1) To review the common cyber security threats.
(2) To review the cyber security attacks.
(3) To classify the cyber security threats of mobile devices and applications.

In this work, first, we conducted a systematic review analysis of cyber security threats related to mobile applications and other technologies such as IOT and cloud computing. Second, we classified and identified the major cyber threats for mobile application platforms and third we provided some countermeasures in a light of mobile technology. Where, mobile technology provides hardware and software solutions to tackle security challenges of mobile applications, which is a novel approach. In order to fill this research gap, this paper aims to present a comprehensive view of cyber security attacks, threats in mobile application platforms and the suggested solutions.

2 Security in Mobile Devices and Applications

Protecting both mobile devices and applications from cyber attacks is one of the critical challenging that ensure the security and privacy of mobile users. Security countermeasures should protect the data, information, hardware and applications in all stages. La Polla et al. [31], determined three main security problems of mobile platforms domain including data confidentiality, privacy and trust.

Data confidentiality is considered one of the fundamentals problem in mobile devices and applications [51]. In mobile platforms context, if the user need to access data should take authorization first in order to prevent attackers to access sensitive data stored on the mobile devices. To achieve that, there is needed to focus on two important aspects are (1) authorization and access control and (2) identity authentication. Mobile devices and applications need to be able to verify the user or device identity is authorized to access the data or not. Where an authorization mechanism helps mobile devices and application to identify if the mobile users or devices are permitted to access data or service. Access control mechanism also ensures of preventing attackers from access to resources of the system. This will establish a secure connection between mobile devices and thus, transition of data between users in a safety way. Another important issue that should be consider in mobile devices and applications is identity authentication. In fact, this issue is very critical in mobile environment, because huge number of users and devices need to authenticate each other through trustable way in a secure manner.

Privacy is an important issue in mobile devices and applications for users, organizations and governments. In the context of mobile platforms, mobile devices are connected, and sensitive data is shared and exchanged over the internet, this make user privacy a sensitive topic in mobile domain. Protecting the privacy of users data in mobile devices from cyber attacks is still a hot topic for many researchers and need to address [2, 29, 63].

3 Research Methodology

In this research, distinct steps were taken to conduct a rigorous systematic review of the literature related to cyber security threats and attacks. The review process was conducted based on existing guidelines established by Kitchenham and Charters [30] which includes (1) identifying the inclusion and exclusion criteria (2) determining the data sources and search strategies and (3) data analysis and coding. This review is considered an essential step before conducting any research paper, and it helps to establish the foundation for knowledge accumulation. It also helps to identify the

Table 1 Inclusion and exclusion criteria for cyber security threats studies

Inclusion criteria	Exclusion criteria
1. The selected studies that include cyber security threats	1. Exclude each study which did not focus on cyber security threats
2. The selected studies that include cyber attack methods and techniques	2. Exclude each study which did not focus on cyber security threats impact
3. The selected studies that include cyber threats impacts	3. Exclude each study which did not focus on cyber security attack techniques.
4. The selected studies that include cyber threats categories	

areas that previous studies have missed. The following subsections describe in detail the steps used for conducting the systematic review in this study.

3.1 Identifying the Inclusion and Exclusion Criteria

In the first step of review, a collection of inclusion and exclusion criteria were determined that were used during the selection of articles. Table 1 show the inclusion and exclusion criteria for cyber security threats studies.

3.2 Determining the Data Sources and Search Strategies

In the second step of a systematic literature review, we collected a large number of studies through a search in the following popular databases: Google Scholar, Wiley, IEEE, ScienceDirect and Springer. The main keywords that were used in the search process are: ("Cyber security" AND "Cyber security threats" AND "Cyber security attacks" AND "Cyber security techniques and methods" AND " Cyber security threats impact" AND "Cyber security categories". We found through the search process more than 1522 articles using the keywords above and classified based on the identified databases as shown in Table 2. Then, we excluded all the items that we found as duplicated, which was 200 articles; thus, the total number of the collected

Table 2 The distribution of papers collected from the top-ranked databases

Database	Number of studies
IEEE	33
Springer	6
Elsevier	22
Google Scholar	30
Total	91

items was reduced to 833. After that, the remaining articles were filtered based on the criteria in Table 1. Finally, 91 studies have met the inclusion criteria and are used in the analysis process. Figure 1 illustrates the systematic review process for this study.

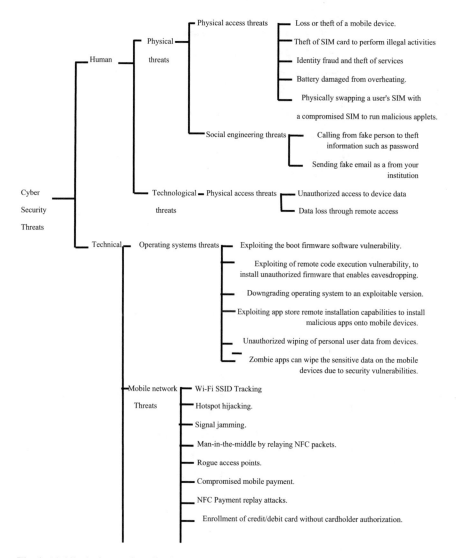

Fig. 1 Mobile devices and applications threats classifications

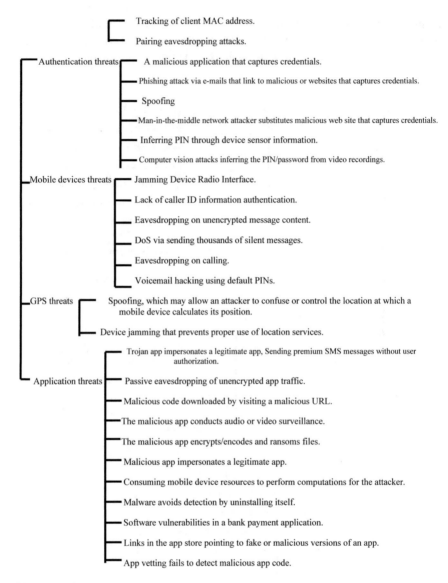

Fig. 1 (continued)

3.3   Data Analysis and Coding

3.3.1   Classification of Cyber Security Threats

Cyber security threat is defined as any action that takes advantage of security weaknesses in a system and has a negative impact on it [16]. As mobile devices and applications become a reality, a growing number of ubiquitous mobile devices have raised the number of the cyber security threats. Unfortunately, mobile devices come with new set of cyber security threats. There is a growing awareness that the new generations of mobile devices could be targeted with malware and vulnerable to attack. In the literature, there are several studies classified the cyber security threats based on attack techniques used by attackers to exploit vulnerabilities [42, 43, 50]. For example, Tomić and McCann [54], categorized the security threats into three levels are: data security level (anonymity and freshness), access security level (accessibility, authorization and authentication) and network security level. The researchers in the same study also mentioned that attacks could occur in all layers from application layer to physical layer. For example, at the application layer level, a malicious attack can be added along the communication link to generate fake messages and data in order to attack the ongoing communication and increase the data collision. The attack in transport layer happen through sending unlimited connection request in order to minimize the node's energy and exhaust its resources and this lead to denial of service. Other attack can be occurred in a network layer in several forms such as spoofing, sinkhole, flooding and replay attack in order to create and send fake messages or causing congestion in the network. Jamming attack at the Data link layer can cause loss of signals and data and destroy the channel and increased interference. At the physical layer level, the attacker can allow unauthorized nodes to access to the network and damage it. Jouini et al. [42] conducted a literature review to classify the security threats in information systems. They established a hybrid model for classifying the security threats for information systems. They classified the security threats into three types: human threats, technological threats and environmental threats. Otuoze et al. [43] proposed a framework for the security threats of smart grid based on threats sources. Where, in the framework, the researchers' classified the security threats into technical and non-technical resource threats. Technical threats was categorized into three types of threats are infrastructure threats, technical operational threats and system data management threats. While non-technical threats were classified into environmental threats and governmental threats. Singh and Shrivastava [52] categorized the security attacks and threats on cloud computing into four levels: authentication Attacks, side Channel Attacks, cloud Malware injection attack and Denial of Service (DoS) attacks. Roman, Lopez, and Mambo [47] classified the security threats of mobile edge computing five assets are: (1) Network infrastructure threats such as man in the middle and denial of service attack, (2) Edge data center threats such as physical damage, privacy leakage, privilege escalation and service manipulation, (3) Virtualization infrastructure threats such as denial of service, misuse of resources, privacy leakage and privilege escalation, (4) core infrastructures threats such as privacy leakage, service manipulation, rogue infrastructure and (5) User devices such as injection of information and service manipulation. Abraham and Chengalur Smith [1] confirmed social engineering malware proliferation through a variety of infiltration channels such as e-mail, social software, websites, and portable media. Mosakheil [40] in his research divided the security threats for blockchain technology into five categories are: (1) Double spending threats, (2) Mining/Pool threats, (3) Wallet threats, (4) Network threats such as DDoS attack, and (5) Smart contracts threats. In the same way, Homayun et al. [26] classified the common cyber security threats by using mapping study, which include phishing, denial of service (DoS), injection attack, man-in-the-middle attacks, session hijacking, SQL injection attack and malware (Table 3).

3.4 Classification of Cyber Security Attacks

Cyber security attack is defined as any activities taken to harm a system or disrupt normal operations through exploiting vulnerabilities using various techniques and tools. Attackers launch attacks to achieve goals either for personal satisfaction or recompense. Cyber attack takes several forms, including (1) passive attack to monitor unprotected network communications in order to decrypt weakly encrypted traffic and getting authentication information; close-in attacks; exploitation by insiders, and so on, and (2) active attack aims to monitor unencrypted traffic to capture sensitive

Table 3 Classification of Cyber security threats

Literature	Target application	Methodology	Findings and Contributions
Tomić and McCann [54]	Wireless sensor networks	Literature survey and analysis	Categorized the security threats into three levels are: data security level (anonymity and freshness), access security level (accessibility, authorization and authentication) and network security level
Jouini et al. [42]	Information systems	Literature review	Established a hybrid model for classifying the security threats for information systems. They classified the security threats into three types: human threats, technological threats and environmental threats
Otuoze et al. [43]	Smart grid	Literature review	Classified the security threats of smart grid into technical and non-technical resource threats. Technical threats was categorized into three types of threats are infrastructure threats, technical operational threats and system data management threats. While non-technical threats were classified into environmental threats and governmental threats
Tomić and McCann [54]	Wireless sensor networks	Literature survey	Classified the attacks that could occur in all layers from application layer to physical layer. For example, at the application layer level, a malicious attack can be added along the communication link to generate fake messages and data in order to attack the ongoing communication and increase the data collision. The attack in transport layer happen through sending unlimited connection request in order to minimize the node's energy and exhaust its resources and this lead to denial of service. Other attack can be occurred in a network layer in several forms such as spoofing, sinkhole, flooding and replay attack in order to create and send fake messages or causing congestion in the network. Jamming attack at the Data link layer can cause loss of signals and data and destroy the channel and increased interference. At the physical layer level, the attacker can allow unauthorized nodes to access to the network and damage it
Singh and Shrivastava [52]	Cloud computing	Literature review	Categorized the security attacks and threats on cloud computing into four levels: authentication Attacks, side Channel Attacks, cloud Malware injection attack and Denial of Service (DoS) attacks

(continued)

Table 3 (continued)

Literature	Target application	Methodology	Findings and Contributions
Roman, Lopez, and Mambo [47]	Mobile edge computing	Developmental	Classified the security threats of mobile edge computing five assets are: (1) Network infrastructure threats such as man in the middle and denial of service attack, (2) Edge data center threats such as physical damage, privacy leakage, privilege escalation and service manipulation, (3) Virtualization infrastructure threats such as denial of service, misuse of resources, privacy leakage and privilege escalation, (4) core infrastructures threats such as privacy leakage, service manipulation, rogue infrastructure and (5) User devices such as injection of information and service manipulation
Mosakheil [40]	Blockchain technology threats	Developmental	Divided the security threats for blockchain technology into five categories are: (1) Double spending threats, (2) Mining/Pool threats, (3) Wallet threats, (4) Network threats such as DDoS attack, and (5) Smart contracts threats
Humayun et al. [38]	Cyber security threats	Mapping study	Classified the common cyber security threats by using mapping study, which include phishing, denial of service (DoS), injection attack, man-in-the-middle attacks, session hijacking, SQL injection attack and malware
Mitrokotsa et al. [37]	Classification of RFID attacks	Literature review and synthesis	Classified threats associated with Radio Frequency Identification systems. They distinguished attacks in the physical layer, network transport layer, application layer, strategic layer, and multilayer
Abraham and Chengalur Smith [1]	Social engineering malware	Literature review and synthesis	Social engineering malware is both pervasive and persistent. Emphasized the importance for organizations to develop a shared social responsibility to combat social engineering malware and not solely on technical solutions. Social engineering malware proliferation through a variety of infiltration channels such as e-mail, social software, websites, and portable media.
Heartfield and Loukas [27]	Social engineering semantic attacks	Survey	Introduced a structured baseline for classifying semantic attacks by breaking down into components and identifying countermeasures

information. In this section, we attempted to review the classifications of cyber security attacks based on the type of behavior used by the attacker. According to the literature, the common cyber security attacks classifications including: (1) Access attacks which allow unauthorized users access to the network or devices such as smart phones with have no right to access [2, 15, 23, 46]. Table 4 summarizes the cyber security attacks classifications, (2) Reconnaissance attack allow attacker to capturing, discovering and mapping of system vulnerabilities such as scanning traffic network, network ports and IP address information [2, 5, 17, 49]. (3) Physical attack this type of attack aims to tamper hardware devices, for example some technologies such as IOT devices operate in outdoor environments may highly susceptible to physical attacks [23, 25, 32, 55]. (4) Denial-of-service (DoS) attack allow the attacker to make the network or device services unavailable to its intended users due to several reasons such as limited computation resources and low memory capabilities. This make mobile platforms are vulnerable to DOS attack [22, 41, 44, 48]. (6) Password-based attack can be done by attackers in two ways: (1) brute force attack by using cracking tools to guess the correct password in order to access valid password, (2) dictionary attack depends on trying several letters and numbers to guess user passwords [53,

Table 4 The common cyber security attacks classifications in different domains

Cyber security attack	Description	Context	Literature
Access attack	Allow unauthorized users access to the network or devices such as smart phones with have no right to access	IOT, Wireless sensor networks, mobile devices	Abomhara and Køien [2]; Ashokkumar, Giri, and Menezes [15]; Damghani et al. [23]; Rahman and Tomar [46]
Reconnaissance attack	Reconnaissance attack allow attacker to capturing, discovering and mapping of system vulnerabilities such as scanning traffic network, network ports and IP address information	Unmanned Aerial Vehicle (UAV), IOT, Network Forensics and mobile applications	Abomhara and Køien [2]; Changsheng [17]; Alabady et al. [5]; Rizal, Mendoza and Gu, [49]
Physical attack	This type of attack aims to tamper with hardware devices, for example some technologies such as IOT devices operate in outdoor environments may highly susceptible to physical attacks	Smart grid, IOT, Mobile networks, Mobile platforms	He and Yan [25]; Damghani [23]; Mavoungou et al. [32]; Van Der Veen [55]
Denial-of-service (DoS) attack	Denial-of-service (DoS) attack allow the attacker to make the network or device services unavailable to its intended users due to several reasons such as limited computation resources and low memory capabilities. This make mobile platforms are vulnerable to DOS attack	Mobile devices, Mobile ad hoc network, IOT	Farina et al. [22]; Paul, Chitodiya, and Vishwakarma [44]; Roland, Langer, and Scharinger [48]; Nawir et al. [41]:
Attack on privacy	Attack on privacy through using remote access methods and malware to spy or stole sensitive information of users or organizations. Privacy protection in mobile devices has become increasingly challenging due to share large amount of information between mobile devices	Cloud storage, IOT, Mobile devices and applications	Abomhara and Køien, [2]; Kang, Wang and Shao [29]; Yu, Chen, and Cai [63]

(continued)

Table 4 (continued)

Cyber security attack	Description	Context	Literature
Password-based attack	Attackers in two ways can do password-based attack: (1) brute force attack by using cracking tools to guess the correct password in order to access valid password, (2) dictionary attack depends on trying several letters and numbers to guess user passwords	IOT, Mobile devices	Shah and Venkatesan [53]; Zhang et al. [62]
Supervisory Control and Data Acquisition attack	Supervisory Control and Data Acquisition attack using malware such as Trojan to take control of the system. Mobile applications are vulnerable to many cyber attacks like Trojan virus	IoT	Abomhara and Køien [2]
Spoofing attack	Spoofing attack is based on obtaining the IP address of the devices to attack the users through enabling attackers to access users' confidential data and use it for malicious purposes	Mobile devices, Unmanned Aerial Vehicle (UAV), Mobile cloud, Mobile ad hoc networks, IOT	Malisa, Kostiainen and Capkun [33]; Huang et al. [28]; Moorthy, Venkataraman, and Rao [39]; Mohammadnia and Slimane [38]; Visalakshi and Prabakaran [58]
Botnet attack	Botnet attack is based on a collection of Internet-connected devices that have been breached and ceded to a malicious device known as botnet controller. The botnet controller able to direct malicious activities in order to damage the network or exploit users and data for materialistic gain (Ali et al. [13, 14]; Moorthy, Venkataraman and Rao [39]	IOT, Mobile clouds	Ali et al. [13, 14]; Moorthy, Venkataraman and Rao [39]

(continued)

Table 4 (continued)

Cyber security attack	Description	Context	Literature
Sybil attack	Sybil attack is a threat in which attacker attempt to obtain identity of honest user and pretend as a distinct user and then attempt to create relationships with an honest users. If the attacker is successful in compromising one of the honest users, he will gain unauthorized privileges that help in the attacking process	Mobile IOT, Wireless ad hoc network, Wireless Sensor Networks (WSN)	Wu and Ma 61; Dong et al. [21]; Vasudeva and Sood [57]

62]. (7) Supervisory Control and Data Acquisition attack using malware such as Trojan to take control of the system. Mobile applications are vulnerable to many cyber-attacks like Trojan virus [2]. (8) Spoofing attack is based on obtaining the IP address of the devices to attack the users through enabling attackers to access users' confidential data and use it for malicious purposes [28, 33, 38, 39, 58]. (9) Botnet attack is based on a collection of Internet-connected devices that have been breached and ceded to a malicious device known as botnet controller. The botnet controller able to direct malicious activities in order to damage the network or exploit users and data for materialistic gain [13, 14, 39]. (10) Sybil attack is a threat in which attacker attempt to obtain identity of honest user and pretend as a distinct user and then attempt to create relationships with an honest users. If the attacker is successful in compromising one of the honest users, he will gain unauthorized privileges that help in the attacking process [21, 57].

4   The Proposed Framework

In fact, mobile devices and applications are prone to different kinds of cyber threats and attacks that affect their users' privacy and there is critical need to understand all cyber threats characteristics in order to prevent their risks. However, most of cyber threats classifications are usually limited and based on one or two criteria in the classification process of threats. In addition, the current frameworks did not present an exhaustive list of cyber threats on mobile devices and applications. These frameworks may be suitable for stable technologies, but in the constantly changing technologies like mobile technology and IOT, it is very different and need to protect against cyber threats continuously [64].

Classification process of cyber threats is an important step that allows users, organizations and governments to know threats that influence their sensitive information and hence protect their devices and systems in advance. In addition, this step helps companies to develop their mobile applications with less vulnerabilities [35]. These threats can be identified through understanding the behavior of existing threats work. Unfortunately, existing classifications do not present an exhaustive list of cyber threats on mobile devices and applications and do not support the classification principles [35, 56, 61]. At that point, the optimal solution is to propose a hybrid model to combine all different classifications of threats.

According to above reasons, we proposed an exhaustive framework for mobile devices and applications-cyber security threat classifications, which includes most cyber threats classification and principles. The main purpose of our framework is to systematically identify cyber security threats, show their potential impacts, draw the mobile users' attention to those threats, and enable them to take protective actions as appropriate. The cyber threats classification in the proposed framework were developed using a literature review in the sections above as the following:

- Cyber security threat sources: The sources that cause cyber threats and we determined two main sources: human and technological.
- Cyber security threat form: The origin of cyber threat source could be either physical threat or technical threat.
- Cyber security threat motivations: The purpose of attackers on mobile devices and applications can be malicious or non-malicious.
- Cyber security threat intention: This criterion aims to identify the intent of attackers who caused the threat, which can be categorized in two classes: accidental or intentional. In addition, this factor helps to understand the attacker behaviour in order to understand its intention, and hence mitigate the risk.
- Cyber security threat type: In our framework, we classified the cyber security threats on mobile devices and applications into nine categories are: physical access threats, operating systems threats, social engineering threats, application threats, authentication threats, network threats, GPS threats and mobile devices threats.
- Cyber threats impact: Threat impact is a malicious action of attackers after violation security of mobile devices and systems. In our framework, we determined the most important cyber threat impacts like corruption of information stored in the mobile devices, destruction of information, theft of information, unauthorized disclosure of sensitive data, malicious versions of applications, theft credit card information and others.

5 Conclusion

Security problems in mobile applications are still a serious concern for many researchers due to the lack security of mobile devices. This makes cyber attackers exploit these vulnerabilities to access the systems illegally. This research aimed to better understanding of the nature of cyber threats in order to establish appropriate countermeasures to prevent or mitigate their effects. We proposed an exhaustive framework for mobile devices and applications-cyber security threat classifications, which includes most cyber threats classification and principles. Our framework is flexible, dynamic and multidimensional and covers most cyber threats on mobile devices and applications.

References

1. Abraham, S., Chengalur-Smith, I.: An overview of social engineering malware: trends, tactics, and implications. Technol. Soc. **32**(3), 183–196 (2010)
2. Abomhara, M., Køien, G.M.: Cyber security and the internet of things: vulnerabilities, threats, intruders and attacks. J. Cyber Secur. Mobil. **22**, 65–88 (2015)
3. Adil, M., Almaiah, M.A., Omar Alsayed, A., Almomani, O.: An anonymous channel categorization scheme of edge nodes to detect jamming attacks in wireless sensor networks. Sensors **20**(8), 2311 (2020)
4. Adil, M., Khan, R., Almaiah, M.A., Binsawad, M., Ali, J., Al Saaidah, A., Ta, Q.T.H.: An efficient load balancing scheme of energy gauge nodes to maximize the lifespan of constraint oriented networks. IEEE Access **8**, 148510–148527 (2020)
5. Alabady, S.A., Al-Turjman, F., Din, S.: A novel security model for cooperative virtual networks in the IoT era. Int. J. Parallel Prog. **48**(2), 280–295 (2020)
6. Adil, M., Khan, R., Almaiah, M.A., Al-Zahrani, M., Zakarya, M., Amjad, M.S., Ahmed, R.: MAC-AODV based mutual authentication scheme for constraint oriented networks. IEEE Access **8**, 44459–44469 (2020)
7. Almaiah, M. A., & Al-Khasawneh, A. (2020). Investigating the main determinants of mobile cloud computing adoption in university campus. Education and Information Technologies, 1–21
8. Khan, M.N., Rahman, H.U., Almaiah, M.A., Khan, M.Z., Khan, A., Raza, M., Khan, R.: Improving energy efficiency with content-based adaptive and dynamic scheduling in wireless sensor networks. IEEE Access **8**, 176495–176520 (2020)
9. Adil, M., Khan, R., Ali, J., Roh, B.H., Ta, Q.T.H., Almaiah, M.A.: An energy proficient load balancing routing scheme for wireless sensor networks to maximize their lifespan in an operational environment. IEEE Access **8**, 163209–163224 (2020)
10. Almaiah, M.A., Dawahdeh, Z., Almomani, O., Alsaaidah, A., Al-khasawneh, A., Khawatreh, S.: A new hybrid text encryption approach over mobile ad hoc network. Int. J. Electric. Comput. Eng. (IJECE) **10**(6), 6461–6471 (2020)
11. Al Hwaitat, A.K., Almaiah, M.A., Almomani, O., Al-Zahrani, M., Al-Sayed, R.M., Asaifi, R.M., Adhim, K.K., Althunibat, A., Alsaaidah, A.: Improved security particle swarm optimization (PSO) algorithm to detect radio jamming attacks in mobile networks. Quintana **11**(4), 614–624 (2020)
12. Almaiah, M.A., Alamri, M.M.: Proposing a new technical quality requirements for mobile learning applications. J. Theoret. Appl. Inf. Technol. **96**, 19 (2018)
13. Ali, I., Ahmed, A.I.A., Almogren, A., Raza, M.A., Shah, S.A., Khan, A., Gani, A.: Systematic literature review on IoT-based Botnet attack. IEEE Access **8**, 212220–212232 (2020)
14. Ali, G., Ally Dida, M., Elikana Sam, A.: Two-factor authentication scheme for mobile money: a review of threat models and countermeasures. Future Internet **12**(10), 160 (2020)
15. Ashokkumar, C., Giri, R.P., Menezes, B.: Highly efficient algorithms for AES key retrieval in cache access attacks. In: 2016 IEEE European Symposium on Security and Privacy (EuroS&P), pp. 261–275. IEEE, March 2016
16. Brauch, H.G.: Concepts of security threats, challenges, vulnerabilities and risks. In: In: Brauch, H. et al. (eds.) Coping with Global Environmental Change, Disasters and Security, pp. 61–106. Springer, Heidelberg (2011)

17. Jiang, C.: Key technologies for integrated reconnaissance and attack system of UAVs. Electron. Opt. Control **2** (2011)

18. Craigen, D., Diakun-Thibault, N., Purse, R.: Defining cybersecurity. Technol. Innov. Manage. Rev. **4**(10), 1–25 (2014)

19. Da Veiga, A.: A cybersecurity culture research philosophy and approach to develop a valid and reliable measuring instrument. In: 2016 SAI Computing Conference (SAI), pp. 1006–1015. IEEE, July 2016

20. Dawson, M., Wright, J., Omar, M.: Mobile devices: the case for cyber security hardened systems. In: New Threats and Countermeasures in Digital Crime and Cyber Terrorism, pp. 8–29. IGI Global (2015)

21. Dong, S., Zhang, X.G., Zhou, W.G.: A security localization algorithm based on DV-hop against sybil attack in wireless sensor networks. J. Electric. Eng. Technol. **15**(2), 919–926 (2020)

22. Farina, P., Cambiaso, E., Papaleo, G., Aiello, M.: Understanding DDoS attacks from mobile devices. In: 2015 3rd International Conference on Future Internet of Things and Cloud, pp. 614–619. IEEE, August 2015

23. Damghani, H., Damghani, L., Hosseinian, H., Sharifi, R.: Classification of attacks on IoT. In: 4th International Conference on Combinatorics, Cryptography, Computer Science and Computation, November 2019

24. Ghadeer, H.: Cybersecurity issues in internet of things and countermeasures. In: 2018 IEEE International Conference on Industrial Internet (ICII), pp. 195–201. IEEE, October 2018

25. He, H., Yan, J.: Cyber-physical attacks and defences in the smart grid: a survey. IET Cyber-Phys. Syst. Theory Appl. **1**(1), 13–27 (2016)

26. Homayoun, S., Dehghantanha, A., Parizi, R.M., Choo, K.K.R.: A blockchain-based framework for detecting malicious mobile applications in app stores. In: 2019 IEEE Canadian Conference of Electrical and Computer Engineering (CCECE), pp. 1–4. IEEE, May 2019

27. Heartfield, R., Loukas, G.: Protection against semantic social engineering attacks. In: Versatile Cybersecurity, pp. 99–140. Springer, Cham (2018)

28. Huang, X., Tian, Y., He, Y., Tong, E., Niu, W., Li, C., Chang, L.: Exposing spoofing attack on flocking-based unmanned aerial vehicle cluster: a threat to swarm intelligence. Secur. Commun. Netw. **2020** (2020)

29. Kang, B., Wang, J., Shao, D.: Attack on privacy-preserving public auditing schemes for cloud storage. Math. Prob. Eng. **2017** (2017)

30. Kitchenham, B., Charters, S.: Guidelines for performing systematic literature reviews in software engineering (2007)

31. La Polla, M., Martinelli, F., Sgandurra, D.: A survey on security for mobile devices. IEEE Commun. Surv. Tutor. **15**(1), 446–471 (2012)

32. Mavoungou, S., Kaddoum, G., Taha, M., Matar, G.: Survey on threats and attacks on mobile networks. IEEE Access **4**, 4543–4572 (2016)

33. Malisa, L., Kostiainen, K., Capkun, S.: Detecting mobile application spoofing attacks by leveraging user visual similarity perception. In: Proceedings of the Seventh ACM on Conference on Data and Application Security and Privacy, pp. 289–300, March 2017

34. Mendoza, A., Gu, G.: Mobile application web app reconnaissance: web-to-mobile inconsistencies & vulnerabilities. In: 2018 IEEE Symposium on Security and Privacy (SP), pp. 756–769. IEEE, May 2018

35. Mylavarapu, R.M., Nigam, A., Hegde, V.B.: U.S. Patent No. 10,686,819. U.S. Patent and Trademark Office, Washington, DC (2020)

36. Mikhaylov, D., Zhukov, I., Starikovskiy, A., Kharkov, S., Tolstaya, A., Zuykov, A.: Review of malicious mobile applications, phone bugs and other cyber threats to mobile devices. In: 2013 5th IEEE International Conference on Broadband Network & Multimedia Technology, pp. 302–305. IEEE, November 2013

37. Mitrokotsa, A., Rieback, M.R., Tanenbaum, A.S.: Classifying RFID attacks and defenses. Inf. Syst. Front. **12**(5), 491–505 (2010)

38. Mohammadnia, H., Slimane, S.B.: IoT-NETZ: practical spoofing attack mitigation approach in SDWN network. In: 2020 Seventh International Conference on Software Defined Systems (SDS), pp. 5–13. IEEE, April 2020

39. Moorthy, V., Venkataraman, R., Rao, T.R.: Security and privacy attacks during data communication in software defined mobile clouds. Comput. Commun. **153**, 515–526 (2020)
40. Mosakheil, J. H.: Security threats classification in blockchains (2018)
41. Nawir, M., Amir, A., Yaakob, N., Lynn, O.B.: Internet of Things (IoT): taxonomy of security attacks. In: 3rd International Conference on Electronic Design (ICED), pp. 321–326. IEEE, August 2016
42. Jouini, M., Rabai, L.B.A., Aissa, A.B.: Classification of security threats in information systems. Procedia Comput. Sci. **32**, 489–496 (2014)
43. Otuoze, A.O., Mustafa, M.W., Larik, R.M.: Smart grids security challenges: classification by sources of threats. J. Electric. Syst. Inf. Technol. **5**(3), 468–483 (2018)
44. Paul, S., Chitodiya, A., Vishwakarma, D.: Detection and prevention methodology for DoS attack in mobile ad-hoc networks. Int. Res. J. Eng. Technol. **6**(5), 6313–6317 (2019)
45. Rabai, L.B.A., Jouini, M., Aissa, A.B., Mili, A.: A cybersecurity model in cloud computing environments. J. King Saud Univ.-Comput. Inf. Sci. **25**(1), 63–75 (2013)
46. Rahman, R.U., Tomar, D.S.: Security attacks on wireless networks and their detection techniques. In: Emerging Wireless Communication and Network Technologies, pp. 241–270. Springer, Singapore (2018)
47. Roman, R., Lopez, J., Mambo, M.: Mobile edge computing, fog et al.: a survey and analysis of security threats and challenges. Future Gener. Comput. Syst. **78**, 680–698 (2018)
48. Roland, M., Langer, J., Scharinger, J.: Practical attack scenarios on secure element-enabled mobile devices. In: 4th International Workshop on Near Field Communication, pp. 19–24. IEEE, March 2012
49. Rizal, R., Riadi, I., Prayudi, Y.: Network forensics for detecting flooding attack on internet of things (IoT) device. Int. J. Cyber-Security Digit. Forensics **7**(4), 382–390 (2018)
50. Sadqi, Y., Maleh, Y.: A systematic review and taxonomy of web applications threats. Inf. Secur. J. Global Persp. 1–27 (2021)
51. Souppaya, M., Scarfone, K.: Guidelines for managing the security of mobile devices in the enterprise. NIST Spec. Publ. **800**, 124 (2013)
52. Singh, A., Shrivastava, D.M.: Overview of attacks on cloud computing. Int. J. Eng. Innov. Technol. (IJEIT), **1**(4) (2012)
53. Shah, T., Venkatesan, S.: Authentication of IoT device and IoT server using secure vaults. In: 17th IEEE International Conference on Trust, Security and Privacy in Computing and Communications/12th IEEE International Conference on Big Data Science and Engineering (TrustCom/BigDataSE), pp. 66–90. IEEE, August 2018
54. Tomić, I., McCann, J.A.: A survey of potential security issues in existing wireless sensor network protocols. IEEE Internet Things J. **4**(6), 1910–1923 (2017)
55. Van Der Veen, V., Fratantonio, Y., Lindorfer, M., Gruss, D., Maurice, C., Vigna, G., Giuffrida, C.: Drammer: deterministic Rowhammer attacks on mobile platforms. In: Proceedings of the 2016 ACM SIGSAC Conference on Computer and Communications Security, pp. 1675–1689, October 2016
56. Varma, P.R.K., Raj, K.P., Raju, K.S.: Android mobile security by detecting and classification of malware based on permissions using machine learning algorithms. In: 2017 International Conference on I-SMAC (IoT in Social, Mobile, Analytics and Cloud) (I-SMAC), pp. 294–299. IEEE, February 2017
57. Vasudeva, A., Sood, M.: Survey on sybil attack defense mechanisms in wireless ad hoc networks. J. Netw. Comput. Appl. **120**, 78–118 (2018)
58. Visalakshi, P., Prabakaran, S.: Detection and prevention of spoofing attacks in mobile adhoc networks using hybrid optimization algorithm. J. Intell. Fuzzy Syst. 1–14 (2020, preprint)
59. Watson, B., Zheng, J.: On the user awareness of mobile security recommendations. In: Proceedings of the SouthEast Conference, pp. 120–127, April 2017
60. Wu, Z., Ma, R.: A novel sybil attack detection scheme based on edge computing for mobile iot environment. arXiv preprint arXiv:1911.03129 (2019)
61. Yan, P., Yan, Z.: A survey on dynamic mobile malware detection. Software Qual. J. **26**(3), 891–919 (2018)

62. Zhang, Y., Xu, C., Li, H., Yang, K., Cheng, N., Shen, X.S.: PROTECT: efficient password-based threshold single-sign-on authentication for mobile users against perpetual leakage. IEEE Trans. Mob. Comput. (2020)
63. Yu, C., Chen, S., Cai, Z.: LTE phone number catcher: a practical attack against mobile privacy. Secur. Commun. Netw. **2019** (2019)
64. Yesilyurt, M., Yalman, Y.: Security threats on mobile devices and their effects: estimations for the future. Int. J. Secur. Appl. **10**(2), 210–235 (2016)

Revisiting the Approaches, Datasets and Evaluation Parameters to Detect Android Malware: A Comparative Study from State-of-Art

Abu Bakkar Siddikk, Md. Fahim Muntasir, Rifat Jahan Lia, Sheikh Shah Mohammad Motiur Rahman, Takia Islam, and Mamoun Alazab

Abstract Alongside the recognition of the android operating system (OS), android malware is on the increase. Cybercriminals are using different techniques to develop malware for android devices. In addition, malware authors are trying to make malicious android applications that severely undermine the potential of traditional malware detectors. The key purpose of the chapter is to analyze and have a different appearance at various techniques of Android malware detection in a variety of research articles. However, this chapter presents an analysis of varied android malware detection approaches and comparing them to supported various parameters like detection technique, analysis method, features extracted and so on. The experiments are based on substantial malware datasets, evaluation parameters and this study employ a wide variety of machine learning techniques, including decision trees and random forests, support vector machines, logistic model trees, and artificial neural networks, also Deep learning techniques. It is a comparative analysis that should be useful in this field for researchers. The analysis shows, based on simple criteria, the

A. B. Siddikk (✉) · Md. F. Muntasir · R. J. Lia · S. S. M. M. Rahman (✉) · T. Islam
Department of Software Engineering, Daffodil International University, Dhaka, Bangladesh
e-mail: abu35-1994@diu.edu.bd

S. S. M. M. Rahman
e-mail: motiur.swe@diu.edu.bd

Md. F. Muntasir
e-mail: fahim35-1900@diu.edu.bd

R. J. Lia
e-mail: rifat35-1845@diu.edu.bd

T. Islam
e-mail: takia35-1014@diu.edu.bd

M. Alazab
College of Engineering, IT and Environment, Charles Darwin University, Darwin, Australia
e-mail: alazab.m@ieee.org

A. B. Siddikk · Md. F. Muntasir · R. J. Lia · S. S. M. M. Rahman · T. Islam
nFuture Research Lab, Dhaka, Bangladesh

Y. Maleh et al. (eds.), *Artificial Intelligence and Blockchain for Future Cybersecurity Applications*, Studies in Big Data 90,
https://doi.org/10.1007/978-3-030-74575-2_7

similarities and differences in essential published research in addition to the accuracy. Thus, this chapter aims to study various android malware detection techniques and to identify plausible research directions. The findings showed that machine learning, with greater detection accuracy, is a more promising method. In order to achieve improved accuracy, future researchers can pursue a deep learning approach with the use of a large dataset.

Keywords Android malware · Deep learning · Machine learning · Malware detection · State-of-art

1 Introduction

The term malware is a contraction of malicious software. Malware is any piece of software that was written with the intent of damaging devices, stealing data, and generally causing a mess. The Android operating system tells the user what systems and data an app will access, but the OS won't block any app activity after installation. Therefore, every Android device should have an Android malware protection program. The risks that an app brings to a device depend on its origins. According to Android Malware Detection Model Based on LightGBM from Wang et al. [7], there are proposed a model this is LightGBM for Machine Learning and they are run this model on a laptop and found 2000 benign samples are downloaded from Baidu app store and Google app store again 2000 malware samples are downloaded from VirusShare (www.virusshare.com). They found the accuracy of LightGBM is 96.4%. Therefore, LightGBM can reduce the time of execution. Danish et al. [1] proposed the IMCFN method. It is mainly divided into two parts: malware image generation and CNN fine-tuning via backpropagation technique. For evaluations, they used two Datasets to find out the android malware which is the Malimg malware dataset (9,435 samples), and IoTandroid mobile dataset (14,733 malware and 2,486 benign samples) they mainly focused on deep learning.

Malware detection approaches can be either static or dynamic [33, 37, 41] Static malware detection relies on features extracted from executable artifacts such as opcodes, bytecodes, byte level, or strings, while dynamic techniques [12, 24] are based on behavioral features from system calls and sandbox. For Example, Baoguo et al. [8] use two datasets one in Microsoft and another Drebin dataset. In the Drebin dataset, the top 10 malware families are selected and a total of 4020 android malware samples are used for experiments based on static and dynamic features. Moreover, many researchers are struggling to detect android malware using many tools and techniques for their purpose as follows.

There are several methods proposed by the researcher to detect android malware as follows:- VizMal [9] operates on an execution trace of an Android application and visualizes it as a sequence of colored boxes, one box for every second of the duration of the execution. Concededly, it is often wont to debug a malware detection method by performing a fine-grained analysis of misclassified applications.

MaMaDroid [36] builds a model of the sequence of API calls as Markov chains, which are successively wont to extract features for machine learning algorithms to classify apps as benign or malicious. Markov Chains are memoryless models where the probability of transitioning from a state to a different one only depends on the present state. This is often represented as a group of nodes, each like a special state, and a group of edges connecting one node to a different label with the probability of that transition.

DroidSieve [34] relies on several features known to be characteristic of Android malware, including API calls. It performs a completely unique deep inspection of the app to spot discriminating features missed by existing techniques, including native components, obfuscation artifacts, and features that are invariant under obfuscation. Evaluate its robustness on a group of over 100K benign and malicious Android apps. For detection, they achieve up to 99.82 curacies with zero false positives. Equivalent features allow family identification with an accuracy of 99.26%. They evaluate its robustness on a set of over 100K benign and malicious Android apps. For detection, they achieve up to 99.82% accuracy with zero false positives. The same features allow family identification with an accuracy of 99.26%.

The main goal of this chapter can be described as follows:

- The most widely used malware detection techniques (machine learning & deep learning) has been analyzed from different top ranking publisher.
- Identify the most used dataset to detect Android malware with an integrated solution.
- To figure out the most optimized algorithm that is effective for detecting android malware.
- Detecting the optimized parameter that is sufficient for generating the malware detection outcome.
- The top ranking publisher (IEEE, Springer etc.) that published the most android-related malware paper has been analyzed.

The rest of the chapter is organized as follows: In this chapter, it has tried to show a statistical analysis of Android malware detection from 2015 to 2020 best-published paper among all best publishers (IEEE, Springer, Elsevier, etc.). In Sect. 2, have briefly reviewed the Statistical, Dynamical analysis on malware Android detection. In Sect. 3, try to explain the process of this work using the proposed methodology. In Sect. 4, show a report on the results of an analysis based on this study for understanding the Android malware from the Analysis and so on. Finally, this paper ends with Sect. 5 presenting the final result.

2   The Most Popular Methods for Detecting Android Malware

For obtaining features there are some methods named static analysis, dynamic analysis and hybrid analysis in android malware detection.

2.1   Static Analysis Approach

The static analysis relies on all the features which are collected without the execution of the code. Several studies including [45–50] have performed static analysis among the 150 papers have been used in this study all of which have been gathered at the reputed publisher sites like IEEE, Springer etc. Mahindru et al. [10]executed an approximate pattern for inspecting any program's attributes for making a balance between correct examining accuracy and methodical number crunching. Alazab et al. [10] had discussed the static analysis of the papers from 2011 to 2016. It showed the detection accuracy rate on different algorithms and also the application classification rate. Taheri et al. [11] showed and analyzed different techniques using static approaches, such as Drebin, StormDroid, and DroidSIFT, ANASTASIA, AndroSimilar, SDHash which are applied on Android apps. Lopes et al. [16] mentioned some static malware detection methods such as Android Asset Packaging Tool (AAPT). It retrieves files from APK. It also mentioned a malware detection framework that used permissions and API calls as features. Yen et al. [17] said about some basic knowledge of static analysis and also some approaches and techniques of it. Zero-day attacks and logic make the static method harder to do its job. Nowadays, most static malware detection techniques are based on content signatures. Call graphs, system calls, dendroid are some of the techniques. A. Saracino et al. [18] mentioned a tool named Alterdroid that compares the differences in behavior between an original app. MADAM is the first system that aims at detecting and stopping at run-time any kind of malware, without focusing on a specific security threat, using a behavior-based and multi-level approach. Not only the accuracy of the runtime detection of MADAM is very high, but it also achieves low performance (1.4%) and energy overhead (4%). An API level of Android application security authentication mechanism (ASCAA), Drebin and a model based on API calls and the use of permissions available in various Android applications for capturing features related to malware behavior are mentioned in Zhao et al. [20]. Wang et al. [39] said about the two main techniques of static analysis named Data flow tracking and Decompiling and also discussed some static analysis tools such as Smali and Apktool.

2.2 Dynamic Analysis Approach

At runtime, dynamic analysis will recognize application behavior and it is often performed in a sandbox environment.

Dynamic analysis is often progressively instructive since it just dissects code that basically executes. Dynamic analysis-based malware detection methods acquire high performance. As well but longer and resource-consuming than their static embodiment. However, dynamic analysis methods are simpler on the detection of the latest malware even as sorts of existing malware. Lopes et al. [16] expressed con to static analysis, the dynamic analysis consists of the execution of a given application during a sandboxed environment, so on monitor its conduct. It's being detected to unknown malware and also developed an android malware detection that uses the frequency of invoked system calls at runtime as features using dataset compromised malicious samples and benign samples in an emulator employing a tool named Monkey. Yene et al. [17] said this sort of method picks out android malware executant the whole apk file, it requires some content to possess associated with the appliance, and await the trigger moment. Zhao et al. [20] mentions that behavior-based detection technology is its superb achievement in dealing with code obfuscation encryption. The feature databases are small and don't require frequent updates. Consequently it's more wont to detect unknown applications almost like known behavior patterns.

The method of anomaly behavior and the use of device calls log is commonly used in the approach to dynamic analysis. The identification of patterns in a specific dataset that do not adhere to a deep-rooted lawful activity is based on anomaly behavior. Although this technique is capable of detecting unknown applications effectively, the false positive rate is high. System Call Log is a process where a software request is serviced from the kernel of the basic operating system. At the kernel level, malicious detection is conducted at high detection precision with few false positive rates.

3 Methodology

A methodology for completing the integrate technology has been used in this article. Initialization, preprocessing, Final Selected Manuscript, Extracting Details, Comparative Analysis, and Findings are six stages. Both of these are carried out step by step. This paper completed all operations smoothly and within a short time by following these measures. It also allows this analysis to gather and extract from the papers all the important material. This research work is also able to collect the data in a simple way through this technique and that will be secure.

3.1 Initialization

The keywords that are used to search for the necessary information were defined at the very first. The keywords identified for this study are Android malware detection, deep learning, and machine learning. Then the search process began on the basis of the keywords. This paper gathers 250 papers very quickly based on the keywords that are listed.

3.2 Preprocessing

Irrelevant papers have been excluded from the 250 papers in preprocessing and the list is reduced by 210 as several papers have been replicated and out of the keywords and goals. Then, the recent publications were classified and the names were listed. Springer, IEEE, ACM, Elsevier, etc. are some of the famous publications that have been established in preprocessing. Finally, only the papers from 2015 to 2020 focused on the identification of android malware with deep learning or machine learning are chosen and the paper numbers have again been reduced to 150.

3.3 Final Selected Manuscript

Consequently, android malware detection with deep learning and machine learning is focused on keywords after searching articles. Then the number of papers was reduced after recognizing common publications and deleting obsolete papers and even taking only papers from 2015 to 2020. Therefore, the number of final manuscripts chosen is 150, which covers papers from 2015 to 2020.

Here goes the diagram of the methodology (Fig. 1)-

3.4 Extract Information

In this step, all the essential data are collected from all 150 papers. The collected essential data are publisher's name, type of the paper, publication's year, evaluation parameters, used datasets, used algorithms and main contribution. Thus, the publishers' names for the 150 papers are mostly IEEE, Springer, ACM, and Elsevier. Three types of papers are found among all the papers named journals, conferences and book chapters. The papers from 2015 to 2020 have already been collected safely and arranged according to the year. According to the collected data, it proves that different papers have used different evaluation parameters.Some of the evaluation parameters are recall, accuracy, precision, f-measure, TPR, FPR, ROC etc. Some of

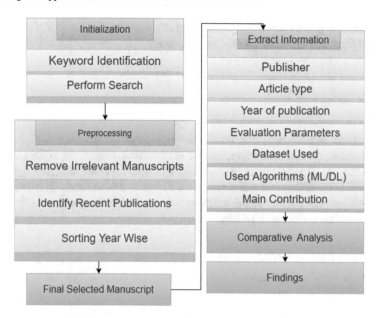

Fig. 1 Diagram of the methodology

the datasets are VirusTotal, Drebin, AMD, ImageNet, AndroZoo, etc. Those papers which used Deep learning algorithms and Machine learning algorithms are collected for this paper. Some deep learning algorithms that are used in the collected papers are Convolutional Neural Network (CNN), Deep Belief Network (DBN), Restricted Boltzmann Machine (RBM), Long short-term memory (LSTM), Deep Neural Nets (DNN), Recurrent Neural Network (RNN), Multimodal Neural Network (MNN), etc. And some machine learning algorithms are used called Support Vector Machine (SVM), K-nearest neighbors (K-NN), K-means, Random Forest(RF), Naïve Bayes (NB), Partial Decision Trees (PART), Neural Network Algorithm (NNA), Multilayer Perceptron (MLP), Decision Tree (DT), etc. This paper also extracted every paper's additional techniques like Data Flow Graph, Control Flow Graph, N-grams, etc.

3.5 Comparative Analysis

Comparative analysis was conducted between them after collecting data from all the documents. For example, all the names of the algorithms and their kind have been specified (deep learning and machine learning). After analyzing the list, the outcome states that Support Vector Machine (SVM) is the most used algorithm in android malware detection and Machine Learning is the most used algorithm sort. Thus, it is obvious from the comparative study that Machine Learning is more used for the detection of Android malware than Deep Learning. For this research work, the assessment parameters of all papers have also been identified. And it means that the most used measurement parameter is accuracy.

3.6 Findings

This analysis would verify the results of the previous steps after completing all the previous steps. For this article, the names of publishers and types of papers were collected from 150 papers from 2015 to 2020. From the previous steps, the algorithms and their forms and the ones that are most used have also been found. The dataset and evaluation parameters, along with the most used dataset and the most used evaluation matrix, are also in the list of results.

4 Result and Discussion

In this section, this proper study was tried to provide comparative analysis based on a comparative study, and the results were finally given in this report. The aim of the study, this article is presented by analyzing various important terms that are currently relevant to the recognition or detection of android malware comprising more than 150 papers. Such pieces of information about our keywords consider's articles available **https://doi.org/10.6084/m9.figshare.12520007.v2**. In this research paper, it showed that most researchers have used a Machine Learning Approach (ML) of the comparison study not only to detect Android Malware, but also to classify several forms of android dataset [10, 42, 43] to find out android malware.

4.1 The Most Applicable Technique

According to the comparative study, A great number of machine learning-based Android malware detection techniques have been proposed in the past few years and till now it has been proposed. Many researchers including [2, 6, 13–15, 19, 21–23, 25–32, 35, 38, 40] used machine learning approaches within 2015-2020 which are covered during the study. Thus, this study found that machine learning techniques in android malware detection have significant contribution and which is increasing day by day.

Machine-learning technique has several false positives applied to real-world data. Most researchers therefore investigated the use of malware machine learning methods [33]. Figure 2 shows that many researchers have used machine learning (ML) techniques over many years to identify Android Malware and researchers expect this to continue in the future.

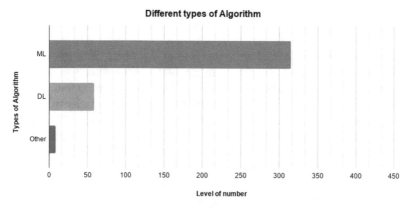

Fig. 2 ML:-Machine Learning approaches; DL:-Deep Learning approaches

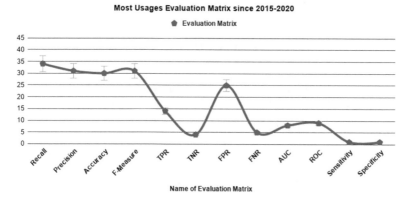

Fig. 3 TPR:-True Positive Rate; TNR:-True Negative Rate; FPR:-False Positive Rate; FNR:-False Negative Rate; AUC:-Area Under the Curve; ROC:-Receiver operating characteristic

4.2 The Most Uses Evaluation Parameters

As shown in Fig. 3 and Fig. 6, most researchers obtained results from the assessment parameters by defining the recall percentage in the assessment matrix. Overall, the model, algorithm and many methods perform very well as a consequence of the performance of the evaluation.

Based on a comparative analysis from 2015 to 2020, all researchers in the paper provided maximum enhancement among all evaluation matrices reached by the Recall to demonstrate and enhance their model or approaches. Alzaylaee et al. [3] suggested one and more methods based on recall, consistency, accuracy, etc. to assess device performance. Since 2015–2020 from this evaluation have found Recall where it is most used to perform their proposed method to evaluate dataset and find out the performance of dataset to detect android malware.

4.3   Analysis of Algorithms

Different researchers have used different types of algorithms at different times for detecting Android malware as shown in the below illustration Fig. 4. A search of Android malware detection-related papers from 2015 to 2020 found that most researchers gave SVM (Support Vector Machine) the highest priority in their paper. Secondly, they have given priority to RF (Random Forest) algorithm and they have used significantly NB (Naive Bayes), FNN (Feedforward Neural Network), DBN (Deep Belief Network), CNN (Convolutional Neural Network), DT (Decision Tree), MLP (Multiple- Path Learning), DNN (Deep Neural Network), NNA (Nearest Neighbour Algorithm). Researchers have used two methods for Android malware detection, one through machine learning and the other through deep learning. A search of Android malware detection related papers from 2015 to 2020 found that most researchers preferred machine learning algorithms for the purpose to detecting Android malware.

From the discussion of the stimulus at last this paper concludes that SVM is a widely used machine learning algorithm for Android malware detection.

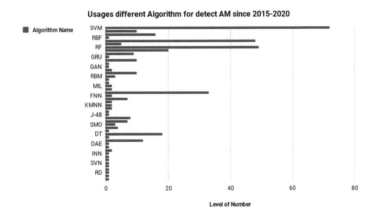

Fig. 4 AM:-Android Malware; SVM:-Support Vector Machines; CNN:-Convolutional Neural Network; NB:-Naive Bayes; RF:-Random Forests; DBN:-Deep Belief Network; DNN:-Deep Neural Network; GAN:-Generative Adversarial Network; NNA:-Nearest Neighbour Algorithm; RBM:-Restricted Boltzmann Machine; MPL:-Multiple- Path Learning; RT:-Real Time; FNN:-Feedforward Neural Network; WANN:-Weight Agnostic Neural Networks; RNN:-Recurrent Neural Network; J-48:-Class of Decision Tree; BN:-Bayesian Network; IBK:-Class of k-nearest Neighbor Algorithm; DT:-Decision Tree; MLP:-Multilayer Perceptron; RR:-Round-Robin; INN:-Incremental Nearest Neighbor; SDG:-Stochastic Gradient Descent; NLP:-Natural Language Processing; RD:-Real Road Networking; BRF:-Balanced Random Forest

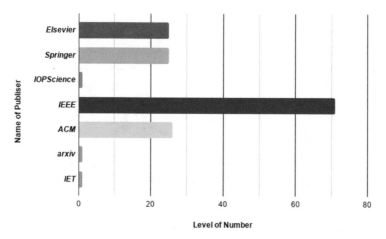

Fig. 5 A comparative analysis of different type of publication

4.4 Publisher

According to the comparative analysis, it stands that IEEE, Elsevier, Springer, ACM, IOPScience, Arxiv, IET are the publishers for all those papers. Figure 5 showed that IEEE had published the highest number of papers among all prominent publishers. Between 2015 and 2020, 72% of papers were published in the IEEEE. Therefore other publishers had published a few papers. Thus, the comparative analysis shows that IEEE has published the highest number of papers on android malware detection based on deep learning and machine learning for the years 2015 to 2020.

4.5 Dataset

From this comparative analysis there have found many datasets where many researchers used different datasets for their purpose to detect intrusion android malware and also showed performance of dataset to how given perfect or accurate result to detect android malware As illustrated in Fig. 6 have analysed many paper since 2015 to 2020. Among all 150 papers there have seen most of all used Drebin dataset to find out the malware and benign. Thus, now the trend goes to the Drebin dataset to detect malware or benign. The Drebin dataset has a lot information which helps to detect intrusion of android malware.

Alazab et al. [44] the book of Deep Learning Applications for Cyber Security in various chapters proposed the Drebin Dataset which contains maximum android applications and malwares using various deep learning applications. Apart from [3–5] used drebin dataset. Drebin is well-known for its scalable and explainable detection.

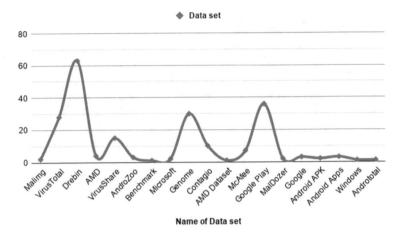

Fig. 6 Classification on different dataset and dataset classes

Table 1 Statistics of most frequently used publishers, algorithms, data sets, classifiers, performance metrics and year-wise distribution of selected studies in Android malware detection.

Top Publishers (2015-2020)			Articles Published by Year			Most used Algorithm(2015-2020) based on classifiers		
Name	No	%	Name	No	%	Name	No	%
IEEE	72	47.9	2015	27	18.0	ML	315	82.7
Elsevier	25	16.7	2016	37	24.6	DL	58	15.2
Springer	25	16.7	2017	32	21.3	Others	08	2.1
ACM	26	17.3	2018	25	16.6			
IET	01	0.7	2019	11	7.3			
Others	01	0.7	2020	18	12.0			
Most used Datasets(2015-2020)			Most used classifiers(2015-2020)			Most used Evaluation Matrix(2015-2020)		
Name	No	%	Name	No	%	Name	No	%
Drebin	63	33.0	SVM	72	18.9	Recall	34	17.6
Google Play	36	18.8	KNN	33	8.7	Precision	31	16.1
Genome	30	15.7	CNN	16	4.2	Accuracy	30	15.5
Virus Total	28	14.7	LSTM	10	2.6	F-Measure	31	16.1
Virus Share	15	7.9	NB	48	12.6	TPR	14	7.3
Contagio	10	5.2	RF	49	12.9	TNR	04	2.1
Android APK	05	2.6	ANN	07	1.8	FPR	25	13.0
MalDozer	02	1.0	NNA	10	2.6	FNR	05	2.6
Malimg	02	1.0	DNN	10	2.6	AUC	08	4.1
Others	04	2.2	Others	07	1.8	ROC	09	4.7

ML:-Machine learning; DL:-Deep learning;SVM:-Support Vector Machine;KNN:-K-nearest Neighbors;CNN:-Convolutional Neural Network;LSTM:-Long Short-term Memory;NB:-Naive Bayes;RF:-Random Forests;ANN:-Artificial neural network;NNA:-Nearest Neighbour Algorithm;DNN:-Deep Neural Network;TPR:-True Positive Rate;TNR:-True Negative Rate; FPR:-False Positive Rate;FNR:-False Negative Rate; AUC:-Area Under the Curve; ROC:-Receiver operating characteristic.

Table 1 Interpretation basically has been used by publishers, articles published by year, algorithms based on Classifiers, datasets, most used classifiers, and performance metrics of envisage studies by the years in 2015 to 2020. This table data has been partially presented previously using several graphs in above. In this table, 'Name', 'No', and '%' refer to the name of the component, the total number of the article those have used the component, and their percentage. Remember and note that all given information has been collected from **150** research articles in **2015–2020**. Firstly, Table 1 interprets the most used publisher android malware detection in research. **IEEE** is the top publisher based on this study among 150 papers almost 72 papers have found where android malware detection papers have been published and Elsevier, Springer respectably. For this study the title of this paper Android malware detection has found maximum in **2016** where total number of paper 37. Secondly has found from 2017 and Thirdly 2015 where have found 27 related papers for that comparative study. This paper based on machine learning and deep learning algorithms Table 1 interprets that the maximum number of algorithms have used **machine learning** for detected android malware where the percentage is 83.7% and second most usages deep learning algorithm where percentage is 15.2% respectively. After that, the list of most used data sets in android malware detection research has been presented. **Drebin** data sets are the most used datasets in android malware detection where considered among all the paper 33.0% used drebin dataset to detect android malware and google play (18.8%) is the second highest used dataset respectively. Next the most used classifiers in android malware detection research. **Suppose vector machine** (SVM) is the most used classifier in the field, it has been considered in 18.9% articles by the respective authors. Second and third used classifiers are Random Forest (12.9%) and Naive Bayes (12.6%) respectively. Finally the most used performance metric in the Android malware detection articles. Most three performance metrics are **Recall** (17.6%), Precision and F-Measure(16.1%) and Accuracy (15.5%) respectively. Moreover, this is the platform which has a huge scope to research android malware detection using machine learning and deep learning algorithms.

5 Conclusion

In this paper, a comparative study of the Approaches, Datasets, Evaluation Parameters and the Trends of Android Malware has been conducted and also study on deep learning and machine learning approaches for intrusion detection. Specifically, this research work analyzed Seven deep learning approaches apart from this paper also analyzed Seven machine learning approaches. This paper has studied keeping a dataset including Drebin, virus total, virus share www.virusshare.com, Genome, Contagio and so on.

From this study and analysis Machine learning is the most uses technique to detect android malware on the other site Drebin is the most used dataset help for intrusion detection of android malware, for this reason, it can be said that the Drebin database and machine learning is the trend for now. As a result, most of the researcher's nowadays used the Drebin dataset and machine learning method to collect malware samples with some important performance indicators, namely, recall, accuracy, precision, false alarm rate, detection rate and so on. Finally, it was found that maximum android malware detection related paper was published in IEEE publication.

For this purpose, researchers need to look at the creation of an improved mechanism in the field of machine learning by exploring more in-depth learning techniques in the detection of Android malware and training the algorithm with large Drebin datasets to fully exploit the model.

References

1. Vasan, D., Alazab, M., Wassan, S., Naeem, H., Safaei, B., Zheng, Q.: IMCFN: image-based malware classification using finetuned convolutional neural network architecture. Comput. Netw. **171**, 107138 (2020)
2. Darabian, H., Homayounoot, S., Dehghantanha, A., et al.: Detecting cryptomining malware: a deep learning approach for static and dynamic analysis. J Grid Comput. (2020)
3. Alzaylaee, M.K., Yerima, S.Y., Sezer, S.: DL-Droid: deep learning based android malware detection using real devices. Comput. Secur. **89**, 101663 (2020)
4. Zou, K., Luo, X., Liu, P., Wang, W., Wang, H.: ByteDroid: android malware detection using deep learning on bytecode sequences. In: Chinese Conference on Trusted Computing and Information Security, pp. 159–176. Springer, Singapore, October 2019
5. Ren, Z., Wu, H., Ning, Q., Hussain, I., Chen, B.: End-to-end malware detection for android IoT devices using deep learning. Ad Hoc Netw. **101**, 102098 (2020)
6. Mercaldo, F., Santone, A.: Deep learning for image-based mobile malware detection. J. Comput. Virol. Hack. Tech. (2020)
7. Wang G., Liu Z.: Android malware detection model based on LightGBM. In: Jain, V., Patnaik,, S., Popențiu Vlădicescu, F., Sethi, I. (eds.) Recent Trends in Intelligent Computing, Communication and Devices. Advances in Intelligent Systems and Computing, vol. 1006. Springer, Singapore (2020)
8. Yuan, B., Wang, J., Liu, D., Guo, W., Wu, P., Bao, X.: Byte-level malware classification based on Markov images and deep learning. Comput. Secur. **92**, 101740 (2020)
9. De Lorenzo, A., Martinelli, F., Medvet, E., Mercaldo, F., Santone, A.: Visualizing the outcome of dynamic analysis of Android malware with VizMal. J. Inf. Secur. Appl. **50**, 101740 (2020)
10. Alazab, M., Shalaginov, A., et al.: Intelligent mobile malware detection using permission requests and API calls. Future Gener. Comput. Syst. **107**, 509–521 (2020)
11. Taheri, R., Ghahramani, M., Javidan, R., et al.: Similarity-based Android malware detection using Hamming distance of static binary features. Future Gener. Comput. Syst. **105**, 230–247 (2019)
12. Abdullah, Z., Muhadi, F.W., Saudi, M.M., Hamid, I.R.A., Foozy, C.F.M.: Android ransomware detection based on dynamic obtained features. In: Ghazali, R., Nawi, N., Deris, M., Abawajy, J. (eds.) Recent Advances on Soft Computing and Data Mining. SCDM: Advances in Intelligent Systems and Computing, vol. 978. Springer, Cham (2020)
13. Lee, W.Y., Saxe, J., Harang, R.: SeqDroid: obfuscated android malware detection using stacked convolutional and recurrent neural networks. In: Deep Learning Applications for Cyber Security, pp. 197–210. Springer, Cham (2019)

14. Ma, Z., Ge, H., Liu, Y., Zhao, M., Ma, J.: A combination method for android malware detection based on control flow graphs and machine learning algorithms. IEEE Access **7**, 21235–21245 (2019)

15. Lou, S., Cheng, S., Huang, J., Jiang, F.: TFDroid: android malware detection by topics and sensitive data flows using machine learning techniques. In: 2019 IEEE 2nd International Conference on Information and Computer Technologies (ICICT), pp. 30–36. IEEE, March 2019

16. Lopes, J., Serrão, C., Nunes, L., Almeida, A., Oliveira, J.: Overview of machine learning methods for Android malware identification. In: 2019 7th International Symposium on Digital Forensics and Security (ISDFS), pp. 1–6. IEEE, June 2019

17. Yen, Y.S., Sun, H.M.: An android mutation malware detection based on deep learning using visualization of importance from codes. Microelectron. Reliab. **93**, 109–114 (2019)

18. Saracino, A., Sgandurra, D., Dini, G., Martinelli, F.: MADAM: effective and efficient behavior-based android malware detection and prevention. IEEE Trans. Dependable Secure Comput. **15**(1), 83–97 (2018)

19. Li, J., Sun, L., Yan, Q., Li, Z., Srisa-an, W., Ye, H.: Significant permission identification for machine-learning-based android malware detection. IEEE Trans. Industr. Inf. **14**(7), 3216–3225 (2018)

20. Zhao, C., Zheng, W., Gong, L., Zhang, M., Wang, C.: Quick and accurate android malware detection based on sensitive APIs. In: 2018 IEEE International Conference on Smart Internet of Things (SmartIoT), Xi'an, pp. 143–148 (2018)

21. Kakavand, M., Dabbagh, M., Dehghantanha, A.: Application of machine learning algorithms for Android malware detection. In: Proceedings of the 2018 International Conference on Computational Intelligence and Intelligent Systems, pp. 32–36, November 2018

22. Rana, M.S., Gudla, C. and Sung, A.H.: Evaluating machine learning models for Android malware detection: a comparison study. In: Proceedings of the 2018 VII International Conference on Network, Communication and Computing, pp. 17–21, December 2018

23. Kim, T., Kang, B., Rho, M., Sezer, S., Im, E.G.: A multimodal deep learning method for Android malware detection using various features. IEEE Trans. Inf. Forensics Secur. **14**(3), 773–788 (2018)

24. Onwuzurike, L., Almeida, M., Mariconti, E., Blackburn, J., Stringhini, G., De Cristofaro, E.: A family of droids-Android malware detection via behavioral modeling: static vs dynamic analysis. In: 2018 16th Annual Conference on Privacy, Security and Trust (PST) (2018). https://doi.org/10.1109/pst.2018.8514191

25. Zhang, Y., Yang, Y., Wang, X.: A novel Android malware detection approach based on convolutional neural network. In: Proceedings of the 2nd International Conference on Cryptography, Security and Privacy (ICCSP 2018), pp. 144–149. Association for Computing Machinery, New York (2018). https://doi.org/10.1145/3199478.3199492

26. Li, W., Wang, Z., Cai, J., Cheng, S.: An Android malware detection approach using weight-adjusted deep learning. In: 2018 International Conference on Computing, Networking and Communications (ICNC) (2018). https://doi.org/10.1109/iccnc.2018.8390391

27. Wang, W., Zhao, M., Wang, J.: Effective Android malware detection with a hybrid model based on deep autoencoder and convolutional neural network. J. Ambient Intell. Humaniz. Comput. (2018). https://doi.org/10.1007/s12652-018-0803-6

28. Varma, P.R.K., Raj, K.P., Raju, K.S.: Android mobile security by detecting and classification of malware based on permissions using machine learning algorithms. In: 2017 International Conference on I-SMAC (IoT in Social, Mobile, Analytics and Cloud) (I-SMAC), pp. 294–299. IEEE, February 2017

29. Milosevic, N., Dehghantanha, A., Choo, K.K.R.: Machine learning aided Android malware classification. Comput. Electr. Eng. **61**, 266–274 (2017)

30. Painter, N., Kadhiwala, B.: Machine-learning-based Android malware detection techniques–a comparative analysis. In: Mishra, D., Nayak, M., Joshi, A. (eds.) Information and Communication Technology for Sustainable Development. Lecture Notes in Networks and Systems, vol. 9. Springer, Singapore (2018)

31. Mahindru, A., Singh, P.: Dynamic permissions based Android malware detection using machine learning techniques. In Proceedings of the 10th Innovations in Software Engineering Conference (ISEC 2017), pp. 202–210. Association for Computing Machinery, New York (2017). https://doi.org/10.1145/3021460.3021485

32. Pektaş, A., Acarman, T.: Ensemble machine learning approach for Android malware classification using hybrid features. In: International Conference on Computer Recognition Systems, pp. 191–200. Springer, Cham, May 2017

33. Varma, P.R.K., Raj, K.P., Raju, K.V.S.: Android mobile security by detecting and classification of malware based on permissions using machine learning algorithms. In: 2017 International Conference on I-SMAC (IoT in Social, Mobile, Analytics and Cloud) (I-SMAC), Palladam, pp. 294–299 (2017)

34. Fereidooni, H., Conti, M., Yao, D., Sperduti, A.: ANASTASIA: ANdroid mAlware detection using STatic analySIs of applications. In: 2016 8th IFIP International Conference on New Technologies, Mobility and Security (NTMS) (2016)

35. Hou, S., Saas, A., Chen, L., Ye, Y.: Deep4MalDroid: a deep learning framework for Android malware detection based on Linux kernel system call graphs. In: 2016 IEEE/WIC/ACM International Conference on Web Intelligence Workshops (WIW) (2016)

36. Chen, S., Xue, M., Tang, Z., Xu, L., Zhu, H.: StormDroid: a streaminglized machine learning-based system for detecting Android malware. In: Proceedings of the 11th ACM on Asia Conference on Computer and Communications Security (ASIA CCS 2016), pp. 377–388. Association for Computing Machinery, New York (2016). https://doi.org/10.1145/2897845.2897860

37. Alzaylaee, M.K., Yerima, S.Y., Sezer, S.: EMULATOR vs REAL PHONE: Android malware detection using machine learning. In: Proceedings of the 3rd ACM on International Workshop on Security And Privacy Analytics (IWSPA 2017), pp. 65–72. Association for Computing Machinery, New York (2017). https://doi.org/10.1145/3041008.3041010

38. Yuan, Z., Lu, Y., Xue, Y.: DroidDetector: Android malware characterization and detection using deep learning. Tsinghua Sci. Technol. **21**(1), 114–123 (2016)

39. Wang, Z., Cai, J., Cheng, S., Li, W.: DroidDeepLearner: identifying Android malware using deep learning. In: 2016 IEEE 37th Sarnoff Symposium, pp. 160–165. IEEE, September 2016

40. Atici, M.A., Sagiroglu, S., Dogru, I.A.: Android malware analysis approach based on control flow graphs and machine learning algorithms. In: 2016 4th International Symposium on Digital Forensic and Security (ISDFS), pp. 26–31. IEEE, April 2016

41. Kate, P.M., Dhavale, S.V.: Two phase static analysis technique for Android malware detection. In Proceedings of the Third International Symposium on Women in Computing and Informatics (WCI 2015), pp. 650–655. Association for Computing Machinery, New York (2015). https://doi.org/10.1145/2791405.279155

42. Wang, X., Yang, Y., Zeng, Y.: Accurate mobile malware detection and classification in the cloud. SpringerPlus **4**(1), 1–23 (2015)

43. Chuang, H., Wang, S.: Machine learning based hybrid behavior models for Android malware analysis. In: 2015 IEEE International Conference on Software Quality, Reliability and Security, Vancouver, BC, pp. 201–206 (2015)

44. Alazab, M., Tang, M. (eds.): Deep Learning Applications for Cyber Security. Springer (2019)

45. Rana, M.S., Rahman, S.S.M.M., Sung, A.H.: Evaluation of tree based machine learning classifiers for android malware detection. In: International Conference on Computational Collective Intelligence, pp. 377–385. Springer, Cham, September 2018

46. Rahman, S.S.M.M., Saha, S.K.: StackDroid: evaluation of a multi-level approach for detecting the malware on Android using stacked generalization. In: International Conference on Recent Trends in Image Processing and Pattern Recognition, pp. 611–623. Springer, Singapore, December 2018

47. Russel, M.O.F.K., Rahman, S.S.M.M., Alazab, M.: AndroShow: a large scale investigation to identify the pattern of obfuscated android malware. In: Machine Intelligence and Big Data Analytics for Cybersecurity Applications, pp. 191–216. Springer, Cham (2021)

48. Russel, M.O.F.K., Rahman, S.S.M.M., Islam, T.: A large-scale investigation to identify the pattern of app component in obfuscated Android malwares. In: International Conference on Machine Learning, Image Processing, Network Security and Data Sciences, pp. 513–526. Springer, Singapore, July 2020
49. Russel, M.O.F.K., Rahman, S.S.M. ., Islam, T.: A large-scale investigation to identify the pattern of permissions in obfuscated Android malwares. In: International Conference on Cyber Security and Computer Science, pp. 85–97. Springer, Cham, February 2020
50. Islam, T., Rahman, S.S.M.M., Hasan, M.A., Rahaman, A.S.M.M., Jabiullah, M.I.: Evaluation of N-gram based multi-layer approach to detect malware in Android. Procedia Comput. Scien. **171**, 1074–1082 (2020)

IFIFDroid: Important Features Identification Framework in Android Malware Detection

Takia Islam, Sheikh Shah Mohammad Motiur Rahman, and Md. Ismail Jabiullah

Abstract Android Malware has grown dramatically day by day because of the rising trends of android operating based smartphones. It has become the main attraction point by attackers now-a-days. Thus, android malware detection has become a major field of investigation among the researchers and academicians who are working with in the field of cyber security. As there are lots of research works have done already, it is still major matter of concern to improve the anti-malware tools. In addition, during the development of anti-malware framework the features of android malware plays the major role. During this study, an important features identification and selection technique has been proposed named IFIFDroid and evaluated which is based on wrapper method. However, the proposed approach can minimize the number of features which helps to machine learning (ML) techniques to learn from less features but perform better. It's found that IFIFDroid can ranking features based on the capacity of individual ML algorithms and comparatively provide better result than existing wrapper method. IFIFDroid proves that there is still way to improve the features selection scheme and provide a strong basement of minimizing the power, execution time during the training by ML algorithms. Though if there is less features to fit without losing accuracy then it will minimize the processing resources as well.

T. Islam (✉) · Md. I. Jabiullah
Department of Computer Science and Engineering, Daffodil International University, Dhaka, Bangladesh
e-mail: takia35-1014@diu.edu.bd

Md. I. Jabiullah
e-mail: drismail.cse@diu.edu.bd

T. Islam · S. S. M. M. Rahman (✉)
Department of Software Engineering, Daffodil International University, Dhaka, Bangladesh
e-mail: motiur.swe@diu.edu.bd

nFuture Research Lab, Dhaka, Bangladesh

Keywords Android malware detection · Features ranking scheme · Android malware analysis · Static analysis

1   Introduction

The usage of mobile devices has been rapidly increasing day by day which is also getting attracted in terms of basic need for its end-user. One report mentioned that android which is a Linux kernel-based operating system (OS) developed by Google, was in the leading position with 82% of total mobile OS in 2016 [1]. Besides that, Android is dominating the mobile market with 85% of the share and has become top positioned in smartphone platforms in 2017 [2, 3] whereas 74% of the universal mobile OS market share is in August 2020 according to StatCounter [4]. Even, Google play contains around 3 million applications that have more than 65 billion downloads [5]. However, because of the vast popularity of android devices, they are being targeted by attackers. In some cases, android devices allow the installation of third-party apps from unknown sources which is also a possible risk to get attacked. In 2016, the rate of attacks in android increased to 40% of total attacks by attackers [1]. In 2017, there was one claim that a total of 316 weaknesses they found on only android operating systems [6]. The statistics from literature and various reports are clearly shown that the popularity of Android OS is growing to customers as well as to the attackers. Attackers are targeting android devices by spreading malware to users. Because of the peak trends of spreading android malwares, it's been a gigantic area of concern among the information security researchers to detect and prevent the malwares in android devices.

To perform a particular task on the device, for instance, sending a text message, each application has to request permission from the user during the installation. However, the majority number of users tend to blindly grant permissions to exotic applications and thereby undermine the purpose of the verification system. As a consequence, malicious applications are hardly enforced by the Android permission system in practice. Android malware detection technology can be divided into three categories: static detection, dynamic detection, and hybrid detection which are found from the state-of-art. Static detection is found on the analysis of defendant code without running the android application. That can obtain high system coverage but faces several countermeasures like code obfuscation and dynamic code loading. As an alternative, dynamic detection contains the analysis of the Android application by running the code. Those can prove compromises that are not easy to explore by static analysis, but the computational assets and time cost of dynamic disclosure are almost high. Hybrid detection is the approach that connects static detection and dynamic detection to obtain an equal between detection effectiveness and efficiency.

Machine learning concept is extensively applied in the detection of Android malware, even based on static, dynamic, or hybrid analysis approaches. The malware detection method which is based on reverse engineering means a classification of general static detection technology. The approach of reverses the implementation based on the semantic features of malicious applications. To decide whether the sample to be detected is a malicious application, it pairs by the specific properties of the recognized malicious applications.

Android malicious applications can execute similar malicious behaviour it called by the APIs [7–9]. It's been identified that there are lots of research works have already done by world renowned researchers but with different features set where different features had influence on the learning base of different machine learning techniques. Thus, the following research questions are considered during this study:

- How can be identified the important features set from an android application for every specific machine learning technique?
- How much influence the features set has on any specific machine learning algorithms?
- How can make a uniform framework for identifying the features set in a random state as it has changed its ranking on every training phase for randomly picking the train set?

However, the major contribution of this research project is proposed a uniform framework to identify important features set before training with machine learning techniques. The framework will help researchers or anti-malware system developers to obtain minimum set of features with maximum detection accuracy. The contributions also include the following:

- It's been found that it is possible to minimize the features set to reach maximum accuracy of any model with minimum features training.
- For producing or generating a model with machine learning algorithms needed more execution time and processing power. Thus, it also can be claimed that as the feature set is less than the learning will take less time and power.

The structure of this chapter is organized as follows. The background and related works are broadly described at Sect. 2. Section 3 represents the proposed framework and research methodology in details. In Sect. 4, evaluation parameters and the Machine Learning (ML) techniques which are used during the implementation and assessment are described. Experimental results and discussion of proposed approach are described with the evaluation of the proposed framework in Sect. 5. Finally, Sect. 6 concludes the chapter with and future directions.

2 Background Study

In this chapter, background study and related works will be discussed and broadly debated.

Alzaylaee [9] proposed a framework based on the deep learning algorithm named DL-Droid. They considered both dynamic and static features for developing their approach. The experiments with more than 30,000 android applications have been performed by them. They also used InfoGain feature ranking algorithm for selecting the important features. Their approach outperformed the combination of dynamic and static features (99.6% accuracy) whereas only the dynamic features provide 97.8% of accuracy. It is mentioned that they performed dynamic analysis using stateful input generation.

Four malware detection methods based on entropy (PDME) and the FalDroid algorithms by using Hamming distance to find similarities between samples proposed by Tehari [10]. They considered their experiments in a different type of features such as API, intent, and permission features on these three datasets. Based on three datasets, including benign and malware Android apps like Drebin, Contagio, and Genome have performed their experiments. The experiment outcomes ensure that their verification accuracy rates of proposed algorithms are more than 90% whereas in some cases, accuracy rates are above 99%.

Ma proposes a combination method for Android malware detection based on the machine learning algorithm and constructed by three detection models for Android malware detection concerning API calls, API frequency, and API sequence. They compared the accuracy and stability of their detection models through a large number of examinations and their experiment's outcome acquired that high accuracy and clearness rate is 98.98% [11].

An anti-malware system that uses customized learning models proposed by Amin which is based on End-to-End deep learning architectures. On that system, operational codes extracted from application attributes of android malware. They have selected to work with independent deep learning models leveraging sequence specialists like recurrent neural networks, Long Short-Term Memory networks, and its Bidirectional variation for static malware analysis on Android. A large number of datasets over 1.8 million android applications show their report an accuracy of 0.999 and F1-score of 0.996 on whereas it can lead to better design of malware detectors [12].

Another android malware detection tool is proposed by McLaughlin using the deep convolutional neural network (CNN) technique. The raw operational code sequences have been extracted from reverse engineering and counted as features during their study. They performed static analysis during the feature's extraction. Though their primary goal was to scan numerous files quickly, they claimed their model to perform on large data with better accuracy [13].

Li introduce Significant Permission IDentification (SigPID), a malware detection system that stands on permission usage analysis to survive the rapidly growing number of Android malware. They proposed three levels of permission data to identify the most significant permissions. Finally, their evaluation finds that their assessment that only 22 permissions are significant and compared another performance of their approach, using only 22 permissions, against a baseline approach that examines all permissions. It is mentioned that they achieve over 90% precision, recall, accuracy, and F-measure, and the analysis times are 4–32 times less than those of using all permissions [2].

MalDozer, a family attribution framework that depends on a sequence classification and automatic Android malware detection using deep learning techniques proposed by Karbab. Based on deep learning techniques they select various malware datasets ranging from 1,000 to 33,000 malware application, and 38,000 benign apps by MalDozer. The solution mentioned that MalDozer accurately detects malware with a false positive rate of 0.06–2, under their all evaluation with multiple datasets, and attributes them to their real families with the F1-Score of 96–99 in percentage [14].

Kim proposes a novel framework for Android malware detection and uses various kinds of features. Those features are clarified using their existence-based extraction method for successful feature representation on malware detection. As a malware detection model, they worked as a multi-modal deep learning technique. Besides, to estimate the performance, they execute several experiments based on 41,260 samples and then compare the accuracy of their model with other deep neural network models. They also evaluated their approach in various aspects between their feature representation method and the usefulness of several features' efficiency in model updates [15].

Based on deep learning algorithms, Ren proposed two end-to-end methods for Android malware detection which have the advantage of their continuous learning activity. They claimed that their proposed methods have the benefit of their continuous learning activity and they evaluated by comparing with some existing detection methods. A dataset containing 8,000 benign and the same number of malicious applications in total 16000 applications used to evaluate their performed. They achieved the detection validity of 93.4 and 95.8 in percentage [16].

Wu introduced an Android malicious application detection structured name called multiview information integration technology (MVIIDroid). On the other hand, their approach extracts applications' multiple components, transforms them into embedding feature vectors and trains a multiple Kernel learning model as the classifier. To describe the effectiveness of their representation, they assess MVIIDroid on two Android malware datasets of 6820 benign applications and 6820 malwares. Besides separating malware from benign applications that they have to achieve superior classification performances [17].

Hou illustrated the Android applications, concerned APIs, and their rich connections as a structured heterogeneous information network (HIN). Instead of using Application Programming Interface (API) calls only, it detects Android malware and further examines it shows that the several connections between them, and create higher-level semantics that requires more effort for attackers to evade the detection. It performed their experimental results to exhibit that their developed system HinDroid outperforms other replacements for Android malware detection techniques [18].

Innovative detection models, proposed by Arora named PermPair, establish and contrast the graphs for malware. Besides extracting a standard sample with the permission pairs from the manifest file of an application. They analyze mainly the pairs of permissions that can be dangerous. It mentioned that they implemented an efficient edge elimination algorithm that was 41% from the normal graph and removed 7% of the useless edges from the malware graph. In addition, the 28% number of decreases in the detection time and shows minimum space utility [19].

Xu performed a detection of DroidEvolver that evaluated on a dataset of 34,722 malicious applications developed over six years and 33,294 benign applications. Based on using the online learning technique, it evolves with feature sets and pseudo label that DroidEvolver makes necessary and lightweight updates. DroidEvolver obtains high detection feature measure (95.27%), which only declines by 1.06% on average per year by the next five years for classifying 57,539 newly presented applications. Their performance ability of DroidEvolver is 28.58 times higher than MAMADROID by malware detection and then compared with the state-of-the-art extra time malware detection method MAMADROID. Finally, the F-measure of DroidEvolver is 2.19 times higher on average [20].

A hybrid model based on deep autoencoder (DAE) proposed by Wang where convolutional neural network (CNN) is used. They recreate the high-dimensional features of Android applications and employ multiple CNN to detect Android malware. They analyzed 13,000 malicious applications and 10,000 benign applications. It mentioned that the accuracy with the CNN-S model is improved by 5%, compared with SVM, while the training time using the DAE-CNN model is reduced by 83% compared with the CNN-S model [21].

Rana assessed four tree-based machine learning algorithms for detecting Android malware in conjunction with a substring-based feature selection approach for the classifiers. For research, they contain 5,560 malware samples where they used the DREBIN dataset with 11,120 applications. Based on machine learning algorithms, they established their performed results While being the Random Forest classifier outperforms the best previously reported solutions. It provides a strong basis for building efficient tools for Android malware detection [22].

Rahman performed a multi-level architecture using stacking concept StackDroid and evaluate which minimizes the error rate. They used the Stacked Generalization process. They used machine learning algorithms and Extreme Gradient Boosting used in level 2 as the final predictor. It mentioned that 97% detection accuracy on the DREBIN dataset and provides an energetic basement for the development of an android malware scanner whichever they obtained 99% of AUC (Area Under Curve), 1.67% of FPR (False Positive Rate) [23].

Russel determined the pattern that is used by attackers to distract malware. They proposed python scripts to extract the pattern of Application (App) components from an obfuscated android malware dataset. Based on the App component pattern, they initiated a matrix form that amassed in a Comma Separated Values (CSV) file. It will conduct to the primary basis of detecting the obfuscated malware [24].

A simulation-based investigation of permissions in obfuscated android malware that was proposed. Based on python scripts to extract the pattern of permissions from an obfuscated malware dataset named Android PRAGuard Dataset. The experimented result shows that the patterns in a matrix form have been found and reserved in a Comma Separated Values (CSV) file which will lead to the fundamental basis of detecting the obfuscated malware [25].

Islam classified the effectiveness of unigram, bigram, and trigram with stacked generalization and unigram provide more than 97% accuracy which is the highest detection rate against bigram and trigram. It mentioned that they were used as a final predictor and meta estimator eXtreme Gradient Boosting (XGBoost). They proved an active foundation to use n-gram techniques in developing android malware detection has been determined from this experiment [26].

Learning-based Android malware detection methods (TLAMD) for IoT Devices was a testing framework proposed by Liu. The proposed framework used on Machine learning techniques. It can perform black-box testing on the system and the evaluation framework can develop adversarial samples for the IoT Android application with a profit rate of nearly 100% [27].

Millar establish three contributions also experimentally exhibit strong against a selection of four prevalent and real-world obfuscation techniques. They propose DANdroid, an innovative Android malware detection that using a deep learning Discriminative Adversarial Network (DAN). It categorizes both obfuscated and unobfuscated applications as each of two malicious or benign. It mentioned that they used three feature sets such as raw opcodes, permissions, and API calls, that are combined in a multi-view deep learning architecture to rise this obfuscation resilience. They performed the dataset of 68,880 obfuscated and unobfuscated malicious and benign samples and multi-view DAN model obtains an F-score of 0.973 and contrast enthusiastically with the state-of-the-art, despite being exposed to the selected obfuscation approach tested both individually and in combination [28].

EveDroid, a scalable and event-aware Android malware detection system, utilizes the behavioral patterns in several cases to effectively detect recent malware based on the observation proposed by Lei. Their events can also reflect apps' possible running activities. On the other hand, they also mention using event groups to describe apps' behaviors at the event level, which can capture a higher level of semantics than in API level and their approaches using API calls as features directly. The performance was based on a dataset that was 14 956 benign and 28 848 malicious Android applications [29].

The used features in the literature are tabulated in Table 1 which prove that the features have significant influence on android malware detection.

3 IFIFDroid: The Proposed Approach

The Proposed Framework or proposed methodology will be described step by step in details in this section.

3.1 Dataset Description

There are lots of public datasets [52–59] available publicly to conduct or experiment during research works. 'DREBIN' dataset is one of the most used datasets among them which is used during the validation or test of the proposed framework. This dataset consists of 123,453 real android applications including 5,560 malware

Table 1 Used features in literature

S/L	Reference	Used features
1	[28]	Raw opcodes, permissions and API calls
2	[29]	Applications behaviors in event level
3	[30]	Static features, dynamic features, and hybrid features
4	[31]	API call graph embedding
5	[32]	URL feature mining
6	[33]	Content-based features, runtime API sequences
7	[34]	System call sequences
8	[35]	Manifest properties, API calls, opcode sequences
9	[36]	Discussed about various features of static analysis including opcode
10	[37]	Permissions, API calls, intents, network traffic, Java classes, and inter-process communication
11	[38]	Call graphs
12	[39]	Network flows and API calls
13	[40]	n-gram features from App's smali code
14	[41]	Permissions, API calls, Network Address and so on
15	[42]	Static features, API package call features and Dynamic Features
16	[43]	Dangerous permissions and components
17	[44]	Opcode sequences
18	[45]	System call
19	[46]	Discuss about permission, intent, uses-feature, application and API including kernel level features
20	[47]	Permission requests and API calls

applications with 179 malware families. From early days of android malware analysis, this dataset performed as a strong basement to study different types of malwares as those malware samples were collected from August 2010 to October 2012 [7, 8].

3.2 Test Bed Setup

The test bed setup is about an experimental environment which includes a Processor of Intel(R) Core (TM) i5-6500 CPU @ 3.20 GHz, 64-bit PC with 16 GB RAM. Linux Mint 18.3 Sylvia was the operating system. Scikit-learn, NumPy, panda and so on which are the packages of python have been used during this study where Python was the programming language.

Fig. 1 Proposed framework

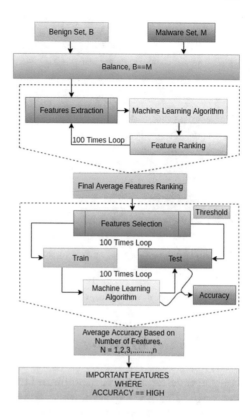

3.3 Pre-processing

It has been mentioned that the dataset has an imbalance in the number of data where only 5560 are malware. Thus, it is necessary to make it balance where there will be equal numbers of malware and benign applications. In this stage, balance operation has been performed using the following formula:

$$f(dataset) = \sigma \begin{cases} \frac{n}{2}, & \text{if } count(malware) = count(benign) \\ 0, & \text{otherwise} \end{cases}$$

After preprocess final dataset has selected based on the equal number of malware and benign. Where (Fig. 1),

$$\sigma = Select,$$
$$0...n = \text{All samples},$$
$$count = \text{Calculate number of sample.}$$

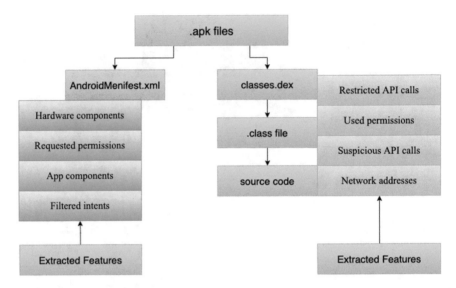

Fig. 2 Reverse engineering process

3.4 Features Extraction

Reverse engineering has been performed to extract the features from the APK file in this stage using Androguard Reverse Engineering Tools [60]. There are two major parts from which the features can be extracted based on the dataset: one is the Manifest.xml (all the permissions are listed there) file and another is the classes.dex (the main source of codes to execute) file. Based on the used dataset, there are total eight features set from those sources depicted in Fig. 2 explained below:

Hardware Components (HC): Android HC also supports VideoCamera, GPS, 3d-accelerometer, compass and provides rich APIs for location and map related functions as well as users can flexibly access, control and process the free Google map. Hardware components implement location based mobile service at low rate cost in mobile systems.

Requested Permissions (RP): The list of Android permissions which are asked to get permitted from users. Android permissions play an important role in the security mechanism allowed by users the installation of an application. Each Malicious Application has run Android 6.0 to request dangerous permission by which it can get access to essential information. For Example: Request CAMERA access permission which is a hardware related permission, for instance Google Play store assumes that the underlying hardware features are required by user's application and filters the application from devices that do not offer it.

App Components (AC): Android app has an app component it is the essential building block. Every component is an entry point through which the system or a user can enter an application. App components are 4 types of such as services, activities, broadcast receivers and content providers.

Filtered Intents (FI): An Intent is a messaging object user can use to request an action from another app component and during inter process and intra process communication in android, intents are performed. Number of malicious applications or malicious activities after rebooting the android phone using BOOT_COMPLETED.

Restricted API Calls (RAC): RAC is performed depend on the allowed permissions of android during the application installation. Malicious activities such as root exploits are indicated by the usage of RAC where the permissions in manifest.xml file aren't requested.

Used Permissions (UP): Whether any application directed to malicious activities or not, it can be identified initially from UP and RAC. Android can define new permissions that are distinct from the pre-installed system permissions and are used to regulate access its.

Suspicious API Calls (SAC): Suspicious API calls means getDeviceId(), Cipher.getInstance(), Runtime.exec() and so on which allows to get access sensitive information about device related to malicious API calls and some of those are used for obfuscation.

Network Addresses (NA): Network addresses which are regularly used by malware to execute external commands or pass data and it minimizes the amount of personal or sensitive data that anyone can transmit over the network.

3.5   Feature Ranking

In this stage, from the rankings list of the features, a high ranked features set has been selected to train a machine learning model and test. The number of high ranked features set such as 1, 2, 3, 4, 5, 6, 7 and 8 features set has been selected sequentially. Initially, Features ranking calculate by CART algorithm and based on coefficient value of each features during training the machine learning techniques. Then, calculate that score for every features 100 times to get more stable scoring from making an average score.

3.6   Features Performance Checking

In this stage, the performance of the selected features with respective machine learning techniques will be evaluated. It's been noticed that here also the machine learning techniques are providing different results for every simulation as the test train was random. To make it reasonable, 100 times loop have been applied and from that the average accuracy for every features set has been calculated.

3.7   Final Selection Based on Performance

From the feature performance, it can be easily selected for which machine learning techniques which features influenced more during the training of that model which will lead a strong basement to develop anti-malware tools.

It's been mentioned that for feature ranking and for training, same ML techniques are applied and evaluated.

4   Evaluation Parameters and Used Machine Learning Techniques

4.1   Evaluation Matrices

Binary classification such as the data as either negative or positive labels has been performed during this study labeled as malware or benign. The decision of this classification has been represented by a structure. Confusion matrix is that structure by which the decision of classifiers can be evaluated [48] Townsend. It consists of with 4 attributes: True positives (TP), True negatives (TN), False positives (FP) and False Negative (FN). True positives (TP) mean correctly identified the benign applications as benign. Identified the malware as benign is defined as False positives (FP). To identify malware as malware correctly referred to True negatives (TN). False negatives (FN) mean the benign one is identified as malware Davis [49]. F1-Score, Precision, Recall, ROC curve, Precision-Recall Curve, Confusion Matrix, False Positive Rate and AUC Sokolova and Boyd. [50, 51] are used during this study to evaluate the effectiveness of IFIFDroid.

ROC Curve: A curve with two plots where False Positive Rate is on the x-axis and True Positive Rate in on the y-axis. The ratio of malware or the fraction of negative values those get wrongly classified as benign or positive is False Positive Rate - FPR. Whereas, TPR - True Positive Rate is the opposite of FPR.

Precision-Recall Curve: A curve with two plots where recall is on the x-axis and precision in on the y-axis. Recall is exactly same as TPR. The rate of correctly identified the true value as true is referred to precision.

Accuracy: The ratio of correctly identified data according to the total amount of data is known as accuracy defined as follows.

$$\text{Accuracy} = (TP + TN)/(TP + TN + FP + FN)$$

Table 2 Representation of features set

S/L	Feature set name	Short form	Set format
1	Hardware components	HC	$\{HC_1, HC_2,\ldots\ldots\ldots, HC_n\}$
2	Requested permissions	RP	$\{RP_1, RP_2,\ldots\ldots\ldots, RP_n\}$
3	App components	AC	$\{AC_1, AC_2,\ldots\ldots\ldots, AC_n\}$
4	Filtered intents	FI	$\{FI_1, FI_2,\ldots\ldots\ldots, FI_n\}$
5	Restricted API calls	RAC	$\{RAC_1, RAC_2,\ldots\ldots, RAC_n\}$
6	Used permissions	UP	$\{UP_1, UP_2,\ldots\ldots\ldots, UP_n\}$
7	Suspicious API calls	SAC	$\{SAC_1, SAC_2,\ldots\ldots, SAC_n\}$
8	Network addresses	NA	$\{NA_1, NA_2,\ldots\ldots\ldots, NA_n\}$

4.2   Machine Learning Algorithms

Machine learning can be defined as a ponder of making machines obtain modern information, unused abilities, reorganize current information. It is utilized in an awfully common way and it alludes to common strategies to extrapolate patterns from large sets or to the capacity to create predictions on new records primarily based on what is learned with the aid of inspecting accessible recognized data. Machine learning techniques can be generally partitioned into two classes: supervised and unsupervised learning. The following are some of the algorithms used for machine learning during this study as followed:

– Extremely Randomized Tree – Extra Tree (ET)
– Random Forest (RF)
– Decision Tree (DT)
– Ada Boost (ADA)
– Gradient Boost (GB)

These algorithms have different categories including: Machine Learning, Ensemble (Bagging Classifiers) and Boosting tree which are used during this study.

5   Experimental Results Analysis and Discussion

In this section, the result obtained from the implementation and assessment of the proposed approach will be described and evaluated briefly. For making the clear representation, the features are labeled as Table 2.

Table 2 represents the features labeling with short form of all features set and the constructions of features in a set. For instance, hardware components may be a set of multiple components like {GPS, camera,.....,touchscreen}. The representation of this set in the dataset is {1, 0, ..., 1} where 1 represents that the components are used in that application if not used then labeled as 0.

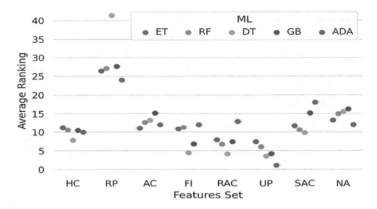

Fig. 3 Comparison of features set in average ranking of all implemented ML techniques

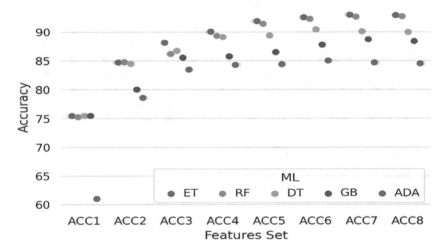

Fig. 4 Comparison of accuracy based on features ranking of all implemented ML techniques

In comparison of all algorithms it's been found that 7 features set on an average from total 8 features set provide maximum accuracy. The top most influenced number of features and accuracy of different machine learning techniques are depicted in Figs. 3 and 4 respectively.

It's been claimed that UP has lower influence on an average for all classifiers from Table 3. On the other hand, RP which is also a set of permissions stand in the first position as a feature set to detect android malwares. Furthermore, RAC only get third position with ADA boost algorithm ranking whereas SAC has fluctuation of ranking by each algorithm. In sum, the feature set of RP, NA, SAC, AC, HC, FI and RAC are the most influenced feature set and top ranked. Finally, traditional wrapper method has been implemented to validate the proposed model and found that proposed approach can improve the accuracy of detection rather than the traditional wrapper method. It's been also obtained that the accuracy difference is not that much

Table 3 Features ranking by each machine learning technique

Feature set	ET	RF	DT	GB	ADA	Average ranking
RP	1	1	1	1	1	1
NA	2	2	2	2	4	2.4
HC	4	6	5	5	7	5.4
AC	5	3	3	4	5	4
SAC	3	5	4	3	2	3.4
RAC	7	7	7	6	3	6
UP	8	8	8	8	8	8
FI	6	4	6	7	6	5.8

Table 4 Comparison between the performance of proposed approach and wrapper method

ML techniques	Wrapper method - accuracy	Proposed method - accuracy
ET	92.87%	93% with 7 feature set
RF	92.15%	92.73% with 7 features set
DT	90.12%	90.49% with 6 features set
GB	88.43%	88.73% with 7 features set
ADA	84.56%	84.68% with 7 feature set

compare to the traditional wrapper method. However, this model indicates that it is possible to improve the traditional wrapper method. The overall comparison of proposed method with traditional wrapper method has been tabulated in Table 4.

6 Conclusion

A feature selection framework named IFIFDroid is proposed and which performed by machine learning methods with multiple algorithms such as Decision Tree, Random Forest, Extremely Randomized Tree, and Gradient Tree Boosting to detect malware on Android by performing static analysis on the DREBIN dataset. Whereas, a static analysis of Android malware applications has been performed considering the features including permission, API call, Intent filter, App component, and System call features are analyzed. In this paper, only eight types of features set are considered. However, there are more features set which will be examined and evaluated with IFIF-Droid. There is also a major point to mention that the difference with the accuracy still not significant with existing wrapper method and in future the framework will be improve by changing some parameters to gain more accuracy in feature ranking, selection and then detection.

References

1. Koli, J.D.: RanDroid: Android malware detection using random machine learning classifiers. In: 2018 Technologies for Smart-City Energy Security and Power (ICSESP), pp. 1–6. IEEE, March 2018
2. Li, J., Sun, L., Yan, Q., Li, Z., Srisa-An, W., Ye, H.: Significant permission identification for machine-learning-based Android malware detection. IEEE Trans. Ind. Inf. **14**(7), 3216–3225 (2018)
3. IDC: Smartphone OS market share, Q1 (2017). https://www.idc.com/promo/smartphone-market-share/os
4. StatCounter. https://www.androidauthority.com/what-is-android-328076/
5. Statista: cumulative number of apps downloaded from Google play as of May 2016. https://www.statista.com/statistics/281106/number-of-android-app-downloads-from-google-play/
6. Agrawal, P., Trivedi, B.: A survey on Android malware and their detection techniques. In: 2019 IEEE International Conference on Electrical, Computer and Communication Technologies (ICECCT), pp. 1–6. IEEE, February 2019
7. Arp, D., Spreitzenbarth, M., Huebner, M., Gascon, H., Rieck, K.: Drebin: efficient and explainable detection of Android malware in your pocket. In: 21th Annual Network and Distributed System Security Symposium (NDSS), February 2014
8. Spreitzenbarth, M., Echtler, F., Schreck, T., Freling, F.C., Hoffmann, J.: MobileSandbox: looking deeper into Android applications. In: 28th International ACM Symposium on Applied Computing (SAC), March 2013
9. Alzaylaee, M.K., Yerima, S.Y., Sezer, S.: DL-Droid: deep learning based Android malware detection using real devices. Comput. Secur. **89**, 101663 (2020)
10. Taheri, R., Ghahramani, M., Javidan, R., Shojafar, M., Pooranian, Z., Conti, M.: Similarity-based Android malware detection using Hamming distance of static binary features. Future Gener. Comput. Syst. **105**, 230–247 (2020)
11. Ma, Z., Ge, H., Liu, Y., Zhao, M., Ma, J.: A combination method for Android malware detection based on control flow graphs and machine learning algorithms. IEEE Access **7**, 21235–21245 (2019)
12. Amin, M., Tanveer, T.A., Tehseen, M., Khan, M., Khan, F.A., Anwar, S.: Static malware detection and attribution in Android byte-code through an end-to-end deep system. Future Gener. Comput. Syst. **102**, 112–126 (2020)
13. McLaughlin, N., Martinez del Rincon, J., Kang, B., Yerima, S., Miller, P., Sezer, S., Safaei, Y., Trickel, E., Zhao, Z., Doupé, A., Joon Ahn, G.: Deep Android malware detection. In: Proceedings of the Seventh ACM on Conference on Data and Application Security and Privacy, pp. 301–308, March 2017
14. Karbab, E.B., Debbabi, M., Derhab, A., Mouheb, D.: MalDozer: automatic framework for Android malware detection using deep learning. Digit. Invest. **24**, S48–S59 (2018)
15. Kim, T., Kang, B., Rho, M., Sezer, S., Im, E.G.: A multimodal deep learning method for Android malware detection using various features. IEEE Trans. Inf. Forensics Secur. **14**(3), 773–788 (2018)
16. Ren, Z., Wu, H., Ning, Q., Hussain, I., Chen, B.: End-to-end malware detection for Android IoT devices using deep learning. Ad Hoc Netw. **101**, 102098 (2020)
17. Wu, Q., Li, M., Zhu, X., Liu, B.: MVIIDroid: a multiple view information integration approach for Android malware detection and family identification. IEEE MultiMedia **27**(4), 48–57 (2020)
18. Rodríguez-Mota, A., Escamilla-Ambrosio, P.J., Salinas-Rosales, M.: Malware analysis and detection on Android: the big challenge. https://www.intechopen.com/books/smartphones-from-an-applied-research-perspective/malware-analysis-and-detection-on-android-the-big-challenge
19. Arora, A., Peddoju, S.K., Conti, M.: PermPair: Android malware detection using permission pairs. IEEE Trans. Inf. Forensics Secur. **15**, 1968–1982 (2019)

20. Xu, K., Li, Y., Deng, R., Chen, K., Xu, J.: DroidEvolver: self-evolving Android malware detection system. In: 2019 IEEE European Symposium on Security and Privacy (EuroS&P), pp. 47–62. IEEE, June 2019
21. Wang, W., Zhao, M., Wang, J.: Effective Android malware detection with a hybrid model based on deep autoencoder and convolutional neural network. J. Ambient Intell. Humaniz. Comput. **10**(8), 3035–3043 (2019)
22. Rana, M.S., Rahman, S.S.M.M., Sung, A.H.: Evaluation of tree based machine learning classifiers for Android malware detection. In: International Conference on Computational Collective Intelligence, pp. 377–385. Springer, Cham, September 2018
23. Rahman, S.S.M.M., Saha, S.K.: StackDroid: evaluation of a multi-level approach for detecting the malware on Android using stacked generalization. In: International Conference on Recent Trends in Image Processing and Pattern Recognition, pp. 611–623. Springer, Singapore, December 2018
24. Russel, M.O.F.K., Rahman, S.S.M.M., Islam, T.: A large-scale investigation to identify the pattern of app component in obfuscated Android malwares. In: International Conference on Machine Learning, Image Processing, Network Security and Data Sciences, pp. 513–526. Springer, Singapore, July 2020
25. Russel, M.O.F.K., Rahman, S.S.M.M., Islam, T.: A large-scale investigation to identify the pattern of permissions in obfuscated Android malwares. In: International Conference on Cyber Security and Computer Science, pp. 85–97. Springer, Cham, February 2020
26. Islam, T., Rahman, S.S.M.M., Hasan, M.A., Rahaman, A.S.M.M., Jabiullah, M.I.: Evaluation of N-gram based multi-layer approach to detect malware in Android. Procedia Comput. Sci. **171**, 1074–1082 (2020)
27. Liu, X., Du, X., Zhang, X., Zhu, Q., Wang, H., Guizani, M.: Adversarial samples on Android malware detection systems for IoT systems. Sensors **19**(4), 974 (2019)
28. Millar, S., McLaughlin, N., Martinez del Rincon, J., Miller, P., Zhao, Z.: DANdroid: a multi-view discriminative adversarial network for obfuscated Android malware detection. In: Proceedings of the Tenth ACM Conference on Data and Application Security and Privacy, pp. 353–364, March 2020
29. Lei, T., Qin, Z., Wang, Z., Li, Q., Ye, D.: EveDroid: event-aware Android malware detection against model degrading for IoT devices. IEEE Internet Things J. **6**(4), 6668–6680 (2019)
30. Liu, K., Xu, S., Xu, G., Zhang, M., Sun, D., Liu, H.: A review of Android malware detection approaches based on machine learning. IEEE Access **8**, 124579–124607 (2020)
31. Pektaş, A., Acarman, T.: Deep learning for effective Android malware detection using API call graph embeddings. Soft. Comput. **24**(2), 1027–1043 (2020)
32. Wang, S., Chen, Z., Yan, Q., Ji, K., Peng, L., Yang, B., Conti, M.: Deep and broad URL feature mining for Android malware detection. Inf. Sci. **513**, 600–613 (2020)
33. Hou, S., Fan, Y., Zhang, Y., Ye, Y., Lei, J., Wan, W., Wang, J., Xiong, Q., Shao, F.: αCyber: enhancing robustness of Android malware detection system against adversarial attacks on heterogeneous graph based model. In: Proceedings of the 28th ACM International Conference on Information and Knowledge Management, pp. 609–618, November 2019
34. Xiao, X., Zhang, S., Mercaldo, F., Hu, G., Sangaiah, A.K.: Android malware detection based on system call sequences and LSTM. Multimedia Tools Appl. **78**(4), 3979–3999 (2019)
35. Feng, R., Chen, S., Xie, X., Ma, L., Meng, G., Liu, Y., Lin, S.W.: MobiDroid: a performance-sensitive malware detection system on mobile platform. In: 2019 24th International Conference on Engineering of Complex Computer Systems (ICECCS), pp. 61–70. IEEE, November 2019
36. Pan, Y., Ge, X., Fang, C., Fan, Y.: A systematic literature review of Android malware detection using static analysis. IEEE Access **8**, 116363–116379 (2020)
37. Kouliaridis, V., Kambourakis, G., Geneiatakis, D., Potha, N.: Two anatomists are better than one-dual-level Android malware detection. Symmetry **12**(7), 1128 (2020)
38. Zhang, H., Luo, S., Zhang, Y., Pan, L.: An efficient Android malware detection system based on method-level behavioral semantic analysis. IEEE Access **7**, 69246–69256 (2019)
39. Taheri, L., Kadir, A.F.A., Lashkari, A.H.: Extensible Android malware detection and family classification using network-flows and API-calls. In: 2019 International Carnahan Conference on Security Technology (ICCST), pp. 1–8. IEEE, October 2019

40. Zhang, Y., Ren, W., Zhu, T., Ren, Y.: SaaS: a situational awareness and analysis system for massive Android malware detection. Future Gener. Comput. Syst. **95**, 548–559 (2019)
41. Zhang, L., Thing, V.L., Cheng, Y.: A scalable and extensible framework for Android malware detection and family attribution. Comput. Secur. **80**, 120–133 (2019)
42. Han, Q., Subrahmanian, V.S., Xiong, Y.: Android malware detection via (somewhat) robust irreversible feature transformations. IEEE Trans. Inf. Forensics Secur. **15**, 3511–3525 (2020)
43. Jiang, X., Mao, B., Guan, J., Huang, X.: Android malware detection using fine-grained features. Sci. Program. **2020**, article ID: 5190138 (2020). https://doi.org/10.1155/2020/5190138
44. Pektaş, A., Acarman, T.: Learning to detect Android malware via opcode sequences. Neuro-computing **396**, 599–608 (2020)
45. Surendran, R., Thomas, T., Emmanuel, S.: GSDroid: graph signal based compact feature representation for Android malware detection. Expert Syst. Appl. **159**, 113581 (2020)
46. Alqahtani, E.J., Zagrouba, R., Almuhaideb, A.: A survey on Android malware detection techniques using machine learning algorithms. In: 2019 Sixth International Conference on Software Defined Systems (SDS), pp. 110–117. IEEE, June 2019
47. Alazab, M., Alazab, M., Shalaginov, A., Mesleh, A., Awajan, A.: Intelligent mobile malware detection using permission requests and API calls. Future Gener. Comput. Syst. **107**, 509–521 (2020)
48. Townsend, J.T.: Theoretical analysis of an alphabetic confusion matrix. Percept. Psychophys. **9**(1), 40–50 (1971)
49. Davis, J., Goadrich, M.: The relationship between Precision-Recall and ROC curves. In: Proceedings of the 23rd International Conference on Machine Learning, pp. 233–240. ACM, June 2006
50. Sokolova, M., Japkowicz, N., Szpakowicz, S.: Beyond accuracy, F-score and ROC: a family of discriminant measures for performance evaluation. In: Australasian Joint Conference on Artificial Intelligence, pp. 1015–1021. Springer, Heidelberg, December 2006
51. Boyd, K., Eng, K.H., Page, C.D.: Area under the precision-recall curve: point estimates and confidence intervals. In: Joint European Conference on Machine Learning and Knowledge Discovery in Databases, pp. 451–466. Springer, Heidelberg, September 2013
52. Zhou, Y., Jiang, X.: Dissecting Android malware: characterization and evolution. In: 2012 IEEE Symposium on Security and Privacy, pp. 95–109. IEEE, May 2012
53. Damshenas, M., Dehghantanha, A., Choo, K.K.R., Mahmud, R.: M0Droid: an Android behavioral-based malware detection model. J. Inf. Priv. Secur. **11**(3), 141–157 (2015)
54. Kiss, N., Lalande, J.F., Leslous, M., Tong, V.V.T.: Kharon dataset: Android malware under a microscope. In: The LASER Workshop: Learning from Authoritative Security Experiment Results (LASER 2016), pp. 1–12 (2016)
55. Li, Y., Jang, J., Hu, X., Ou, X.: Android malware clustering through malicious payload mining. In: International Symposium on Research in Attacks, Intrusions, and Defenses, pp. 192–214. Springer, Cham, September 2017
56. Wei, F., Li, Y., Roy, S., Ou, X., Zhou, W.: Deep ground truth analysis of current Android malware. In: International Conference on Detection of Intrusions and Malware, and Vulnerability Assessment, pp. 252–276. Springer, Cham, July 2017
57. Lashkari, A.H., Kadir, A.F.A., Gonzalez, H., Mbah, K.F., Ghorbani, A.A.: Towards a network-based framework for Android malware detection and characterization. In: 2017 15th Annual Conference on Privacy, Security and Trust (PST), p. 233-23309. IEEE, August 2017
58. Maiorca, D., Ariu, D., Corona, I., Aresu, M., Giacinto, G.: Stealth attacks: an extended insight into the obfuscation effects on Android malware. Comput. Secur. **51**, 16–31 (2015)
59. Allix, K., Bissyandé, T.F., Klein, J., Le Traon, Y.: AndroZoo: collecting millions of Android apps for the research community. In: 2016 IEEE/ACM 13th Working Conference on Mining Software Repositories (MSR), pp. 468–471. IEEE, May 2016
60. Androguard. https://github.com/androguard/androguard

AntiPhishTuner: Multi-level Approaches Focusing on Optimization by Parameters Tuning in Phishing URLs Detection

Md. Fahim Muntasir, Sheikh Shah Mohammad Motiur Rahman, Nusrat Jahan, Abu Bakkar Siddikk, and Takia Islam

Abstract Phishing is an alarming issue among the cybercriminals. In the last decade, online services have revolutionized the world. Due to the revolutionary transformations of web service, the reliance on the web has increased day by day. Security threats have emerged due to the increasing reliance on online orientation. There are many types of anti-phishing solutions available that have been proposed by many researchers. However, this chapter is to propose an intelligent framework to detect phishing URLs based on the optimized learning architecture scheme. Multi-layer based structures have been implemented to detect phishing URLs using Deep Neural Network (DNN), Neural Network (NN) and Stacking. These architectures are evaluated with various tuning hyper-parameters to obtain the optimized output named AntiPhishTuner. As a result, five-layer based DNN can provide accuracy of 0.95 with the minimum mean squared error (MSE) 0.30, and also a mean absolute error (MAE) 0.074 where the number of epochs was 50 and Adam optimizer as an optimizer. Using two-layer NN with AdaGard optimizer can provide accuracy of 0.95, with MSE 0.30 and MAE 0.074. NN provides these results with 150 epochs. Stack generalization can reach maximum accuracy 0.97 in binary classification with MAE 2.1. This chapter can provide a better lead to researchers and anti-phishing tools

Md. F. Muntasir (✉) · S. S. M. M. Rahman (✉) · N. Jahan · A. B. Siddikk · T. Islam
Department of Software Engineering, Daffodil International University, Dhaka, Bangladesh
e-mail: fahim35-1900@diu.edu.bd

S. S. M. M. Rahman
e-mail: motiur.swe@diu.edu.bd

N. Jahan
e-mail: nusrat.swe@diu.edu.bd

A. B. Siddikk
e-mail: abu35-1994@diu.edu.bd

T. Islam
e-mail: takia35-1014@diu.edu.bd

Md. F. Muntasir · S. S. M. M. Rahman · A. B. Siddikk · T. Islam
nFuture Research Lab, Dhaka, Bangladesh

Y. Maleh et al. (eds.), *Artificial Intelligence and Blockchain for Future Cybersecurity Applications*, Studies in Big Data 90,
https://doi.org/10.1007/978-3-030-74575-2_9

developers to make an initial decision about the approach that should be followed for further extension.

Keywords Uniform resource Locator (URL) · Phishing · Deep Neural Network (DNN) · Neural Network (NN) · Stacking

1 Introduction

Phishing is a fraudulent technique used by both social and technological engineering for the purpose of stealing user identities and personal account information and credentials from financial accounts Huang [22]. There are a broad variety of phishing forms, including algorithms, link handling, email phishing, domain spoofing, phishing using HTTPS, SMS, pop-ups. prefix, suffix, subdomain, IP address, URL-length, '@' symbol, spear phishing, dual-slash attributes, port, https token, request URL, URL-anchor, tag-links, domain age are phishing attributes Rahman [13].

The elements of a phishing platform are typically equivalent to a few legitimate websites literally and externally. Today's security concerns are increasingly rising due to phishing. According to an eminent Washington-based cyber security company F5 Systems, Inc., which stamped its target choice, sociology and technological infiltration, a phishers strategy combines three special tasks Pompon [23]. According to the Anti-Phishing Working organization, there were 18,480 momentous phishing attacks and 9666 curiously phishing regions in March2006. It impacts billions of site clients and enormous costing boundaries to businesses Viktorov [24]. The prospective expenditure of computerized offense to the around the global network could be a phenomenal 500 billion USD and a clue break will fetch the ordinary organization around 3.8 million USD expenditure, considering that evidence by Microsoft, in 2018.

There are several proposed solutions that researchers have provided. For example, detect phishing websites through a hierarchical clustering approach which bunches the vectors produced from DOMs together concurring to their corresponding distance Cui [25]. A few considers centered on detecting phishing URLs by using the potential characteristics of URLs. One to two hidden layers are usually used for neural networks. In some cases of deep learning, the number of layers varies. But it requires nearly more than 150 layers Le [11]. There are a few rules to decide the number of layers that incorporate two or less layers for basic data sets and for computer vision, time series, or with intricate datasets extra layers can give way better results Rahman [13]. Mostly classification the data patterns are accessible in a structured way. But the URL information isn't accessible in a settled pattern. Applying the classification methods or machine learning techniques in URL data. In this way additional approaches ought to be utilized for overseeing the URLs Woogue [26]. Phishing could be a pivotal issue in web security. Phishing detection technique Enables URLs recognition through Various URLs evaluations. Apropos assess the URLs, a number of procedures are accessible. Among the accessible techniques the machine learning techniques are more compelling and precise. Such techniques the malicious URLs patterns become acquainted by classification algorithm and when requisite. It distinguishes the URLs sorts that are phishing or legitimate Dong [6]. Phish tank database is a norm assortment that keeps track of phishing reported URLs by various web security organizations. This database stores a variety of features Mohammad [27].

Phishing is the most known online security threat and it can be called fraudulent practice on criminal activities. Which is the main concern of phishing attackers. Usually, phishing attackers mimic legitimate websites for credential information such as online banking, e-commerce websites so that user's expose their sensitive information such as name, password, login credential, credit card information, health-related information etc. refers to mimic sites. Attackers collect user's information and carry out various fraudulent activities by phishing attacks Abutair [1].

URLs play a significant role in phishing attacks, where attackers send malicious URLs to users through various communication channels such as emails, social media, etc., and sending URLs look like a valid URLs Shirazi [17].

Typically, three ways are used to take advantage of phishing attacks Hutchinson [9]. First of all, mimic the legitimate web interface which looks exactly like a legitimate interface is called web-based phishing. Considering it valid, Phishers fool user provides credential information. Secondly, attackers use web-based techniques to send phishing content via email. The third one is which phishing attack also occurred by malware-based where attackers inject malicious code to user's system Dong [6]. In any case, why machine learning-based anti-phishing framework is used for phishing detection? Because to detect those phishing attacks some traditional approaches like Blacklisting, regular expression, and signature matching are used, however those approach fail to detect unknown URLs Rahman [13].

Detecting the unknown pattern of malicious URLs database signatures have always remains updated. However, by the expansion of research in the number of machine learning-based research for malicious URLs detection, it's observed that deep learning-based architecture provides better performance than existing machine learning algorithms Harikrishnan [10].

The principal objectives of this chapter can be stated as follows:

– Assessment of AntiPhishTuner with tuning optimizer for Neural Network as well as Deep Learning (Deep Neural Network).
– Phishing URLs detection has been implemented to improve the accuracy by the stacking concept.
– Combining all types of classification can perform phish stack, like machine learning, ensemble learning and neural network based approach as base classification.
– Expressing intelligent Anti-phishing architectures with optimization tuning.
– Effect of learning rate in neural network-based technique.
– Appraise of training accuracy with regard to mutate in learning rate.
– Detecting the optimized parameter that are suitable to develop the result for DNN and NN.
– Detecting the combination of adaptive learning optimization algorithm with DNN and NN.

The remainder of the paper is organized as follows: in Sect. 2, represents, Literary Review. In Sect. 3, represents the methodology of phishing URLs detection using multilayer approaches and the dataset information which are used to experiment and evaluate has been described. Experiments, evaluation parameters along with obtained results have been identified and analyzed in Sect. 4. Finally, Sect. 5 concludes the paper.

2 Literature Review

Adebowale [2] proposed an ordinary technique that there are some users who steal confidential information from websites and call those users are phishing users. This activity commonly happens by fake websites or malicious URLs that are called fraudulent ventures. Cybercriminals use fraudulent activities to create a well-designed phishing attack. Gaining access to the victim's systems the cybercriminals could install malware or inapt protected user systems.

Acquisti [4] suggested that reduce the threat of phishing assaults, indicating at directing the hazard of phishing attacks, various strategies are recommended to get ready and instruct end-users to recognize phishing URLs.

Wang [20] suggested ensemble classifiers for e-mail filtering that excluded five algorithms that are Support Vector Machines, K-Nearest Neighbor, Gaussian Naive Bayes, Bernoulli Naive Bayes, and Random Forest Classifier.

Ultimately random forest was improved accuracy 94.09% to 98.02%. Gupta, S. and Singhal, A. [8] proposed that approximately for minimum execution time random forest tree is an admirable strategy to detect malicious URLs.

Vrbančič [19] recommended setting parameters of deep learning neural networks that are swarm intelligence-based techniques. After that the proposed technique applied to the classification of phishing website and capable of better detection by comparing to the existing algorithm.

El-Alfy, E. S. M. [7] recommended for training the nodes framework that connected unsupervised and supervised algorithms. Phishing sites depend on feasibility neural networks and clustering K medoids. Feature selection and module is used to reduce space capacity is used by K-medoid technique. Thirty features are achieved 96.79% accuracy by the desired technology.

Le [11], recommended to DNN, are trained with implied deep stacking. The evaluated covers of the past outlines are upgraded as they were at the conclusion of each DNN preparing epoch, and after that, the upgraded evaluated veils give extra inputs to train the DNN within the other epoch. At the test period, the DNN makes expectations successively in a repetitive manner. In expansion, we propose to utilize the L1 loss for training. Implicit.

Winterrose [21] claimed that exploring distinctive properties of veritable oversees methodologies for recognizing phishing web goals. Phishing URLs utilizing significant learning strategies, for the case, profound Boltzmann machine (DBM), stacked auto-encoder (SAE), and profound neural organization (DNN). DBM and SAE are utilized for pre-preparing the show with a predominant depiction of information for attribute assurance. DNN is utilized for twofold gathering in recognizing darken URL as either a phishing URL or a genuine URL. The proposed system fulfills a higher area rate of 94% with an under most false-positive rate than other machine learning procedures.

Rahman [30] suggested that to detect phishing attack in several anti-phishing systems for that reason used six machine learning classifiers (KNN, DT, SVM, RF, ERT, and GBT) and three publicly accessible datasets with multidimensional attributes could be used due to a lack of proper selection of machine learning classifiers. Using confusion matrix, precision, recall, F1-score, accuracy and misclassification rate to evaluate the performance of the classifiers. Find better performance that obtained from Random Forest and Exceptionally Randomized Tree of 97% and 98% accuracy rate for detection of phishing URLs respectively. Gradient Boosting Tree offers the best performance with 92% accuracy for multiclass feature set.

Sahingoz [31] proposed a real-time anti-phishing process that combines seven different classification algorithms also with different feature sets. Through using NLP-based Random Forest algorithm, 97.98% accuracy was observed.

3 AntiPhishTuner: Proposed Approach

The proposed approach has been depicted in Fig. 1 and described in details step by step in this section.

3.1 Dataset

A publicly accessible dataset has been used for training or creating the architecture. The initial part of this model is to collect data and analyze the datasets. This dataset was collected from the UCI repository. It has a total of 11055 different types of URLs. It has a total 30 features used to train the model Rahman [14]. Table 1 represents the various aspects of the used dataset.

Table 1 Dataset information

Total features of dataset	30 features
Total URLs	11,055 URLs
Phishing URLs	4898
Legitimate URLs	6157

3.2 Feature Description

Before analyzing the features selection part, features and the ability to use these features need to be evaluated. Basically, there are four primary features and a total of 30 sub-features. Based on the details, each feature offers details as to whether the website may be phishing, legitimate or suspicious. This segment provides the planning to point up the features.

1. Address bar-based features: The address bar that means URL bar or location bar could be a GUI gadget that appears in an ongoing URL. According to the dataset it has 12 sub-features. That is appeared on the Table 2 below.

Table 2 Address bar-based features

Name of the features	Explanation
Ip Address	In the event that IP address is utilized as an elective of a domain name within the URL that is a phishing website and client can almost be sure somebody is attempting to take his credential data. From this dataset, discover 570 URLs having an IP address which add up to 22.8% of the dataset and proposed a rule IP address is in URL that called Phishing, otherwise its Legitimate
Length of URLs	Long URLs are mostly utilized to cover up the dubious portion within the address bar because it contains malicious content. Deductively, no well-founded length that recognizes phishing URLs from legitimate ones. For that legitimate URLs proposed length of the URLs is 75. In this study to guarantee the accuracy measured the length of URLS is suspicious, legitimate or a phishing site in this dataset and proposes an average length. From this proposed condition the URL length is less than or equal 54 and it is classified as legitimate, if the URL is larger than 74 then it is phishing. According to the dataset found 1220 URLs that's length greater than or equal 54
TinyURLs	For shortening the URL length tinyURL is used. It diverts to the most page to click the shorter URL. This interface is like a phishing site since rather than an authentic site it diverts the end client to fake sites
Operate the @ Symbol	Web browsers mostly ignore the segment that is attached with @ symbol. Because it is kept away from real addresses. According to the dataset, finding 90 URLs that have the '@' symbol will add up to only 3.6%
Operate the "//" symbol	After HTTP or HTTPS the "//" symbol is used as legitimate URLs. On the off chance that after the initial protocol statement that's considered phishing URLs. "//" symbol is utilized for diverting to other sites
Domain names prefix or suffix separated by "-" symbol	If any URL contains the "-" symbol in its domain name then consider it's a phishing URLs. Generally validated URLs don't contain the "-" symbol
Operate the "." symbol in domain	Operate the "." symbol in domain When a sub-domain with the domain name is added, it has to include dot. Considering suspicious in case drop out more than one subdomain and larger than that will point it like a phishing
HTTPS with secure socket layer	Most of the legitimate site HTTPS protocol and the age of certificate is exceptionally vital for using HTTPS. For this that's need a trusted certificate
Expiry date of domain	Principally domain name have longer expiry date for legitimate sites

(continued)

Table 2 (*continued*)

Name of the features	Explanation
Favicon	Favicon can divert clients to suspicious sites, when it is stacked from outside space. It's by and large utilized in websites and it's a graphic image
Utilizing insignificant ports	Phishers continuously discover defenselessness and attempt to require an advantage on the off chance that any URLs has some open ports that's superfluous
HyperText Transfer Protocol in domain	The phishing websites are considered if any URLs of this website have HTTPS on domain name

Table 3 Abnormal based features

Name of the features	Explanation
Request URL	From another domain on the off chance that a page contains larger amount of outside URLS that's considered it suspicious or phishing
Having URL of anchor	Comparable to the request URL features, the chance of phishing increases, more \<a\> tags utilized inside the site
Link among (Meta, script, Link) tag	It is calculated as either suspicious or phishing formed on their proportion if the tag contains large number of outer links
Server form handler	Phishing is considered in case the Server shape handler is blank or empty. Server frame handler diverts to a distinctive domain It's checked as suspicious
Having an email to submitting information	It is considered as phishing, rather than a server, web form coordinated to an individual email is submitted the information
Abnormal URLs	It considered as phishing, In case the character isn't included within the URLs

2. Abnormal Based Features: It for the most part centers on abnormal exercises on the site. According to the dataset it has 5 sub-features. That appears on the Table 3 below.

3. HTML and JavaScript based features: According to the dataset it has 5 sub-features. That appears on Table 4 below.

4. Domain based features: Using domain names prepares effortlessly identifiable and unforgettable names numerically. According to the dataset it has 7 sub-features. That appears on Table 5 below.

Table 4 HTML and JavaScript based features

Name of the features	Explanation
Forwarding website	It can be frightening, on the off chance that diverting is happened different times
Customization of status bar	To alter the status bar of the URLs can be utilized on "Mouseover" occasion. It continuously appears off genuine URLs and stows away the fake URLs. at a time When it's connected on the site that's obliging as phishing
Right click disabled	Users can't check the source code; right-click functions are impaired mainly by Phishers. When the framework is debilitated within the site that's obliging as phishing
Having Pop-up Window	Pop-up window with a text field is consisted by a web page that's obliging as phishing
Custom IFrame	Stowing them away within the website phisher could be utilized IFrame. In for the most part Connect outside substance to appear in a domain utilized by IFrame

Table 5 Domain based features

Name of the features	Explanation
Age of domain	Obliging a authentic site as phishing site tend to live for shorter period of time in the event that the age of domain is longer than six month
Record of DNS	It is exceedingly recommended as phishing site within the event DNS record isn't contained by website
Traffic of website	Colossal amount of individuals visit websites for the most part because it would have higher positioning. Positioning can distinguish on the off chance that a location is phishing or not. A phishing site is being tends to have a lower chance by the next ranked site
Ranking of page	In most time that phishing websites have no PageRank value since this value is allotted on its importance
Indexing of Google	A legitimate site can be accepted by a site that has a title on the google index
Reports statistical	Guessing it as phishing webpage within the event the have of the webpage has a place in any beat phishing IP's or domains
Joins indicating to page	Phishing site prohibiting have much links indicating apropos it since it has shorter lifetime

3.3   Deep Learning Algorithm

This study has considered five adaptive optimizer such as Stochastic Gradient Descent (SGD), ADAM, ADADELTA, ADAGARD, and RMSPROP used for evaluation of NN and DNN.

3.4   Machine Leraning Algorithm

This study has considered some machine learning algorithms for stacking such as Support Vector Machine (SVM), Decision Tree (DT), Naive Bayes (GNB), Linear Discriminant Analysis (LDA), Random Forest (RF), Multilayer Perceptron (MLP), Stochastic Gradient Descent (SGD), Logistic Regression (LR), k nearest neighbors (KNN) and Gaussian are used for Stacked Generalization as a base classifiers in first step, and 10-fold cross validation Adebowale [3] has been used. Here, XGBoost classifier is being used as meta estimator for final prediction in second step.

3.5   Model Generation Phase

The above methodology indicates three types of multilayer approaches: NN, DNN and stack generalization respectively. The main purpose of this model is to determine the best output through evaluation by applying stacking technique and neural network and deep neural network on the processed data set and to propose an optimized model based on that output.

Now an optimized output will be provided by applying neural network and deep neural network technique on this dataset.

After loading the features from the dataset, the data set is split into two parts, test and train. The train segment is applied to a two-layer neural network architecture and Somesha [16] a five-layer deep neural network architecture, respectively.

Since the data set is of binary type, for binary classification problem non-linear activation function ReLU have been used for hidden layers of neurons and sigmoid function have been used for output layers of neurons Vrbančič [19]. According to this architecture, five types of adaptive optimizer have been used here.

The next step is to compile the model using these adaptive optimizers. It is then divided into two parts, train and validation, by splitting the train set. The model is fitted using a number of epochs and early stopping techniques, to prevent overfitting. Now two outputs are available by evaluating the two models using the test set. After applying the approach, stack generalization technique has been applied in the dataset.

The evaluation technique of stack generalization has been described in figure. It's a multilevel approach. Stacking is usually done in two steps. In the first level stacking provides transitory prediction using based on classifiers with k-fold cross validation

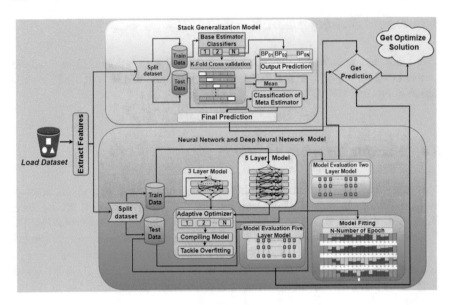

Fig. 1 Methodology for phishing URLs detection

and output probability prediction are revealed. During the system formation the output prediction and transitory prediction of step one are used in second steps. The estimate theory of phish stack are described below Rahman [14]:

- In the first step of stacking by using base classifiers to predict train and test set according to the second step the desired predictions are being acquired then that are considered as features.
- Stacking is a multilevel approach so any kind of algorithm can be used to predict it in two steps.
- This proposed system used k-fold cross-validation so that it eluded overfitting for this training set and each fold of the train portion it may predict using out-of-fold. According to this proposed model the value three to ten is used for k-fold cross validation after all provides output using a test set.

In the first step at the end of training the data the output is predicted using the test set. This time it's complete with all folds technique that's needed to mean for estimating all values from all folds that are used.

In the second step connected to another classifier that's called a meta-estimator on the train set, from the test set it performs terminal prediction. This approach takes extra time because it again adds a classifier for its performances. When the k-fold cross validation done in the first step then prediction is not completed these are completed on the second step.

Three outputs are obtained from the above multilayer techniques then a model is selected based on the decision, according to the value of the output. An optimized architecture is proposed based on that model.

Table 6 Evaluation parameters

Assessment parameter	Assessment parameters formula	Statement of the assessment parameter
Mean Absolute Error (MAE)	$\text{MAE} = \frac{\sum_{i=1}^{n} |y_i - x_i|}{n}$	It is the average value of all absolute errors [5]
Mean Square Error (MSE)	$\text{MSE} = \frac{1}{n} \sum_{i=1}^{n} \left(Y_i - \hat{Y}_i\right)^2$	It is the average value of all squares errors
AUC-ROC curve	For Positive Recall TRP = TP/(TP + FN). For Negative Recall FPR = 1 − Specificity = 1 − TN/(TN+FP) = FP/TN+FP	AUC - ROC curve is intrigued with True Positive Rate that belongs on y-axis, in opposition to the False Positive Rate that belongs on x-axis [1]
Precision - Recall Curve	For Positive Precision P = TP/(TP + FP) For Negative Precision N = TN/(TN+FN) For Positive Recall PR = TP/(TP + FN) For Negative Recall NR = TN/(TN+FP)	According to the precision-recall curve for a single classifier, estimating and intrigued the precision in opposition to the recall [12]
Accuracy	Accuracy = (TP + TN)/(TP + TN + FP + FN)	Accuracy means the rate of prediction that model executes [15]
Misclassification rate	Error Rate = 1 − Accuracy	The failings of identify value that is not appropriate for classification

4 Result and Discussion

4.1 Environment Setup

The experiment that has been conducted is Intel(R) Core(TM) i3-7100 CPU @2.40 GHz processor, 64-bit PC with 4 GB RAM. The operating system is Windows 10 pro Education and python has been used to implement the architecture To Detect Phishing URLs with the packages of python such that TensorFlow, scikit-learn, Keras, Pandas, and NumPy.

4.2 Evaluation Parameters

The system was mainly focused on evaluation based on data phishing or legitimate that's identified by binary classification. Confusion matrix, Accuracy, Precision-Recall Curve, Classification report, AUC-ROC Curve, Mean Absolute Error (MAE), Mean square Error (MSE) used to evaluate the performance of this system. The evaluation parameters [14] for assessment are described in the Table 6:

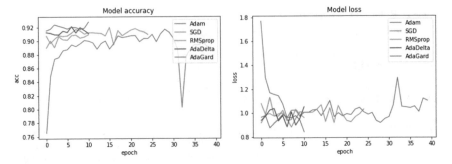

Fig. 2 Accuracy and loss DNN five optimizer

4.3 Experiment Result

The representation of the various optimizers of DNN and NN was shown in Tables 7 and 8. Five separate adaptive optimizers have been used for this experiment, the number of hidden layers, the learning rate and the epoch size are considered HL, LR and EPS respectively. According to this condition, HL5 means the number of hidden layers is 5 and the number of hidden layer 2 is HL2. For this evaluation, 15 types of learning rate and 10 types of epoch size were used for 20 times iteration for these five optimizer's. After 20-fold iteration, have chosen a better combination of epoch size and learning rate to achieve optimized performance so that this model is more accurate.

4.3.1   Case Study #1

Evaluation rate of five Adaptive Optimizer with accuracy and loss for DNN.

As illustrate in Fig. 2 have shown that used different deep learning adaptive optimizer to take the decision which optimizer would be the best for anti-phishing proposed model. In this case Adam optimizer given the highest accuracy among all the optimizer where SGD optimizer given slightly low performance. On the contrary, model optimizer loss their performance while tuning the model for the prediction of proposed model with the selected optimizer. Where every optimizer loss their performance based on their adaptive quality. In this case being understand to take the optimizer based on their performance and loss accuracy for the shake which optimizer will be the best fit for anti-phishing proposed model.

The ROC curve and precision-Recall curve the have been shown in Fig. 3. Maximum accuracy 0.955 attained from Adam individually. In case of precision-recall curve and the AUC-ROC curve SGD and AdaGard do better provides 0.96. SGD and AdaGard perform better in ROC curve and precision-Recall curve than others.

The analysis shows clearly in Table 7 that the learning rate has an essential contribution to the success of profound neural systems among all the measurement or

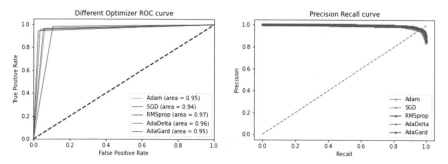

Fig. 3 Different optimizer for ROC curve and Precision-Recall curve for DNN

Table 7 Evaluation table for DNN

Serial	Optimizer	Label	Learning rate	Epochs	Accuracy	MSE	MAE
1	Adam	HL5	0.01	50	0.955	0.030	0.074
2	SGD	HL5	0.001	100	0.951	0.021	0.049
3	RMSprop	HL5	0.0003	150	0.953	0.028	0.076
4	AdaDelta	HL5	0.0027570	250	0.954	0.023	0.078
5	AdaGard	HL5	0.0017470	150	0.953	0.018	0.049

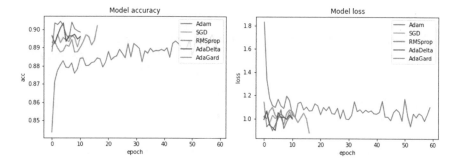

Fig. 4 Accuracy and loss NN five optimizer

appraisal parameters. It was found that the maximum accuracy of 0.955, MSE 0.030 and MAE 0.074 of the hidden five layers using Adam optimizer, along with the 50 epochs and 0.01 learning rate (HL5 EPs50). Observing all the outcomes from Table 7 from above, it can be observe that all the optimizer provides 95% accuracy of which Adam pays a little more, Adam is the top scorer Vrbančič [19].

4.3.2   Case Study #2

Evaluation rate of five Adaptive Optimizer with accuracy and loss for NN

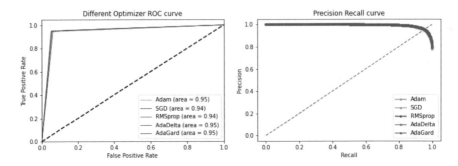

Fig. 5 Different optimizer ROC curve and Precision-Recall curve for NN

Table 8 Evaluation table for NN

Serial	Optimizer	Label	Learning rate	Epochs	Accuracy	MSE	MAE
1	Adam	HL2	0.0017470	150	0.948	0.014	0.058
2	SGD	HL2	0.001	128	0.945	0.026	0.086
3	RMSprop	HL2	0.0003	200	0.948	0.026	0.080
4	AdaDelta	HL2	0.0027570	250	0.949	0.016	0.067
5	AdaGard	HL2	0.0017470	150	0.955	0.030	0.074

As discussed before in Fig. 2 similarly in this phase according to the illustration in Fig. 4 used different deep learning adaptive optimizer where AdaGard optimizer given the highest accuracy among all the optimizer for two layer NN where both Adam and RMSprop optimizer given slightly low performance for the NN two layer. According to their performance, model have been loss their performance while evaluated the model to find the best optimizer if two NN layer used for all the adaptive optimizer.

The ROC curve and precision-Recall curve the have been shown in Fig. 5. Maximum accuracy 0.955 attained from AdaGard individually. In case of precision-recall curve and the AUC-ROC curve Adam, SGD, AdaDelta and AdaGard do better provides 0.95, expect RMSprop.

From the experiment it's clearly shown in Table 8 that the learning rate has an essential contribution to the success of profound neural systems among all the measurement or appraisal parameters. It was found that the maximum accuracy of 0.955, MSE 0.030 and MAE 0.074 of the hidden five layers using AdaGard optimizer, along with the 150 epochs and 0.0017470 learning rate (HL2, EPs150) which is slightly near with the DNN.

Table 9 Accuracy of machine learning classifier algorithm

LR	LDA	KNN	DT	GNB	SVM
0.927	0.921	0.936	0.955	0.593	0.944

Table 10 Build model stack and the increased accuracy of Machine learning Algorithm

LR	LDA	KNN	DT	GNB	SVM
0.966	0.965	0.965	0.966	0.965	0.966

Table 11 Misclassification rate and accuracy of temporary prediction

Algorithm	Accuracy	Misclassifation rate
RF	0.96	0.0047
DT	0.95	0.0063
MLP	0.96	0.0022
SVM	0.94	0.0072
SGD	0.91	0.0073
GNB	0.59	0.012

4.3.3   Case Study #3

The main purpose of stacked generalization is used a higher grade model to combine low grade models to achieve higher predictive accuracy. Stacking combines multiple model and learns it up for classification task.

According to Table 9, first of all here 6 machine learning algorithms are used on the data of the desired dataset then some accuracy is found on the basis of that algorithm. These algorithm are used to build a stack model.

In this step a stack model is generated by applying these algorithm. According to Table 10 this algorithm have changed in their accuracy after generating a stack model. The stack stipulates that it combines multiple models and learns for classification task. So purpose of this step is to stack learn stack.

The stack has already been learned, now it knows how to process a model. Table 11 represent the final accuracy and misclassification rate for first step. This work is done by two steps. So this table's value indicates the first step's prediction or temporary prediction because after second step prediction will find final prediction. The next step is to build a model, according to the study a model has been created XGBClassifier and through that model fitted the previous trained data and predict the final results.

The precision-Recall curve and the ROC curve have been shown in Fig. 6. The first step shows that the maximum accuracy 0.96 with minimum error rate. RF and MLP do better individually where precision-recall curve and the AUC-ROC curve,

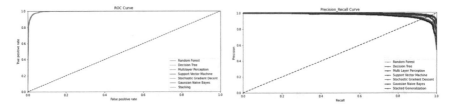

Fig. 6 Different algorithms ROC curve and Precision-Recall curve

Table 12 Evaluation rate of stacking

Algorithm	Accuracy	Accuracy of stack generalization
LR	0.927	0.966
LDA	0.921	0.965
KNN	0.936	0.965
DT	0.955	0.966
GNB	0.953	0.965
SVM	0.944	0.966

stacked generalization performs low. However in the time of final prediction stack generalization provides accuracy 0.97.

4.3.4 Discussion on the Difference Among the Three Multilayer Approaches

Stack Generalization

In this study binary classification type dataset has been selected for evaluation of three multilayer approaches. It is a well-known fact that machine learning algorithms provide good results for binary classification type datasets. The main features of stack generalization is that it integrates with low grade models using high grade models and also known as ensemble algorithm that basically works in two layers. According to the stacking concept, it learns multiple machine learning algorithms in the first layer, and then gives the predictions as an output which is used as the learning of another algorithm in the second layer. The machine learning algorithm used as the final predictor, this learning is more error-free. The main target of stack generalization is to develop the result of low grade models Li [28] New model is trained by other models that are already trained from a dataset. Most commonly stacking uses simple linear function (mean, median, average etc.) to assemble the prediction for other models. According to the stacking techniqe for binary classification type data sets will provide more accuracy than NN and DNN. So optimized output can be obtained using this stacking concept (Table 12).

Table 13 Highest rate for NN

Optimizer	Label	Learning rate	Epochs	Accuracy	MSE	MAE
AdaGard	HL2	0.0017470	150	0.955	0.030	0.074

Table 14 Highest rate for DNN

Optimizer	Label	Learning rate	Epochs	Accuracy	MSE	MAE
Adam	HL5	0.01	50	0.955	0.030	0.074

Neural Network

According to the determination rule, the number of hidden layer two is used for arbitrary decisions with rational activation functions Rahman [13]. Therefore two hidden layer neural network is the best approach for a given data dataset. The first layer is called the input layer according to the structure of the neuron. Previous layer outcome obtained to be the weighted input to the following layer, there is no correlation among each layer but NN shows craved conduct is made finding the correct weight by knowing NN Fister [29]. The structure of the neural network is similar to tree structure. There are units in each layer of the neural network. These units indicate how deep these layers can go. The value of unit basically indicates how depth the data will go and how many combinations will be tree based. A complete accurate outcome is obtained from multiple averages of a value. Stacking provides more accuracy than a neural network for a given data set, therefore neural networks provide more optimized results than stacking. Stacking is done in two layers whereas neural networks provide optimized output from many tree base combinations. Stacking is done in two layers so in case of complex data it will performed low whereas each hidden layer of neural network have units which indicates how deep it will go in tree base structures, so that it will provide an optimized output from multiple combinations (Table 13).

Deep Neural Network

This study ultimately dictates three multi-layer NN, DNN, and stacking strategies. Evaluating their output reveals that their outcomes among these stacks are virtually the same. Here two layers are used for NN, 5 for DNN, and two steps have been used for stack generalization. Stacking technique and NN have decent results for a basic dataset, such that the output of the dataset is lowered whether it has complicated or complex values. According to DNN, it works with a large number of layers and uses the value of the unit as needed. The DNN model-based architecture from this study offers good results for every form of dataset much of the time (Table 14).

5   Conclusion

Though phishing is a sensational phenomenon in today's cyber space, it is a matter of concern to investigate for securing the future. In this study, anti-phishing techniques have been developed based on NN, DNN and stacking concept. Parameter adjustment plays a vital role for these techniques. Among those parameters, learning rate is one of them. This is an unimaginable footstep of increasing the performance of NN and DNN based on systems. Here is an assessment of the effect of parameters that will be an evidence in the development of NN and DNN based system. The amount of data in the data set affect the system learning base. In the case of stacking, Random Forest and Multilayer perception provides better results for precision and recall. However stack generalization helps better to enhance the overall accuracy.

Basically, this chapter indicates three multilayer techniques that are NN, DNN, stacking along with the parameter tuning for neural network based architectures. Evaluating their performance shows that the results they provide almost same outcome. Apart from those, stacking provides better accuracy with less complex dataset. Here 2 layers are used for NN, 5 layers for DNN and stack generalization has used two layer. DNN and NN layers have units which indicate how deep these layers can go. Fundamental difference between NN and DNN is that NN works with two layers on the behalf of DNN works with more than two layers.

In recapitulate, the final outcome is obtained from the averages of multiple outputs. Stacking technique and NN provide better results for the dataset with simplicity, if the dataset holds complex or more complicated values then performance is getting decreased. According to DNN, it works with a large number of layers and uses the value of the unit as needed. From this study, DNN model based architecture provides better results on an average for any type of dataset.

References

1. Abutair, H.Y., Belghith, A.: Using case-based reasoning for phishing detection. Procedia Comput. Sci. **109**, 281–288 (2017)
2. Adebowale, M.A., Lwin, K.T., Hossain, M.A.: Deep learning with convolutional neural network and long short-term memory for phishing detection. In: 2019 13th International Conference on Software, Knowledge, Information Management and Applications (SKIMA), pp. 1–8. IEEE, August 2019
3. Adebowale, M.A., Lwin, K.T., Sanchez, E., Hossain, M.A.: Intelligent web-phishing detection and protection scheme using integrated features of Images, frames and text. Expert Syst. Appl. **115**, 300–313 (2019)
4. Acquisti, A., Adjerid, I., Balebako, R., Brandimarte, L., Cranor, L.F., Komanduri, S., Wilson, S.: Nudges for privacy and security: understanding and assisting users' choices online. ACM Comput. Surv. (CSUR) **50**(3), 1–41 (2017)
5. Absolute Error. https://www.statisticshowto.datasciencecentral.com/absolute-error/
6. Dong, Z., Kapadia, A., Blythe, J., Camp, L.J.: Beyond the lock icon: real-time detection of phishing websites using public key certificates. In: 2015 APWG Symposium on Electronic Crime Research (eCrime), pp. 1–12. IEEE, May 2015

7. El-Alfy, E.S.M.: Detection of phishing websites based on probabilistic neural networks and K-medoids clustering. Comput. J. **60**(12), 1745–1759 (2017)
8. Gupta, S., Singhal, A.: Dynamic classification mining techniques for predicting phishing URL. In: Soft Computing: Theories and Applications, pp. 537–546. Springer, Singapore (2018)
9. Hutchinson, S., Zhang, Z., Liu, Q.: Detecting phishing websites with random forest. In: International Conference on Machine Learning and Intelligent Communications, pp. 470–479. Springer, Cham, July 2018
10. Harikrishnan, N.B., Vinayakumar, R., Soman, K.P., Poornachandran, P., Annappa, B., Alazab, M.: Deep learning architecture for big data analytics in detecting intrusions and malicious URL. Big Data Recommender Syst. Algorithms, Architectures, Big Data, Secur. Trust **303** (2019)
11. Le, H., Pham, Q., Sahoo, D., Hoi, S.C.: URLNet: learning a URL representation with deep learning for malicious URL detection. arXiv preprint arXiv:1802.03162 (2018)
12. Precision Recall Curve and what are they, Available from: https://acutecaretesting.org/en/articles/precision-recall-curves-what-are-theyand-how-are-they-used
13. Rahman, S.S.M.M., Gope, L., Islam, T., Alazab, M.: IntAnti-Phish: an intelligent anti-phishing framework using backpropagation neural network. In: Machine Intelligence and Big Data Analytics for Cybersecurity Applications, pp. 217–230. Springer, Cham (2021)
14. Rahman, S.S.M.M., Islam, T., Jabiullah, M.I.: PhishStack: evaluation of stacked generalization in phishing URLs detection. Procedia Comput. Sci. **167**, 2410–2418 (2020)
15. Rana, M.S., Rahman, S.S.M.M., Sung, A.H.: Evaluation of tree based machine learning classifiers for android malware detection. In: International Conference on Computational Collective Intelligence, pp. 377–385. Springer, Cham, September 2018
16. Somesha, M., Pais, A.R., Rao, R.S., Rathour, V.S.: Efficient deep learning techniques for the detection of phishing websites. Sādhanā, **45**(1), 1–18 (2020)
17. Shirazi, H., Bezawada, B., Ray, I.: "Kn0w Thy Doma1n Name" unbiased phishing detection using domain name based features. In: Proceedings of the 23nd ACM on Symposium on Access Control Models and Technologies, pp. 69–75, June 2018
18. Understanding AUC-ROC Curve. https://towardsdatascience.com/understanding-auc-roc-curve-68b2303cc9c5
19. Vrbančič, G., Fister Jr, I., Podgorelec, V.: Swarm intelligence approaches for parameter setting of deep learning neural network: case study on phishing websites classification. In: Proceedings of the 8th International Conference on Web Intelligence, Mining and Semantics, pp. 1–8, June 2018
20. Wang, Z.Q., Wang, D.: Recurrent deep stacking networks for supervised speech separation. In: 2017 IEEE International Conference on Acoustics, Speech and Signal Processing (ICASSP), pp. 71–75. IEEE, March 2017
21. Winterrose, M.L., Carter, K.M., Wagner, N., Streilein, W.W.: Adaptive attacker strategy development against moving target cyber defenses. In: Advances in Cyber Security Analytics and Decision Systems, pp. 1–14. Springer, Cham (2020)
22. Huang, Y., Yang, Q., Qin, J., Wen, W.: Phishing URL detection via CNN and attention-based hierarchical RNN. In: 2019 18th IEEE International Conference On Trust, Security And Privacy In Computing And Communications/13th IEEE International Conference On Big Data Science And Engineering (TrustCom/BigDataSE), pp. 112–119. IEEE, August 2019
23. Pompon, R., Walkowski, D., Boddy, S., Levin, M.: 2018 phishing and fraud report: attacks peak during the holidays. F5 LABS (2018)
24. Viktorov, O.: Detecting phishing emails using machine learning techniques (Doctoral dissertation, Middle East University) (2017)
25. Cui, Q., Jourdan, G.V., Bochmann, G.V., Couturier, R., Onut, I.V.: Tracking phishing attacks over time. In: Proceedings of the 26th International Conference on World Wide Web, pp. 667–676, April 2017
26. Woogue, P.D.P., Pineda, G.A.A., Maderazo, C.V.: Automatic web page categorization using machine learning and educational-based corpus. Int. J. Comput. Theory Eng. **9**(6), 427–432 (2017)

27. Mohammad, R.M.A.: An ensemble self-structuring neural network approach to solving classification problems with virtual concept drift and its application to phishing websites (Doctoral dissertation, University of Huddersfield) (2016)
28. Li, Y., Yang, Z., Chen, X., Yuan, H., Liu, W.: A stacking model using URL and HTML features for phishing webpage detection. Future Gener. Comput. Syst. **94**, 27–39 (2019)
29. Fister, I., Suganthan, P.N., Kamal, S.M., Al-Marzouki, F.M., Perc, M., Strnad, D.: Artificial neural network regression as a local search heuristic for ensemble strategies in differential evolution. Nonlinear Dyn. **84**(2), 895–914 (2016)
30. Rahman, S.S.M.M., Rafiq, F.B., Toma, T.R., Hossain, S.S., Biplob, K.B.B.: Performance assessment of multiple machine learning classifiers for detecting the phishing URLs. In: Data Engineering and Communication Technology, pp. 285–296. Springer, Singapore (2020)
31. Sahingoz, O.K., Buber, E., Demir, O., Diri, B.: Machine learning based phishing detection from URLs. Expert Syst. Appl. **117**, 345–357 (2019)

Improved Secure Intrusion Detection System by User-Defined Socket and Random Forest Classifier

Garima Sardana and Abhishek Kajal

Abstract Research has considered the intrusion Detection system (IDS) to make detection and classification of intrusions, attacks, and different types of data-stealing activities. Existing research in the field of the IDS system has been considered. The model has been developed to send and receive data. The IDS system is proposed to detect, classify the intrusion with the integration of the Random forest algorithm. The socket programming has been used to transfer data from sender to receiver. To secure the transmission used defined port number has been used. Moreover, the data traveling over the network would be in encrypted form to avoid the possibility of data manipulation or access by an unauthentic user. The IDS system is capable to trace attacks of different categories as the random classifier has classified the attacks.

Keywords IDS · Random forest algorithm · Socket programming · Classification

1 Introduction

As the use of Web-Based Services and Applications are increasing day by day, the probability of Cyber Attacks is also increasing. When sensitive and important user data travels over a network, both internal and external intruders may try to attack or hack this data. The attackers can use manual and machine-based method for this purpose. These attackers are becoming more powerful and efficient. It has become a challenge to stop and avoid these attackers or hackers. The data attacks or these types of data-stealing are known as Cyber Crime and these malicious people who perform these types of activities are known as cyber attackers. From time to time researchers or specialized teams are working in this field and proposing innovative, flexible, and more trusted IDS systems [1].

G. Sardana (✉) · A. Kajal (✉)
Department of Computer Science and Engineering, Guru Jambheshwar University of Science and Technology, Hisar, India

Y. Maleh et al. (eds.), *Artificial Intelligence and Blockchain for Future Cybersecurity Applications*, Studies in Big Data 90,
https://doi.org/10.1007/978-3-030-74575-2_10

1.1 Intrusion Detection System

Intrusion Capturing Model systems are used to make detection and classification of intrusions, attacks, and different types of data-stealing activities. This system is used towards the network and host level system and works automatically on time. Based on intrusive behaviors, the intrusion detection systems are differentiated between network-based and host-based systems.

This system is considered in the form of a criminal warning. Let us take an example to understand this point. To protect a home from theft lock systems are used in homes. As soon as it identifies such type of activity it alerts the owner by rings an alarm. However, input congestion coming from the Internet in the direction of the escape firewall is filtered by firewalls.

The stock of IDS hardware is already available. It uses the information which is generated from an individual host-based IDS (HIDS). In addition to this, it also uses those IDSs which take advantage of information gathered from an entire segment of the normal system.

For example, it becomes possible for outer users to make a connection to the intranet by dialing through a modem. This modem is put into operation in a personal system of businesses. A firewall is not able to identify this type of entry. An Intrusion Prevention System becomes a system of protection and threat stoppage technology. It controls system traffic flows to identify and stop the exploitation of weak points.

1.2 Types of IDS

1   Host-based IDS
2   Network-based IDS
3   Hybrid IDS

Host-Based IDS views a sign of intrusion in the local system. For analysis purpose of examination, information related to the host system's logging is used by them. Host manager is considered in the form sensing element. The sensing element which is based on the Host's information, include network and other logs produced by controller treatment and data of objects not reflected in normal controller checking & logging method.

Network-Based IDSs is used to detect network traffic. It requires the implementation of sensing elements throughout the system. This type of system is required for a particular section of the network. It is used to analyze the network and activity of protocols.

Hybrid IDS system tries to integrate the advantages of all IDS while removing its loopholes. In this system, the sensing element and hosts communicate with a core administration or director stage. Introduction of data which is collected from network-based sensing elements and host-based computer program becomes the biggest problem for the supplier of a hybrid instruction detection system.

1.3 Random Forest Classifier

Random forest (RF) [16] has been known as an ensemble classifier (see Fig. 1). It has been developed to enhance the accuracy and performance of IDS systems. This classifier includes several decision trees. After reviewing several classifiers, it is clear that there is less error in the classification of intrusion using RF which shows its efficiency and applicability. This classification is better than other classification classifiers. This classifier considers several trees, minimum node size, and the number of features to split each node. There are many advantages of this classifier such as it saves the generated forests which can be used for future reference. It also resolves the issue related to fitting. Accuracy and variable importance generate without any manual effort. When the individual trees are constructed in a random forest, randomization is used to choose the best node to split on. This value is equal to $\sqrt{A}$, where A is the number of attributes in the dataset. Therefore, RF formulated several noisy trees that influence accuracy and wrong decision.

In this research paper the Sect. 1 is introducing intrusion detection system along with its types. Moreover the role of random forest classification for IDS detection has been presented. Section 2 is presenting existing researches in field of intrusion detection while Sect. 3 is focusing on the problem statement. The Sect. 4 has focused on research methodology where socket programming has been considered along with client server transmission mechanism. The concept of user defined ports and predefined port have been considered in this section. Intrusion Sect. 5 is presenting the tools used for IDS detection and Sect. 6 discussed the proposed work. Section 7 explores the results obtained during simulation and finally Sect. 8 presented the conclusion.

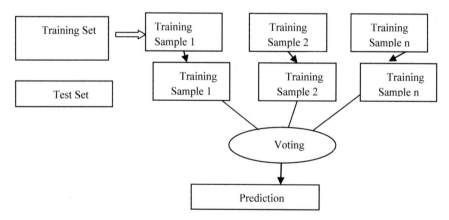

Fig. 1 Random forest classifications

2 Literature Review

In 2020, Y. Zhou et al. [1] developed an Efficient Intrusion Detection System. In their work, they considered Feature Selection and Ensemble Classifier. In 2020, Y. J. Chew et al. [2] applied a Decision Tree in the company of responsive size in the system—dependent upon intrusion detection. The design which was proposed by them has been examined in the company of six proportions Gure KDDCup, NIDS records. In 2020, Song, Yajie, and Bu, Bing et al. [3] proposed a new design for Intrusion Detection. For the achievement of this purpose, they combine Network and hardware. The work is confirmed on equipment on the up and up reenactment foundation of CBTC frameworks. Recreation results demonstrate strategy accomplished ninety-seven point six four percent genuine certain rate. It can essentially increase the level of safety insurance in support of the CBTC structure. In 2019, A. Arul Anitha et al. [4] developed ANNIDS: Artificial neural network dependent intrusion detection system in support of the internet of things (IoT). They discussed that there are several studies on IoT which have revealed that Artificial Neural Network (ANN) is best to acquire accurate detection rate as compared to other approaches. In 2019, A. Khraisat et al. [5] surveyed the system of intrusion detection, various methods, records, and related issues. Their survey work has presented a taxonomy of contemporary IDS. In addition to this, a comprehensive review related to notable recent works is made. In 2019, R. Vinayakumar et al. [6] developed an ethos of deep learning in support of a system that is used for smart intrusion detection. Their research work introduced a new form of deep learning which deep neural network (DNN) becomes. In 2018, Meira, Jorge et al. [7] gave relative Results Unsupervised Techniques in Cyber Attack Novelty Detection. Interruption location has been known as a significant need in current occasions. PC frameworks are continually being casualties of pernicious assaults. In 2018, Kolli, Satish and Lilly et al. [8] considered CSA for PTC with the support of DIDS. Railroads are intending to finish the execution of PTC frameworks by 2020 with the essential security destinations of maintaining a strategic distance from between train crashes, train crashes and guaranteeing railroad specialist wellbeing. In 2018, Clotet et al. [9] talked about an ongoing inconsistency based on IDS for digital assault discovery at the mechanical procedure level of Critical Infrastructures. This work presents a constant abnormality based discovery framework intended during the mechanical procedure of essential structure (CI). In 2018, T. Tian et al. [10] improves the maximization method of ant lion. Its application is also improved. It was done in the system in which hydraulic turbines are dominating to identify its parameter. In 2017, Aleroud [11] utilized circumstantial data for the detection of internet attacks. An ongoing pattern is towards the information based interruption identification frameworks (IDSs). IDSs in which information is used in the form of base saves information regarding digital attacks and probable weak points and utilize this information. In 2017, Al-Dabbagh et al. [12] composed a System of intrusion Detection given internet Attacks in mobile control systems. In this article, a proposed topology for a remote organized control framework has been concentrated under a few digital assault situations. In 2017, S. M. Alqahtani et al. [13] made a comparison in the middle of various organizations' techniques in support of cloud IDS warnings along with fuzzy organizers. In their research work, they utilized general classification algorithms. In 2017, S. Mouassa et al. [14] discussed the Ant lion optimizer to resolve the problems related to the transmission of ideal and sensitive energy in the electric network. In 2017, B. B. Rao et al. [15] explained Fast KNN Classifiers to utilize in developing an efficient IDS system. A couple of KNN grouping methods that are very fast are considered in their work.

3 Problem Statement

Classifiers that are established based on machine learning already exist. When this classifier is implemented it will make the efficiency of intrusion detection systems better. But, at the same time, it also includes some vulnerability. The performing period of the existing system needs improvement. For this purpose, additional nodes are added in the direction of the available group. In addition to this, existing systems are not able to deliver complete details related to malware patterns and quality. In short, for making the efficiency better, a complicated pattern of DNNs are made trained over the latest device with the help of a distributed method.

A complicated pattern of DNNs is not trained in this work using the standard of intrusion detection system records. It happens because the expenses of computerized calculation related to the company of complicated patterns of DNNs are comprehensive. Using the standard of intrusion detection system records, solutions that are established based on machine learning designs raise many challenging concerns which are as follows:

1 Design generates highly incorrect encouraging speed in the company of intrusion large extent.
2 Designs are not capable of being generalized because in the present assessment, for representing design efficiency just an individual record is implemented.
3 Design examines up to this point does not consider present network conjunction
4 To maintain the present fast-growing network dimension and activities, a different type of solution becomes a requirement.

All the above said challenging issues become the primary encouragement in support of this research where the effectiveness of existing machine learning classifiers becomes the main focal point. However existing methods generate highly incorrect encouraging speed by capturing how the supply of networks is used. Within standard conjunction arrangements of attacks are present in the company of highly low description and for a long period.

4 Research Methodology

The proposed work has make using of socket programming for client server model. Transmission requires IP address and port no during transmission. The receiver needs to initialize the connection before receiving data from sender. The data is then transferred from sender to receiver. The intrusion detection model is detecting and classifying the intruder during transmission process.

Client-Server Model
Two requests related to a system can be started at the same time, but practically it is not needed. Because of that, it becomes necessary to form the application of the transmission network in such a way that it can execute compulsory network function in a specified order, in place of parallel function. Primarily, the server performs and stays in this post until it obtains the network packet which is delivered in its direction when the client performs. When the primary contact completes either consumer or server becomes capable of delivering and obtaining information.

IP4 Addresses
These addresses extended up to thirty-two bits. Normally they exist in the market standard form of numbers. All four bytes due to which thirty-two address build exists in the form of whole number (zero to two hundred and fifty-five) and divided through a dot.

Port
For the identification of sockets in an exception manner address related to web service, adjacent rules and numbers linked with the port are used. Because of this reason, whenever a socket is formed, it is compared in the company of internet protocol address and port number. Ports become the objects of a computer program in the middle of different demands. As soon as a host obtains a packet, it moves in the direction of the protocol heap and finally reaches the application layer. A packet during data transmission is consisting of a port number & IP address where data is to be delivered within data. Port number 1 to 1023 are reserved for existing services but 1024 to 65,535 are available for our programs.

Socket Programming
Programming is already used for communication in the middle of two applications. This application runs in the settings of two Java runtime environments. There is the feasibility that the programming related to Java Socket may be connection-oriented. It is also possible that this programming can be done in the absence of connection. The Socket, as well as Server Socket classes, have been used. These are applied for programming related to connection-oriented sockets. Datagram Socket in the company of Datagram Packet classes has been applied. These are used to do programming related to without connection socket. The user in socket programming must know given two points below:

Server IP Address

1. Port number
2. Socket class

The socket is used to communicate among devices. Socket class could be applied to make the socket. Several methods of socket class are used to create connection, close connection, read and write data from on socket to another socket.

Server Socket Class

It becomes possible to use it in the formulation of the server socket. It is applied to make communication with clients. The following table explains those methods which are available inside the Server Socket

Important Methods

To run this code there is a need to open two command prompts. Here execution of each program takes place on every command prompt (see Fig. 2). During the execution of client code, information is shown over the server-side (Table 1).

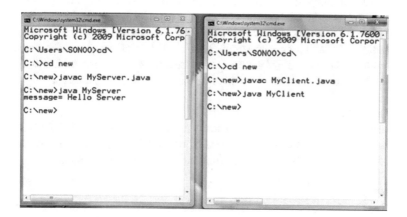

Fig. 2 Output of program

Table 1 Server socket class methods

Method	Description
Public socket accept()	In addition to linking in the middle of server & client, proceeds socket
Public synchronized void close()	Stop the server socket

5   Tools in Intrusion Detection

The products which are made available to the general public after intrusion detection manage a variety of administrative objectives that are related to their safety [2]. Hardware that is used to provide safety is considered here.

SNORT

It is a type of computer program which can be accessed freely. To define congestion, languages which are made based on rules are used by it. This language is adjustable [6]. It files packets out of internet protocol address in a format that can be inspected visually by human beings. It identifies lots of worms; weak points explore attempts, scan ports in addition to other illegal activities. With the help of different pre-processors, content discovery, and the examination of various rules and regulations, all these things are identified.

OSSEC-HIDS

It becomes famous in the form of open source security. It is a computer program that can be assessed freely. It works concerning crucial computer program. The structure which is established based on the Client/Server model is used by it. It is so much capable that it can easily deliver OS logs in the direction of the server in support of examination work and for preservation. It is already implemented in machines which carry out log examination, ISPs, educational institutions & information hub. For the observations and assessment of certified logs and firewalls, HIDS is used.

FRAGROUTE

It becomes famous in the form of a forwarding device that is used for division. Bundles of internet protocol are delivered by the attacker in the direction of the frag router. After that these packets are broken and processed for the party.

HONEYD

It becomes hardware that developed basic moderators into the network [6]. When the facilities are utilized through a host, it permits an individual host to demand various locations on a local area network in support of networks' computerized calculation. It becomes possible to criticize the basic engine or to track their path [6].

KISMET

In support of the intrusion detection system which is mobile, it becomes a benchmark. This system is arranged inside the useful load of packets and happenings of WIDS. It would identify the intruder gateway.

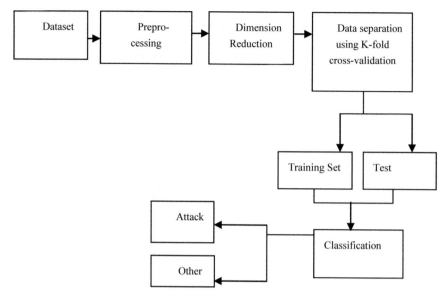

Fig. 3 Process flow of proposed work

6 Proposed Work

The work which has been introduced here contains two phases: (1) Quality Collection (2) Organizations. In this work, one against-all method was included in support of organizing all attacks. For the identification of usual datasets, we established them to class one, and the remaining attacks are established to class two different. After that, qualities are collected and organization is done through the radiofrequency method. The complete arrangement of the system is displayed in the diagram (see Fig. 3).

7 Result and Discussion

The simulation process is presenting the client server transmission in sender and receiver where user defined port is used to transfer information between two nodes. The random forest classifier has been applied to classify the intrusion.

7.1 Client Server Setting in Sender and Receiver

Here, the usage of the network is put forward. Net bean-based Integrated Development Environment. It is already highlighted in the diagram (see Fig. 4).

Fig. 4 On the server-side, we have made designing and written code to enable the download option and disable the download option

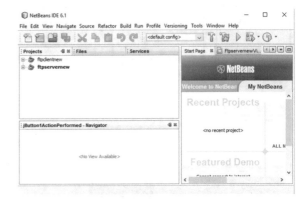

Fig. 5 Design view of receiver application

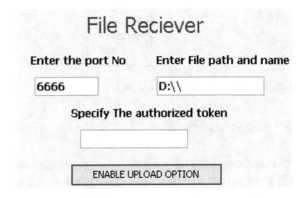

The design view of the file receiver module (see Fig. 5). Here the port no, the path of file, content decoding token has been specified.

7.2   Sender Implementation

The design view of the file sender module (see Fig. 6). Here the port no, the path of the file, IP address of receiver, content-encoding token has been specified.

Running Application
There is a need to upload a text file from sending to the receiver side. In the following figure nn.txt file has been shown (see Fig. 7).

At the time of execution of the file sender code (see Fig. 8), it is essential to clarify whether port no is more than 1023. File path and approval token are also specified.

During the execution of the file sender module, there is a requirement of user-defined port number 6666, file path, and authorization of token. The specification of IP addresses to put target location for file for broadcasting (see Fig. 9).

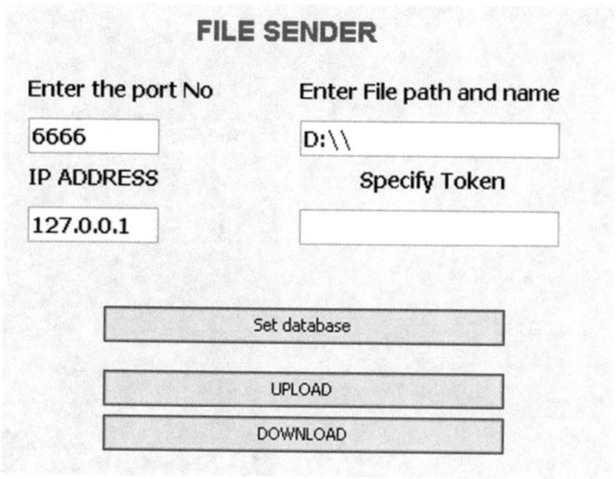

Fig. 6 Code to implement
UPLOAD on the sender side

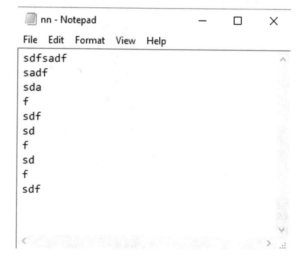

Fig. 7 Running applications

7.3 Random Forest Implementation

The simulation of random forest classifier has been performing using MATLAB and the classification result is shown in Figs. 10 and 11.

　　≫call_generic_random_forests

Confusion Matrix
After training of the dataset testing module is run then the confusion matrix is generated considering various attributes. The true classes are presented on y-axis and predicted classes are presented on the x-axis (Fig. 12).

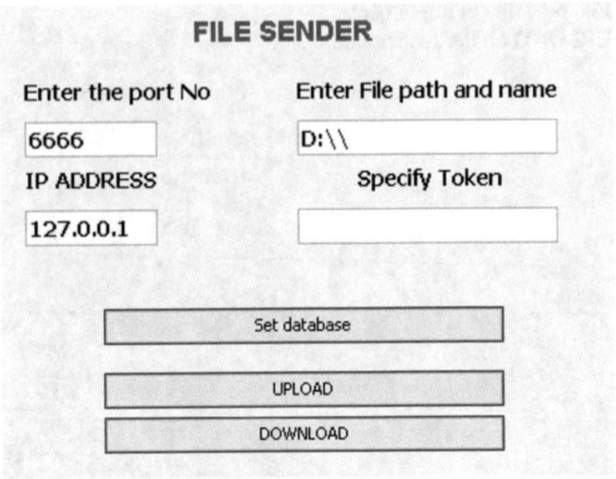

Fig. 8 Running applications

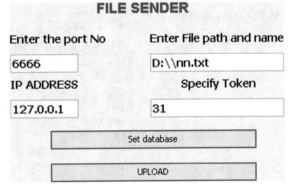

Fig. 9 File sender applications

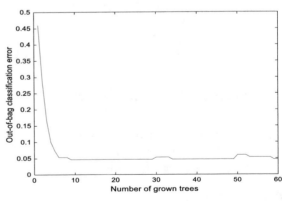

Fig. 10 The classification error according to grown trees

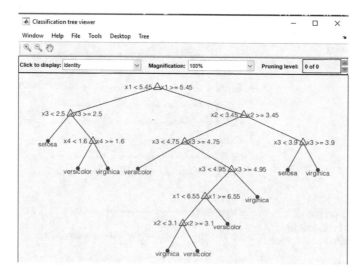

Fig. 11 Classification tree viewer

Fig. 12 Confusion matrix of proposed model

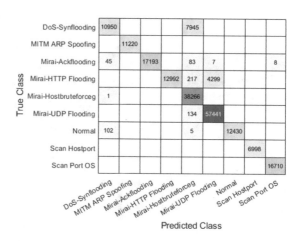

Considering the above confusion matrix chart presenting accuracy, precision, recall value, and f-score is generated. The accuracy chart in the case of existing work is presented below (Figs. 13, 14, 15, 16 and 17).

Fig. 13 Comparison of proposed and previous in case of accuracy

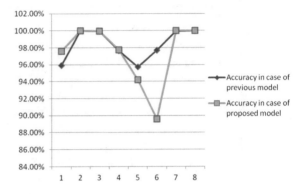

Fig. 14 Comparison of proposed and previous in case of precision

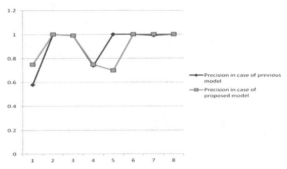

Fig. 15 Comparison of proposed and previous in case of recall

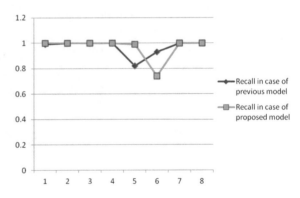

Fig. 16 Comparison of proposed and previous in case of F-score

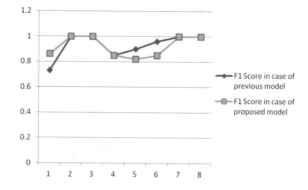

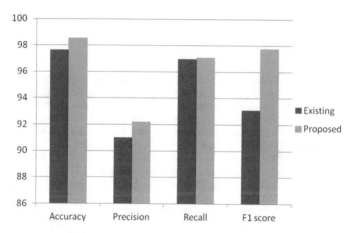

Fig. 17 Comparison of proposed and previous IDS model

8 Conclusion

The research concludes that a random forest is proven as an ensemble classifier. It is capable to enhance the accuracy and performance of IDS systems. Results conclude that this classifier includes many decision trees. After the simulation, it has been concluded that there is less error in the classification of intrusion using RF which is efficient and applicable. The proposed system is more secure as it has made use of a user-defined socket during data transmission. The probability of IDS attacks has been reduced due to the presence of encryption and user-defined socket transmission. The accuracy, precision value, recall value and f-score is found better in case of proposed IDS system.

9 Future Scope

Research has been focused to provide security against different attacks such as DOS attacks and Probe attacks. For improving the efficiency of present IDS systems a new model was added to monitor the DNS and BGP events in the Networks. The future work on relevant research has been analyzing various machine learning classifiers that are utilized to increase performance in the case of IDS. This research work also introduces a method for enhancement of system efficiency which is based upon intrusion Detection. For this using Random Forest Classifier is bringing into use. Future researchers are supposed to evaluate the new technique by calculating the quality of service parameters.

References

1. Zhou, Y., Cheng, G., Jiang, S., Dai, M.: Building an efficient intrusion detection system based on feature selection and ensemble classifier. Comput. Netw. **174**, 107247 (2020)
2. Chew, Y.J., Ooi, S.Y., Wong, K.S., Pang, Y.H.: Decision tree with sensitive pruning in network-based intrusion detection system. Lect. Notes Electr. Eng. **603**, 1–10 (2020)
3. Yajie, S., Bing, B., Li., Z.: A novel intrusion detection model using a fusion of network and device states for communication-based train control systems (2020)
4. Anitha, A.A., Arockiam, L.: ANNIDS: artificial neural network-based intrusion detection system for internet of things. Int. J. Innov. Technol. Explore. Eng. **8**(11), 2583–2588 (2020)
5. Khraisat, A., Gondal, I., Vamplew, P., Kamruzzaman, J.: Survey of intrusion detection systems: techniques, datasets, and challenges. Cybersecurity **2**(1), 1–22 (2019)
6. Vinayakumar, R., Alazab, M., Soman, K.P., Poornachandran, P.: Deep learning approach for intelligent intrusion detection system. IEEE Access **7**, 41525–41550 (2019)
7. Jorge, M.: Comparative results with unsupervised techniques in cyber attack novelty detection. In: Proceedings, vol. 2, pp. 1191 (2018). https://doi.org/10.3390/proceedings2181191
8. Kolli, S., Joshua, L., Wijesekera, D.: Providing Cyber Situational Awareness (CSA) for PTC using a distributed IDS system (DIDS) (2018)
9. Clotet, X., Moyano, J., León, G.: A real-time anomaly-based IDS for cyber-attack detection at the industrial process level of critical infrastructures. Int. J. Crit. Infrastruct. Prot. **23**, 11–20 (2018)
10. Tian, T., Liu, C., Guo, Q., Yuan, Y., Li, W., Yan, Q.: An improved ant lion optimization algorithm and its application in hydraulic turbine governing system parameter identification. Energies **11**(1), 95 (2018)
11. Aleroud, A., Karabatis, G.: Using contextual information to identify cyber-attacks (2017). https://doi.org/10.1007/978-3-319-44257-0_1
12. Al-Dabbagh, A., Li, Y., Chen, T.: An intrusion detection system for cyber attacks in wireless networked control systems. IEEE Trans. Circ. Syst. II Express Briefs (2017)
13. Alqahtan, S.M., John, R.: A comparative analysis of different classification techniques for cloud intrusion detection systems' alerts and fuzzy classifiers. In: Proceeding Computer Conference, pp. 406–415 (2018)
14. Mouassa, S., Bouktir, T., Salhi, A.: Antlion optimizer for solving optimal reactive power dispatch problem in power systems. Eng. Sci. Technol. Int. J. **20**(3), 885–895 (2017)
15. Rao, B.B., Swathi, K.: Fast KNN classifiers for network intrusion detection system. Indian J. Sci. Technol. **10**(14), 1–10 (2017)
16. Farnaaz, N., Abbar, M.A.: Random forest modeling for network intrusion detection system. Procedia Comput. Sci. **89**, 213–217 (2016)

Spark Based Intrusion Detection System Using Practical Swarm Optimization Clustering

Mohamed Aymen Ben HajKacem, Mariem Moslah, and Nadia Essoussi

Abstract Given the availability growth of data in large networks, intrusion detection systems become an important challenge since they require efficient methods to discover attacks from such networks. This paper proposes a new Spark based intrusion detection system using particle swarm optimization clustering, referred to as IDS-SPSO, for large scale data able to provide good tradeoff between scalability and accuracy. The use of Particle swarm optimization clustering is argued to avoid the sensitivity problem of initial cluster centers as well as premature convergence. In addition, we propose in this work to take advantage of parallel processing based on the Spark framework. Experiments performed on several large collections of real intrusion data have shown the effectiveness of the proposed intrusion detection system in terms of scalability and clustering accuracy.

Keywords Intrusion detection · Big data · Clustering · PSO · Spark

1 Introduction

Given the ever increasing growth and popularity of Internet, network intrusion detection becomes an important challenge to provide protection and security for information. This is explained by the large number of users and the large amount of data exchanged which makes it difficult to distinguish between the normal connections and attacks. To this end, intrusion detection systems (IDSs) are designed to deal with large amounts of data in order to protect a system against network attacks.

M. A. B. HajKacem (✉) · M. Moslah · N. Essoussi
LARODEC, Institut Supérieur de Gestion de Tunis, Université de Tunis,
41 Avenue de la liberté, cité Bouchoucha, 2000 Le Bardo, Tunisia
e-mail: nadia.essoussi@isg.rnu.tn

Y. Maleh et al. (eds.), *Artificial Intelligence and Blockchain for Future Cybersecurity Applications*, Studies in Big Data 90,
https://doi.org/10.1007/978-3-030-74575-2_11

197

Several machine learning techniques were applied for IDSs in the literature [1, 6, 11, 23, 26, 33]. Clustering is one of the machine learning techniques that is used to organize data into groups of similar data points called also clusters [18]. Clustering methods can be mainly categorized into five classes namely hierarchical, density-based, grid-based, model-based and partitional methods [39]. K-means [24] as one of the partitional clustering methods, remains the most efficient because of its simplicity and linear time complexity. However, it is sensitive to the selection of initial cluster centers, since it can produce local optimal solutions when the initial cluster centers are not properly selected [10].

To deal with this issue, several optimization algorithms were introduced to solve the data clustering problem [8, 14, 17, 20, 27, 31, 34]. Genetic optimization algorithm which is based on a mutation operator to deal with clustering task was designed in [20]. Simulated annealing optimization was also used for data clustering in [8]. Particle Swarm Optimization (PSO), was proposed to solve the clustering problem, by using multiple search directions with social behavior to enhance the quality of the clustering result [34]. Among these algorithms, Particle Swarm Optimization (PSO) has gain a great popularity because of its efficiency [29].

On the other hands, conventional intrusion detection methods based on clustering fail to scale with larger sizes of network traffic and are computationally expensive in terms of memory. To deal with large scale data, several distributed clustering methods were designed in the literature [2–5, 12, 15, 30, 40, 41]. Most of these methods use the MapReduce framework [13] for data processing. However, MapReduce is unsuitable for iterative algorithms since it requires repeated times of reading and writing to disks. Spark [9, 32] is introduced to overcome the limitations of MapReduce, particular for processing iterative algorithms. It is an in–memory parallel framework for processing Big data using a cluster of machines. Compared with the MapReduce framework, Spark is more efficient and approximately 10 to 100 times faster for data processing task [7].

This paper proposes a new Spark based intrusion detection system (IDS-SPSO). The proposed system builds the intrusion detection model using PSO clustering. To the best of our knowledge, this is the first work that implements parallel intrusion detection system using PSO and Spark framework. The aim is to show how the proposed system takes advantage of Spark and PSO to deal with real large scale intrusion data to achieve high accuracy quality and scalability.

The remainder of this paper is organized as follows: Sect. 1 presents background definitions related to the Particle Swarm Optimization (PSO) and parallel frameworks. Section 2 discusses the related works in the area of intrusion detection methods based on clustering. Then, Sect. 3 describes the proposed parallel intrusion detection system while Sect. 4 presents the experimental results performed on large real intrusion data. Finally, Sect. 5 gives concluding remarks and some future works.

2 Preliminaries

This section first presents background definitions related to the Particle Swarm Optimization (PSO) followed by the parallel frameworks which are used in this work.

2.1 Particle Swarm Optimization

Particle Swarm Optimization (PSO) was introduced by the electrical engineer Eberhart and the social psychologist Kendy [29]. This algorithm was proposed to simulate the social behavior of birds when searching for food. When a bird recognizes a food area, it broadcasts the information to all the swarm. Hence, all the birds follow him and this way they raise the probability of finding the food since it is a collaborative work. So, the behavior of birds within swarms was turned into an intelligent algorithm capable of solving several optimization problems.

PSO is a population based optimization algorithm. It consists of a swarm of particles where each particle represents a potential solution to the optimization problem. Each particle P_i is characterized at the time t, by the current position $x_i(t)$ in the search space, the velocity $v_i(t)$, the personal best position $pbest\,P_i(t)$ and the fitness value $pbest\,F_i(t)$. The personal best position represents the best fitness value the particle has ever seen, which is calculated by:

$$pbest\,P_i(t+1) = \begin{cases} pbest\,P_i(t)\,if\,f(pbest\,P_i(t)) <= f(x_i(t+1)) \\ x_i(t+1)\,if\,f(pbest\,P_i(t)) > f(x_i(t+1)) \end{cases} \qquad (1)$$

The personal best position represents the best fitness value any particle has ever experienced, which is calculated by:

$$gbest\,P(t+1) = min\,(f(y),\,f(gbest\,P(t))) \qquad (2)$$

where $y \in \{pbest\,P_0(t),\,...,\,pbest\,P_S(t)\}$. The following equation is used to update the particles positions within the problem search space.

$$x_i(t+1) \leftarrow x_i(t) + v_i(t) \qquad (3)$$

While the following equation is used to update the particle velocities.

$$v_i(t+1) \leftarrow wv_i(t) + c_1r_1(pbest\,P_i(t) - x_i(t)) + c_2r_2(gbest\,P(t) - x_i(t)) \qquad (4)$$

where w is the inertia weight, $x_i(t)$ is the position of the particle P_i at the time t, $v_i(t)$ is the velocity of the particle P_i at the time t, c_1 and c_2 are two acceleration coefficients, and r_1 and r_2 are two random values in the range [0, 1]. The main algorithm of PSO is outlined in Algorithm 1.

Algorithm 1. The main algorithm of PSO

1: **Input:** Input data set R
2: **Output:** Particles information
3: Create an initial population of particles from R.
4: **while** Convergences not reached **do**
5:    Calculate the fitness value of particles.
6:    Update the personal best position of each particle using Equation 1.
7:    Update the global best position using Equation 2.
8:    Change the velocities and positions using Equation 3 and 4 respectively.
9: **end while**

2.2 MapReduce Framework

MapReduce [13] is a parallel programming framework for data processing. As shown in Fig. 1, MapReduce is composed of three phases namely *map*, *shuffle* and *reduce*. Each phase processes data through $<key/value>$ pairs. The map phase applies the map function by taking in parallel each $<key/value>$ and generates a set of intermediate $<key'/value'>$ pairs. Then, the shuffle phase merges all intermediate values which share the same intermediate key as a list. The reduce phase applies the reduce function to group all intermediate values associated with the same intermediate key. Note the implementation of the MapReduce framework is available in Hadoop [35]. And the inputs and outputs of MapReduce are stored in a distributed file system which is called Hadoop Distributed File System (HDFS). Despite of its performance to deal with Big data, MapReduce framework is unsuitable to fit when executing iterative algorithms [22]. Since, it requires at each iteration reading and writing data from disks, which can increase the running time.

2.3 Spark Framework

Spark framework supports iterative computation and has an improved processing speed compared to MapReduce since it utilizes in-memory computations using the resilient distributed datasets (RDDs). These RDDs can be cached in memory to be used in multiple consecutive operations. Spark [42] is introduced to run with Hadoop [35], especially by reading data from HDFS. Moreover, it provides a set of in-memory operators, beyond the standard MapReduce, with the aim of processing data more rapidly on distributed environments compared to MapReduce [32]. Spark framework proposes two types of operators which can deal with RDD called, transformations and actions. The transformations are designed to execute a function to the whole records and generate new RDD. Map, ReduceBykey and MapPartition are examples of transformations. The actions are designed to return a value to the program and store the final result of the computation in a file system. Filter and Count are examples of actions. The Data flow of Spark framework is shown in Fig. 2.

Fig. 1 Flowchart of
MapReduce framework

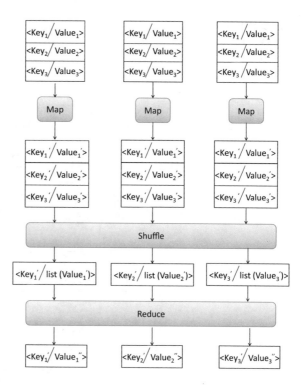

3 Related Works

Several intrusion detection based on machine learning techniques were proposed in
the literature [1, 6, 11, 23, 26, 33]. These methods can be divided into supervised or
unsupervised according the type of the used data during the processing are labelled
or not. Several techniques have been designed for intrusion detection systems using
unsupervised approach such as clustering-based methods [16, 19, 21, 28].

Peng et al. [28] proposed mini batch k-means clustering method for intrusion
detection. They employ the principal component analysis technique to reduce the
number of dimensions of the used data set in order to enhance the clustering efficiency.
However, this method considers only a small sample size of intrusion data set which
can leads a loss of quality.

Leung et al. [21] designed a density-based clustering method which employs the
frequent pattern tree in order to solve the high dimensionality of the used data set.
This method was tested on one million records and achieved a good detection results.

Jiang et al. [19] proposed a fuzzy c-means clustering intrusion detection method
where they employ a weighting strategy for the record membership calculation. This
method was tested with five data samples where each sample having ten thousand
records. The results show high false positives rates with satisfactory detection rates.

Fig. 2 Data flow of Spark framework

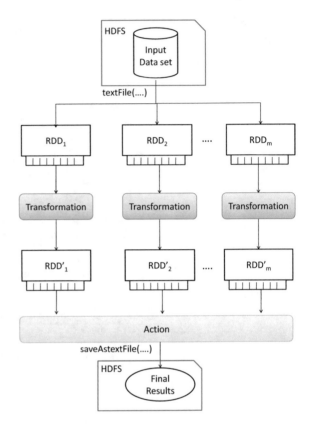

Harish et al. [16] proposed a modified version of fuzzy c-means clustering method for anomaly detection. They employ Principal component analysis (PCA) as feature selection technique to deal with curse of dimensionality. In addition, this methods is based on using gaussian kernel as distance measure to compute the distance between cluster center and samples. The advantage of using gaussian kernel is that it reduces the effect of noise.

Wankhade et al. [36] proposed ensemble clustering method to deal with intrusion detection problem. This method is based on k-means and divide and merge strategy which is used to select the accurate number of cluster centers. The experimental results have shown that this strategy can improve detection rate and lower false alarm rate.

Although the attested performance of existing intrusion detection methods, they fails to organize large network traffic. To solve the large scale intrusion data, parallel methods were proposed to perform distributed computations [2–5, 12, 15, 30, 40, 41]. Most of these methods use the MapReduce as a parallel programming framework.

Aljarah et al. [2] proposed a parallel intrusion detection system through the MapReduce framework referred to as IDS-MRPSO. In addition, they build a clustering model by solving the intrusion detection problem using PSO optimization algorithm. Finally, the proposed system has been tested using real large scale of intrusion data with different training subset sizes to evaluate the scalability and the detection quality. However, MapReduce framework is not appropriate to deal with iterative algorithms since it requires at each iteration reading and writing data from disks.

Wang and Han [37] proposed a network intrusion detection based on parallel DPC clustering. This method is based on cut off distance strategy which reduces the number of comparisons between data points and clusters. Furthermore, they proposed fitting the DPC clustering using Spark framework in order to deal with the scalability. However this method remains sensitive to the random selection of initial cluster centers [10].

It is important to note that our proposed system is the first work which is based on fitting a parallel intrusion detection system through Spark framework. Compared with the MapReduce, Spark is a good in-memory parallel framework for data processing.

4 Proposed Intrusion Detection System (IDS-SPSO)

Large network traffic data needs an efficient intrusion detection system to protect it against attacks. The proposed intrusion detection system incorporates the data clustering process based on the PSO algorithm. Furthermore, PSO clustering is distributed using Spark framework in order to scale with large network traffic. As shown in Fig. 3, the proposed intrusion detection system consists of three main phases: *pre-processing phase*, *data detector modeling phase*, and *validation phase*. The first phase is devoted to apply set of data pre-processing techniques such as missing values removal, categorical feature elimination and data normalization. Once the pre-processing phase is completed, we propose in the second phase to apply Spark based PSO clustering method (S-PSO) [25] on training data in order to generate global best centroids vectors. In the third phase, we evaluate the quality of the detection model by computing distances between the testing data and the final global best centroids vectors.

4.1 Pre-processing Phase

First, we remove the records which contain missing values since we use these records in the distance computation when building clusters. So, we cannot use in the distance computation a record which contains a missing value. Then, we eliminate the categorical features. We propose in this work to consider only numerical features in the distance computation, because we need a special distance metric for the categorical features. After that, we apply the normalization to the obtained data set in order to avoid the bias problem for some features which have a large variability between

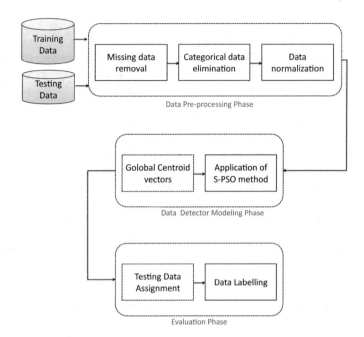

Fig. 3 Flowchart of IDS-SPSO system

minimum and maximum values. The normalization process is performed using the following equation:

$$x_{ij_{new}} = \frac{x_{ij} - x_{j_{min}}}{x_{j_{max}} - x_{j_{min}}} \tag{5}$$

where x_{ij} is the value of record i for feature j, $x_{ij_{new}}$ is the normalized value of record i for feature j, $x_{j_{min}}$ is the minimum value of feature j and $x_{j_{max}}$ is the maximum value of feature j.

4.2 Data Detector Modeling Phase

The data detector modeling phase consists on applying S-PSO method [25] to the data results from the pre-processing phase. The authors proposed an efficient PSO clustering method using Spark. The experimental results on large scale data show that S-PSO scales very well with increasing data and achieved a good clustering accuracy. This method reads the data set only once in contrast to existing MapReduce implementation of PSO clustering. Hence, it exploits the flexibility provided by Spark framework, by using in-memory operations that alleviate the consumption time of existing MapReduce solution [2].

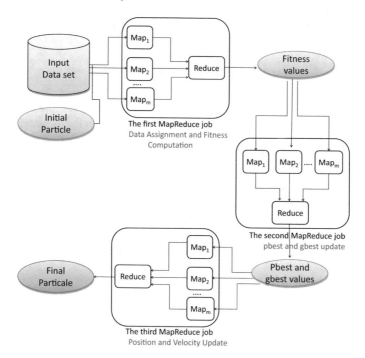

Fig. 4 Flowchart of S-PSO method

S-PSO method is composed of three MapReduce jobs namely, *Data assignment and fitness computation*, *Personal and global best update* and *Position and velocity update*. The main process of the S-PSO method is described in Fig. 4.

4.2.1 Data Assignment and Fitness Computation

In the first MapReduce job, S-PSO starts by creating an initial population which is composed of particle's position, velocity, personal best position and personal best fitness. To this end, the positions of particles are randomly initialized from the input data set and they represent the initial cluster's centroids.

Then, the data set is divided into chunks and each chunk is assigned to a map function. The particle's information are broadcast to all chunks. The map function first assigns each data point to the nearest cluster centroid in each particle by computing distances. Then, the map function generates a key value pair as output where the key represents the couple particleID and centroidID and the value represents the minimum distance between a data point and the centroidID in a particleID.

Once all the data points are affected to the nearest cluster centroid, a reduce function is applied to compute the fitness value by combining merging data from different map functions. The fitness value is computed using the total sum of squares errors by:

$$Fitness = \frac{\sum_{j=1}^{k} \sum_{i=1}^{|C_j|} d(r_i, C_j)}{k} \qquad (6)$$

where $d(r_i, C_j)$ represents the distance between the record r_i and the cluster's centroid C_j, $|C_j|$ represents the number of records assigned to the centroid C_j and k represents the number of clusters. Then, the reduce function generated key value pairs as output where the key represents the particleID and the value represents the fitness value.

Let $R = \{r_1...r_n\}$ the input data set. Let $P(t) = \{P_1(t)...P_S(t)\}$ the set of the particle's information where $P_c(t) = \{x_c(t), v_c(t), pbest P_c(t), pbest F_c(t)\}$ represents the information of particle c in the iteration t where $x_c(t)$ is the position, $v_c(t)$ is the velocity, $pbest P_c(t)$ is the best position and $pbest F_c(t)$ is the best fitness. Let $F = \{F_1...F_S\}$ the set of fitness values where F_c is the fitness value of the particle c. The main steps of Data assignment and fitness computation MapReduce job is described in Algorithm 2.

Algorithm 2. Data assignment and fitness computation MapReduce job

---

1: **Input:** input data set R, Particle information P(t)
2: **Output:** Fitness values F
3: Divide the data set R into m chunks $R = \{R^1...R^m\}$
4: % *Map Phase*
    Let R^p be assigned to map task p.
5: **for** each $r_i \in R^p$ **do**
6:    **for** each $P_c \in P(t)$ **do**
7:       $x_c(t) \leftarrow$ Extract positions from $P_c(t)$
8:       Affect each data point to its nearest cluster centroid by computing distances.
9:       Let mindis the minimum computed distance.
10:     Let CentroidID the index of the cluster centroid where the record data point r_i is affected.
11:     Let ParticleID the index of the particle P_c.
12:    **end for**
13:    Emit (key: ParticleID, CentroidID/value: mindis)
14: **end for**
15: % *Reduce Phase*
16: **for** each $P_i(t) \in P$ **do**
17:    Calculate fitness value F_i using Equation 6.
18:    Emit (key: ParticleID /value: F_i)
19: **end for**

4.2.2 Pbest and Gbest Update

Once the new particle's fitness are computed, they are automatically distributed to RDDs collections. However, the computation of pbest and gbest is not an expensive operation. So, it does not need to be executed in parallel manner. Then, each particle updates its personal best position and the global best position.

Let $pbest F(t) = \{pbest F_1(t)...pbest F_S(t)\}$ is the set of personal best fitness values where $pbest F_i(t)$ is the pbestF of the particle i at iteration t.

Let $pbest P(t) = \{pbest P_1(t)...pbest P_S(t)\}$ is the set of personal best position where $pbest P_1(t)$ is the pbestP of the particle i at iteration t. Let $gbest P$ is the position of the best particle. The main steps of the pbest and gbest update MapReduce job is described in Algorithm 3.

Algorithm 3. Pbest and gbest update MapReduce job

1: **Input:** F, $pbest F(t)$, $pbest P(t)$
2: **Output:** $pbest F(t + 1)$, $pbest P(t + 1)$, $gbest P$
3: $gbest P \leftarrow \emptyset$
4: **for** each $P_i(t) \in P(t)$ **do**
5:    $pbest F_i(t + 1) \leftarrow \emptyset$
6:    $pbest P_i(t + 1) \leftarrow \emptyset$
7:    **if** $(pbest F_i(t) \le F_i)$ **then**
8:       $pbest F_i(t + 1) \leftarrow pbest F_i(t)$
9:       $pbest P_i(t + 1) \leftarrow pbest P_i(t)$
10:   **else**
11:      $pbest F_i(t + 1) \leftarrow F_i$
12:      $pbest P_i(t + 1) \leftarrow x_i(t + 1)$
13:   **end if**
14: **end for**
15: Let i^* is the index of particle having the best fitness value.
16: $gbest P \leftarrow x_{i^*}(t)$

4.2.3   Position and Velocity Update

During this MapReduce job, S-PSO starts by assigning the particles information to different map functions. Then, the map function performs the velocity and position update using the Eqs. 3 and 4. While the reduce function groups all the intermediate key value pairs computed from the different map functions. Once the reduce phase is completed, the data set and particle's information are distributed in RDDs collections which are stored in memory for the next iteration. For more details about the S-PSO method, the readers can refer to [25].

Let $x(t) = \{x_1(t)...x_S(t)\}$ the set of position values where $x_i(t)$ is the position of the particle i at iteration t. Let $v(t) = \{v_1(t)...v_S(t)\}$ the set of velocity values where $v_i(t)$ is the velocity of the particle i at iteration t. The main steps of Position and velocity update MapReduce job is described in Algorithm 4.

Algorithm 4. Position and velocity update MapReduce job

1: **Input:** $gbest\, P$, $P(t)$
2: **Output:** $P(t + 1)$
3: % *Map Phase*
   Let $P_p(t)$ be assigned to a map task p.
4: $x_i(t + 1) \leftarrow \emptyset$
5: $v_i(t + 1) \leftarrow \emptyset$
6: Update the new position value $x_i(t + 1)$ using 4
7: Update the new velocity value $v_i(t + 1)$ using 3
8: Emit(key: 1/value: $P_i(t + 1)$)
9: % *Reduce Phase*
10: Group outputs from the different map functions and update the new particle information $P(t + 1)$.
11: Emit $(P(t + 1))$

4.3 Evaluation Phase

Once the data detector modeling phase is completed, we extract the global best centroid vectors from the final particle's information. During this phase, we evaluate the detection model by computing distances between the testing records and the global best centroids vectors. After that, we affected the testing records to the their nearest clusters by computing distances. The main steps of the evaluation phase is described in Algorithm 5.

Algorithm 5. Evaluation phase

1: **Input:** Testing data T, Final Particle information P
2: **Output:** Assigned data
3: **for** each $t_i \in T$ **do**
4:    Let $C()$ the k centroids extracted from the final particle P.
5:    Compute distances between x_i and C.
6:    Assign x_i to its nearest centroid.
7: **end for**

Finally, the cluster labeling process is applied to predict the correct labels for clusters which are generated in the testing data assignment step. The assignment of cluster labels is performed by retrieving the maximum percentage of intersections between the true labels of the testing data, and the assigned clusters that are generated by applying the testing data assignment step.

Figure 5 illustrates an example to better understanding the cluster labeling process. For cluster C_1, the percentage of the normal records is $\frac{3}{4}$ while the percentage of the attack records is $\frac{1}{4}$. Hence, cluster C_1 is a normal cluster. For cluster C_2, the percentage of the normal records is $\frac{1}{3}$ while the percentage of the attack records is $\frac{2}{3}$. So, cluster

C_2 is a attack cluster. Similarly, for cluster C_3, the percentage of the normal records $\frac{1}{3}$ is while the percentage of the attack records is $\frac{2}{3}$. So, cluster C_3 is a attack cluster.

4.4 Time Complexity Analysis

In order to show the effectiveness of the proposed system, we describe in the following the evaluation of the time complexity of the S-PSO method. Given n is the data set size, k is the number of clusters, c is the number of data chunks, l is the number of iterations and s is the swarm size.

The data assignment step is the most expensive operation in PSO algorithm since it requires computing distances between each record to all the clusters of each particle in the swarm. Then, this step has to be repeated several times until convergences. Thus, the time complexity of PSO is evaluated by $O(n.k.s.l)$. The S-PSO first divides the input data into c chunks that could be executed in parallel manner. So, S-PSO requires processing n/c records for each iteration. Hence, the time complexity of S-PSO is evaluated by $O(n/c.k.s)$.

Fig. 5 An illustrative example of the clusters labeling process

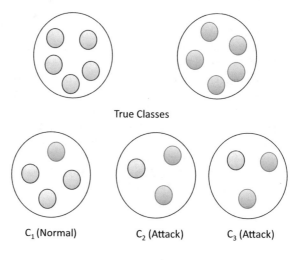

True Classes

C_1 (Normal) C_2 (Attack) C_3 (Attack)

Predicted Clusters

5   Experiments and Results

5.1   Environment

The experiments are realized on a cluster of 4 machines where each node has 2-core 2.30 GHz CPU E5400 and 1 GB of memory. The experiments are performed using Apache Spark version 2.1.1, scala version 2.1.1, Apache Hadoop 2.7.0 and Ubuntu 16.04.

5.2   Data Set Description

In order to evaluate the performance of the proposed system, we used a Big intrusion detection data set[1] which was employed as the benchmark at the Knowledge Discovery and Data Mining in 1999. This data set contains a standard set of data which includes a wide variety of normal and attack connections in a military network environment.

Each record in the collected data set represents a connection between two IP addresses. The data contains 4,898,431 connection records which are classified into normal traffic and four kinds of attacks namely, denial of service (DoS), probe (PRB), remote to local (R2L) and user to root (U2R). Each connection is described by 3 categorical and 38 numerical features for a total of 41 features.

A set of pre-processing techniques were applied on the training and testing data sets. We first start by removing the records that have missing values. Then, we reduce the number of features to 38 by eliminating the 3 categorical features. Finally, we apply the normalization process on the training and testing data sets.

In order to evaluate the impact of the data size on the performance of the detector model, we extract 4 different data samples from the whole training data set. To simplify the names of the data samples, we will use the following notations Train20, Train40, Train80 and Train100 to denote an extracted data set which stores 20%, 40%, 80% and 100% of the whole training data set. Statistics of these data sets are summarized in Table 1.

Table 1 Summary of the data samples

Data set	Number of connections	Normal	Attack
Train20	979,686	194,556	785,130
Train40	1,959,372	389,112	1,570,260
Train80	3,918,745	778,225	3,140,520
Train100	4,898,431	972,781	3,925,650

---

[1]https://archive.ics.uci.edu/ml/machine-learning-databases/kddcup99-mld/.

5.3 Evaluation Measures

In order to evaluate the scalability of the proposed system, we use the Speedup measure [38] which consists on fixing the data set size and varying the number of machines. The Speedup measure is defined as follows:

$$Speedup = \frac{T_1}{T_m},\tag{7}$$

where T_1 is the running time of processing data on 1 machine and T_m is the running time of processing data on m machines.

In order to evaluate the quality of clustering of the proposed system, we used true positives, true negatives, false positives, and false negatives. A true positive (TP) indicates that the intrusion detection system detects precisely a particular attack having occurred. A true negative (TN) indicates that the intrusion detection system has not made a mistake in detecting a normal connection. A false positive (FP) indicates that a particular attack has been detected by the intrusion detection system but that such an attack did not actually occur. A false negative (FN) indicates that the intrusion detection system is unable to detect the intrusion after a particular attack has occurred. We use in this paper the True Positive Rate (TPR) and False Positive Rate (FPR), which are defined in Eq. 8 and 10 respectively.

$$TPR = \frac{TP}{TP + FN}\tag{8}$$

$$FPR = \frac{FP}{FP + TN}\tag{9}$$

Furthermore, we use the Area Under Curve (AUC) measure [43] to combine the TPR and FPR which is considered a good indicator of these rates. The AUC can be defined as follows:

$$AUC = \frac{(1 - FPR) \times (1 + TPR)}{2} + \frac{FPR \times TPR}{2}\tag{10}$$

A greater value of these measures indicates better quality results.

5.4 Results

We use in the experiments the following parameters: the number of particles to 10, the number of iterations to 50, the inertia weight to 0.72 and the acceleration coefficients to 1.49. We first evaluate the accuracy of the proposed IDS-SPSO compared to IDS-MRPSO system. Table 2 reports the TPR, FPR, and AUC values obtained by

Table 2 Comparison of the accuracy of IDS-SPSO versus IDS-MRPSO

Dataset	Method	TPR	FPR	AUC
Train20	IDS-MRPSO	0.903	0.038	0.933
	IDS-SPSO	0.848	0.096	0.875
Train40	IDS-MRPSO	0.911	0.021	0.945
	IDS-SPSO	0.856	0.085	0.888
Train80	IDS-MRPSO	0.935	0.013	0.961
	IDS-SPSO	0.879	0.068	0.904
Train100	IDS-MRPSO	0.939	0.013	0.963
	IDS-SPSO	0.883	0.059	0.905

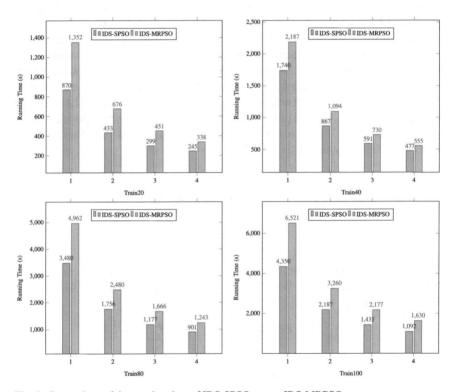

Fig. 6 Comparison of the running time of IDS-SPSO versus IDS-MRPSO

the proposed system using different training data samples sizes compared to IDS-MRPSO system. The obtained results show that the proposed IDS-SPSO gives nearly same results of existing IDS-MRPSO system. In addition, we observed from this table that the TPR value of IDS-SPSO using the whole training data (i.e. Train100) reaches the best value compared to smaller training data sets. Furthermore, Table 2 shows that IDS-SPSO obtains the lowest FPR for Train100 data set. For instance, the IDS-MRCPSO system has a high TPR of 0.883 for Train100, while it has a TPR of 0.848 for Train20. In addition, it has a low FPR of 0.059 for Train100, while it has a PDR of 0.096 for Train20. Hence, we observed that the proposed system can distinguish effectively between the normal and attacks data records. Finally, we concluded that the obtained results show the improvement of accuracy when using larger training data sets.

We then evaluate the running time of the proposed system compared to the IDS-MRPSO system. Figure 6 shows the running time results for the 4 training data samples using different numbers of machines. The obtained results show that the proposed system is faster than existing IDS-MRPSO system. For instance, the IDS-SPO is faster by a factor of 1.52 and 2.66 than IDS-MRPSO respectively for Train20 and Train100 data sets. From this Figure, we can also observe the improvement of running time when the number of machines is increased. For example the running time on 1 machine takes 870, 1740, 3480 and 4350 s for Train20, Train40, Train80, and Train100, respectively, while the running time on 4 machines takes 245, 477, 901 and 1092 s for the same samples respectively.

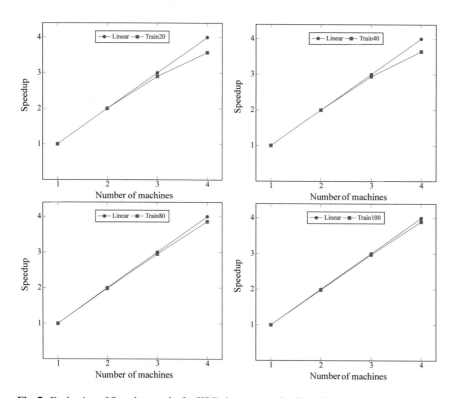

Fig. 7 Evaluation of Speedup results for KDD data set samples from 20% to 100% sizes

We then evaluate the scalability of the proposed system, by running multiple experiments with different number of machines. Figure 7 shows the Speedup results using different training data sizes with different numbers of machines. From this Figure, we observed that Speedup results become important especially when the data size is increased. For example, the Speedup value when running IDS-SPSO using 4 machines for Train20 is 3.57 while it is 3.90 for Train100 data. In addition, the proposed system shows approximately a linear speedup when the number of machines increases. This is explained by the benefits of the in-memory processing of Spark framework, which can significantly reduce the network cost when we increase the number of machines.

6 Conclusion

In this paper, we proposed an intrusion detection system IDS-SPSO for large scale of network traffic. The proposed system incorporates clustering analysis to build the detection model by solving the intrusion detection problem using particle swarm optimization clustering algorithm. We have also shown in this work that the intrusion detection system can be efficiently distributed through Spark framework. Experiments were realized on a real intrusion data set in order to evaluate the scalability of the proposed system. The experimental results show the efficiency of the IDS-SPSO when we increase both the number of machines and the training data size. Furthermore, the experiments results show that using larger training data leads to better detection rates by keeping the false alarm very low.

Our future work is to incorporate algorithms which are capable of providing automatically the number of clusters. Furthermore, we will extend the proposed system by employing feature selection techniques to extract the most important features when building clusters.

References

1. Amini, M., Rezaeenour, J., Hadavandi, E.: A neural network ensemble classifier for effective intrusion detection using fuzzy clustering and radial basis function networks. Int. J. Artif. Intell. Tools 25(02), 1550033 (2016)
2. Aljarah, I., Ludwig, S.A.: Parallel particle swarm optimization clustering algorithm based on MapReduce methodology. In: 2012 Fourth World Congress on Nature and Biologically Inspired Computing (NaBIC), pp. 104–111 (2012)
3. HajKacem, M.A.B., N'cir, C.E.B., Essoussi, N.: MapReduce-based k-prototypes clustering method for big data. In: Proceedings of Data Science and Advanced Analytics, pp. 1–7 (2015)
4. HajKacem, M.A.B., N'cir, C.E.B., Essoussi, N.: STiMR k-means: an efficient clustering method for big data. Int. J. Pattern Recogn. Artif. Intell. 33(08), 195–215 (2019)
5. HajKacem, M.A.B., N'cir, C.E.B., Essoussi, N.: One-pass MapReduce-based clustering method for mixed large scale data. J. Intell. Inf. Syst. 52(3), 619–636 (2019)

6. Bouteraa, I., Derdour, M., Ahmim, A.: Intrusion detection using classification techniques: a comparative study. Int. J. Data Min. Model. Manag. **12**(1), 65–86 (2020)
7. Bhathal, G.S., Singh, A.: Big data: Hadoop framework vulnerabilities, security issues and attacks. Array **1**, 100002 (2019)
8. Babu, G.P., Murty, M.N.: Simulated annealing for selecting optimal initial seeds in the k-means algorithm. Indian J. Pure Appl. Math. **25**(12), 85–94 (1994)
9. Chen, C.P., Zhang, C.-Y.: Data-intensive applications, challenges, techniques and technologies: a survey on big data. Inf. Sci. **275**, 314–347 (2014)
10. Celebi, M.E., Kingravi, H.A., Vela, P.A.: A comparative study of efficient initialization methods for the k-means clustering algorithm. Expert Syst. Appl. **40**(1), 200–210 (2013)
11. Chew, Y.J., Ooi, S.Y., Wong, K.S., Pang, Y.H., Hwang, S.O.: Evaluation of black-marker and bilateral classification with J48 decision tree in anomaly based intrusion detection system. J. Intell. Fuzzy Syst. **35**(6), 5927–5937 (2018)
12. Cui, X., Zhu, P., Yang, X., Li, K., Ji, C.: Optimized big data k-means clustering using MapReduce. J. Supercomput. **70**(3), 1249–1259 (2014)
13. Dean, J., Ghemawat, S.: MapReduce: simplified data processing on large clusters. Commun. ACM **51**(1), 107–113 (2008)
14. Esmin, A.A., Coelho, R.A., Matwin, S.: A review on particle swarm optimization algorithm and its variants to clustering high-dimensional data. Artif. Intell. Rev. **44**(1), 23–45 (2015)
15. Gowanlock, M., Rude, C.M., Blair, D.M., Li, J.D., Pankratius, V.: A hybrid approach for optimizing parallel clustering throughput using the GPU. IEEE Trans. Parallel Distrib. Syst. **30**(4), 766–777 (2018)
16. Harish, B.S., Kumar, S.A.: Anomaly based intrusion detection using modified fuzzy clustering. IJIMAI **4**(6), 54–59 (2017)
17. Ilango, S.S., Vimal, S., Kaliappan, M., Subbulakshmi, P.: Optimization using artificial bee colony based clustering approach for big data. Cluster Comput. **22**(5), 12169–12177 (2019)
18. Jain, A.K.: Data clustering: 50 years beyond k-means. Pattern Recogn. Lett. **31**(8), 651–666 (2010)
19. Jiang, W., Yao, M., Yan, J.: Intrusion detection based on improved fuzzy C-means algorithm. In: 2008 International Symposium on Information Science and Engineering, vol. 2, pp. 326–329. IEEE (2008)
20. Krishna, K., Murty, M.N.: Genetic k-means algorithm. IEEE Trans. Syst. Man Cybern. Part B (Cybern.) **29**(3), 433–439 (1999)
21. Leung, K., Leckie, C.: Unsupervised anomaly detection in network intrusion detection using clusters. In: Proceedings of the Twenty-Eighth Australasian Conference on Computer Science, vol. 38, pp. 333–342 (2005)
22. Lin, J.: MapReduce is good enough? If all you have is a hammer, throw away everything that's not a nail!. Big Data **1**(1), 28–37 (2013)
23. Li, Z.: A neighbor propagation clustering algorithm for intrusion detection. Revue d'Intelligence Artificielle **34**(3), 331–336 (2020)
24. MacQueen, J., et al.: Some methods for classification and analysis of multivariate observations. In: Proceedings of the Fifth Berkeley Symposium on Mathematical Statistics and Probability, vol. 1, pp. 281–297 (1967)
25. Moslah, M., HajKacem, M.A.B., Essoussi, N.: Spark-based design of clustering using particle swarm optimization. In: Clustering Methods for Big Data Analytics, pp. 91–113. Springer, Cham (2019)
26. Maglaras, L.A., Jiang, J.: A novel intrusion detection method based on OCSVM and K-means recursive clustering. EAI Endorsed Trans. Secur. Saf. **2**(3), e5 (2015)
27. Paul, D., Saha, S., Mathew, J.: Improved subspace clustering algorithm using multi-objective framework and subspace optimization. Expert Syst. Appl. **158**, 113487 (2020)
28. Peng, K., Leung, V.C., Huang, Q.: Clustering approach based on mini batch Kmeans for intrusion detection system over big data. IEEE Access **6**, 11897–11906 (2018)
29. Poli, R., Kennedy, J., Blackwell, T.: Particle swarm optimization. Swarm Intell. **1**(1), 33–57 (2007)

30. Shahrivari, S., Jalili, S.: Single-pass and linear-time k-means clustering based on MapReduce. Inf. Syst. **60**, 1–12 (2016)
31. Singh, H., Kumar, Y.: A neighborhood search based cat swarm optimization algorithm for clustering problems. Evol. Intell. **13**, 593–609 (2020)
32. Shyam, R., Bharathi Ganesh, H.B., Kumar, S., Poornachandran, P., Soman, K.: Apache spark a big data analytics platform for smart grid. Procedia Technol. **21**, 171–178 (2015)
33. Taheri, S., Bagirov, A.M., Gondal, I., Brown, S.: Cyberattack triage using incremental clustering for intrusion detection systems. Int. J. Inf. Secur. **19**, 597–607 (2020)
34. Van der Merwe, D., Engelbrecht, A.P.: Data clustering using particle swarm optimization. In: The 2003 Congress on Evolutionary Computation, CEC 2003, vol. 1, pp. 215–220 (2003)
35. White, T.: Hadoop: The Definitive Guide. O'Reilly Media Inc., Sebastopol (2012)
36. Wankhade, K.K., Jondhale, K.C.: An ensemble clustering method for intrusion detection. Int. J. Intell. Eng. Inform. **7**(2–3), 112–140 (2019)
37. Wang, J., Han, D.: Design of network intrusion detection system based on parallel DPC clustering algorithm. Int. J. Embed. Syst. **13**(3), 318–327 (2020)
38. Xu, X., Jager, J., Kriegel, H.-P.: A fast parallel clustering algorithm for large spatial databases. In: High Performance Data Mining, pp. 263–290. Springer (1999)
39. Xu, D., Tian, Y.: A comprehensive survey of clustering algorithms. Ann. Data Sci. **2**(2), 165–193 (2015)
40. Yang, L., Chiu, S.C., Liao, W.K., Thomas, M.A.: High performance data clustering: a comparative analysis of performance for GPU, RASC, MPI, and OpenMP implementations. J. Supercomput. **70**(1), 284–300 (2014)
41. Zhao, W., Ma, H., He, Q.: Parallel k-means clustering based on MapReduce. In: IEEE International Conference on Cloud Computing, pp. 674–679 (2009)
42. Zaharia, M., Chowdhury, M., Franklin, M.J., Shenker, S., Stoica, I.: Spark: cluster computing with working sets. In: HotCloud 2010, vol. 10, p. 95 (2010)
43. Zhu, W., Zeng, N., Wang, N.: Sensitivity, specificity, accuracy, associated confidence interval and ROC analysis with practical SAS implementations. In: NESUG Proceedings: Health Care and Life Sciences, Baltimore, Maryland, vol. 19, p. 67 (2010)

A New Scheme for Detecting Malicious Attacks in Wireless Sensor Networks Based on Blockchain Technology

Mohammed Amin Almaiah

Abstract Wireless sensor networks (WSNs) work in various domains such as smart cities, healthcare domains, smart buildings and transportation. These networks share sensitive data across multiple sensor nodes, smart devices and transceivers. These sensitive data in WSNs environment is susceptible to various cyber-attacks and threats. Therefore, an efficient security mechanism is needed to handle threats, attacks and security challenges in WSNs. This paper proposed a new scheme using Heuristic, Signature and voting detection methods to identify the optimal countermeasures to detect the malicious and security threats using Blockchain technology. In our scheme, the cluster head node (CN) use the three detection systems with Blockchain to detect the malicious sensor nodes. Also, CN uses important parameters such as sensor node-hash value, node-signature and voting degree for malicious to detect malicious nodes in WSNs. The overall results statistic showed that 94.9% of malicious messages were detected and identified successfully during our scheme's simulation.

Keywords Wireless sensor networks · Blockchain technology · Malicious sensor attacks

1 Introduction

Wireless sensor networks (WSNs) have become a well-known and popular source of sensitive data sharing and other human life activities such as smart homes, bank transactions, etc. This huge increase in the use of WSNs has resulted in a significant raising in cybersecurity attacks. Security problems in WSNs are still a serious concern for many researchers due to the infrastructures' heterogeneous nature and its weakness in the operational environment. This makes cyber attackers exploit these vulnerabilities to access the systems illegally [1]. WSNs have several limitations in terms of lower power, computational processing and limited resources [2]. WSNs contains millions of wireless sensor nodes, which collect data according to their

M. A. Almaiah (✉)
Department of Computer Networks and Communications, King Faisal University,
Al-Ahsa 31982, Saudi Arabia
e-mail: malmaiah@kfu.edu.sa

© The Author(s), under exclusive license to Springer Nature Switzerland AG 2021
Y. Maleh et al. (eds.), *Artificial Intelligence and Blockchain for Future Cybersecurity Applications*, Studies in Big Data 90,
https://doi.org/10.1007/978-3-030-74575-2_12

assigned task and share it with other sensor nodes. This interconnection of these devices in a heterogeneous environment makes them more vulnerable to cybersecurity issues and threats. Therefore, cyber-attacks have become a serious concern, which led to many security solutions from the research community.

Cybersecurity is defined as a combination of security procedures, techniques, tools, and guidelines to protect the networks and devices over the internet [3]. Cybersecurity is one of the most critical issues for all countries of the world by protecting their assets and securing their information by detecting and mitigating the various cyber threats and attacks [4]. Many researchers have presented many techniques to address multiple types of security issues and problems in WSNs. However, they are still insufficient to protect the wireless networks from the ever-increasing security vulnerabilities and attacks. As a result, protecting WSNs from cyber attacks and threats has become essential and has prompted many researchers to conduct more research in recent years.

Several mechanisms and approaches have been proposed in the literature for detecting and mitigating cybersecurity attacks and threats for WSNs environment [2]. Each of these approaches has different tools and features to tackle various security attacks and breaches [5]. In this work, we conducted an overview analysis of the leading security issues related to WSNs to identify the significant cyber threats and provided the solutions in light of Blockchain technology. Blockchain technology includes hardware and software solutions to tackle the security challenges of WSNs, which is a novel approach. To fill this research gap, this paper aims to present a new scheme using Heuristic, Signature and voting detection methods to identify the optimal counter measures to detect the malicious and security threats using Blockchain technology.

Blockchain network uses a peer-to-peer network to record the data. In peer-to-peer policy, all wireless sensor nodes could be clients and servers simultaneously, so they communicate between peers. Based on the peer-to-peer approach, in our scheme, all nodes communicate as peers. This means each node can act as sender and receiver. All nodes share information (their neighbors' nodes) with their direct observers' nodes of every activity in the network. All the messages should be signed using asymmetric encryption technique to authenticate the information of the member nodes. The cluster head nodes have the public keys of all sensor nodes of the network to validate the identity of the member sensor nodes. Based on asymmetric encryption principles, each node in the network has its private key, and only this node can sign their sending messages using its ID (identifier). In this way, no one node can send a report of information (block) with a fake identity due to asymmetric encryption, one of the Blockchain principles. Thus, when a node observes an up-normal activity from another node, it directly reports it's observantly. This node reports this message using its signature as a signed block into the Blockchain system and sharing it with other nodes using peer-to-peer network. Once this message is recorded in the Blockchain system is very difficult to modify as it is shared through the permanent peer-to-peer storage. In case one of the nodes is hijacked, the node will start to send fake information, but signed by its correct identifier, and then the voting technique will detect its up-normal activity as shown in the current experiment of our approach in this study.

In this paper, we shed light on the use mechanisms and solutions of Blockchain technology to tackle the security challenges of WSNs, which is a novel approach. This paper, among the first studies that focus on analyzing the security problems of WSNs in light of Blockchain technology, is still a hot topic for many researchers to conduct more research in the future. Specifically, our research aims to address the following research question:

How blockchain technology provides more security options to enhance the security of WSNs from malicious attacks?

The rest of the paper is organized as Sect. 2 of the paper contains the background of the study. Similarly, Sect. 3 overviews the proposed model. The experiment implementation and result statistics for our scheme are discussed in Sect. 4. Section 5 summarizes and concludes the paper.

1.1 Research Motivation and Significance

Despite many benefits related to WSNs, security concerns and vulnerabilities in sensor devices are still the main challenges for WSNs [5]. This problem may allow intruders to breach the security of sensor devices and access the network. This may lead to steal sensitive information or damage the network [6, 7]. Therefore, to prevent such these security problems, there is a need to propose efficient security schemes and mechanisms to handle security breaches in WSNs. The current WSNs security schemes still have complexity in the high cost of energy consumption [8]. Due to the limited resources of the WSNs, incorporating security features to prevent and avoid malicious attacks is a complex challenge [9, 10]. If we exploit the Blockchain technology and its solutions, we can provide high-level security for WSNs and prevent malicious attacks simultaneously. Blockchain could also provide better security for WSNs by detecting malicious sensor nodes, routing attacks and intrusion detection. If Blockchain technology is employed in WSNs, the security advantages will be immense for WSNs.

2 Background of the Study

2.1 Security Issues in WSNs

Wireless sensor nodes are not intelligent enough to handle various cybersecurity threats solitarily. Therefore, a robust security mechanism is needed to help wireless devices to take cyber threats and vulnerabilities. Wireless devices have limited resources in terms of small memory to store security applications. Moreover, wireless devices are susceptible to various threats, due to their limited resources such as onboard power, memory, and processing. Furthermore, these devices' structure is straightforward, consisting of small processing chips, sensors, and transceivers. Due to this fragile and straightforward structure of wireless devices, various cyber-attacks such as DDOS can be propagated, which causes many problems in the network and halting of devices. There are other security issues of WSNs, such as detecting malicious nodes, intrusion detection, and authentication should be taken into consideration. [11] categorized the security issues into three levels: data security level (anonymity and freshness), access security level (accessibility, authorization and authentication) and network security level. The same study [11] also mentioned that attacks in WSNs could occur in all layers from the application layer to the physical layer. For example, at the application layer level, a malicious node can be added along the communication link to generate fake messages and data to attack the ongoing communication and increase the data collision. The transport layer attack happens by sending unlimited connection requests to minimize the node's energy and exhaust its resources, which leads to a denial of service. Another attack can be occurred in a network layer in several forms such as spoofing, sinkhole, flooding and replay attack to create and send fake messages or causing congestion in the network. Jamming attack at the Data link layer can cause loss of signals and data and destroy the channel and increased interference. At the physical layer level, the attacker can allow unauthorized nodes to access the network and damage it. Other researchers also have focused on the security issues in the IoT environment and how to detect security threats and vulnerabilities [12–14]. In a first attempt by [15] to design trust and authenticate scheme for WSNs based on Blockchain technology and to investigate the applicability of Blockchain in WSNs to address the security problems. This paper aimed to propose a new scheme using Heuristic, Signature and voting detection methods to identify the optimal countermeasures to detect the malicious and security threats using Blockchain technology.

2.2 Overview of Blockchain Technology

Blockchain is a robust technology that could be used to improve the security of WSNs by sharing and checking the data by the different sensor nodes deployed in the networks by using blockchain principles [16–18]. In this case, Blockchain can be defined as a distributed and collaborative security mechanism employed to guarantee the integrity, security and safety of information. In the Blockchain system, the data is stored in multiple records, which is called blocks. This information is distributed between all blocks deployed in the network by using links. These links between blocks are secured by using cryptography mechanisms, where each block has a hash of the content of the previous block. Based on that, any record cannot be modified without modifying all the next blocks. An important note, if any block in the chain were modified, the hash of the next block also would require to be modified, and this modification also would need a change in the upcoming block and so on. All of these blocks are saved in a distributed and decentralized manner in different nodes. In this way, no one of these blocks could be changed unless most nodes accept the change and do it. Thus, the data is safely and permanently stored in wireless sensor networks. In this way, cyber-attacks are very difficult to be propagated and implemented.

Blockchain could help improve the WSNs security from malicious attacks, prevent malicious activities through consensus mechanisms, and detect data tampering based on its underlying characteristics including data encryption, transparency, immutability, auditability and operational resilience [19]. In addition, Blockchain characteristics provide an impenetrable platform for cyber-attacks due to include typical network cybersecurity controls, practice and procedures.

2.3 Applicability of Blockchain in WSNs

Blockchain principles inspire the current solution of wireless network security issue. Each wireless sensor node has a list of records of the sensor nodes' identifiers (IDs) that have reported the existence of themselves based on direct observation node. This information is distributed along with the wireless network, so most sensor nodes must have this information permanently and securely. If a malicious node attempted to modify the blocks by inserting fake information such as false hash key, this would be dropped by other sensor nodes due to use the Peer-to-Peer principles of Blockchain, unless the majority of sensor nodes are malicious. Still, this scenario will not be happened according to the robust principles of Blockchain. Another scenario could happen when a sensor node reported to have observed abnormal activity from a node. This information cannot be ignored in wireless networks that used the Blockchain system, since it could have happened later. Thus, this fake information could be introduced in the blocks distributed in sensor nodes. To address this problem, trust policy is employed to cope with these cases by identifying behaviors, actions and activities of a node that are only reported by only one direct observer node repetitively, based on this hypothesis that other several sensor nodes normally observe each activity. Another scenario that can be happened is that a malicious node will ignore the detection of any other compromised malicious node and so on; however, normally these malicious nodes would probably also be observed by other nodes using trust table.

In our scheme, before the detection systems starting, all the sensor nodes (SN) and cluster head nodes (CN) should be registered in the Blockchain system. Each cluster head node has a list of all the neighbours' nodes' public keys in the network. Each node should be signed each message received and send it securely to all neighbours' nodes to avoid malicious-attacks by the compromised malicious node. Thus, all nodes can sign and forward all messages over the network by asymmetric encryption [20]. In this way, each sensor node in the wireless network can know a list of malicious nodes that the other nodes have observed.

In case one of the nodes is hijacked, the node will start to send fake information, but signed by its correct identifier, and then the voting method will detect its upnormal activity. In this way, each cluster node can know the list of malicious nodes that have been detected through a result of voting (malicious or benign) by the other nodes and send the result to the Blockchain system. Finally, according to the voting results on the Blockchain and detection systems on the CN, the detection system decides whether to eliminate the network's suspect node.

2.4   Research Contribution

In our scheme, to exploit the benefits of Blockchain, we designed a private Blockchain. In our model, each cluster head node (CN) is responsible for authenticating the participating sensor nodes belonging to their location and storing the sensor nodes-IDs. The CN node also uses the Heuristic Detection System to calculate the hash value and check it if this value on the Blockchain system or not, so that it becomes easy to detect the malicious sensor nodes in the WSNs. Second, The CN node is responsible for validating the sensor node signature by using Signature-Based System. The Blockchain system has a list of all node signatures in the network and shares it with all CNs. It is assumed that each sensor node should sign each message received and send it securely to all neighbors' nodes to avoid cyber-attacks through cluster head nodes. Thus, all nodes can sign and forward all messages over the network by using asymmetric encryption. In this way, each cluster head node can investigate whether signatures already exist on the Blockchain system or not. Thus, based on the result on the Blockchain, the signature detection system in the cluster head node can decide whether to delete the suspect node or not. Finally, the cluster node can know the list of malicious nodes that have been detected through a result of voting (malicious or benign) by the other nodes and send the result to the Blockchain system. According to the voting results on the Blockchain and detection systems on the CN, the detection system decides whether to eliminate the network's suspect node.

3   Proposed System

In this section, we present an overview of the proposed scheme and some assumptions in the scheme. Then, we describe the detection systems used in the proposed system. Next, we calculate the malicious degree by applying the elimination decision formula. Finally, we present a countermeasure against mass voting by malicious nodes.

3.1 Overview of Proposed Scheme

In this work, a new scheme of assigning three functions for the cluster head nodes (CNs) with communication with the Blockchain system. These three functions based on three detection systems are (1) heuristic-based system, (2) signature-based system and (3) voting-based system to detect the malicious sensor nodes. Figure 1 presents an overview of the proposed scheme in our work. There are three main components in this proposed model:

(1) Sensor Nodes (SNs): each sensor node has low computing power and memory space because batteries power it. (1) Sensor Nodes (SNs): each sensor node has low computing power and memory space because batteries power it.
(2) Cluster head nodes (CNs): Cluster head node has the more computational capability, more storage space, and more communication distance. Also, all nodes' public keys and signatures are preset in CN by Blockchain system to authenticate sensor nodes before laying out the network. It also has lightweight authentication certification (LAC) to verify the authentication and exchange the secret key between the Blockchain system and the cluster head nodes (CN).
(3) Blockchain: This trusted system is used to initialize the sensor node and has a shared secret key (SK) with the CN. Also, Blockchain is used to store the authentication results of sensor nodes in CN in a distributed way.

For this proposed scheme, due to a large number of sensor nodes distributed in the network, we divided them into different groups according to cluster head nodes'

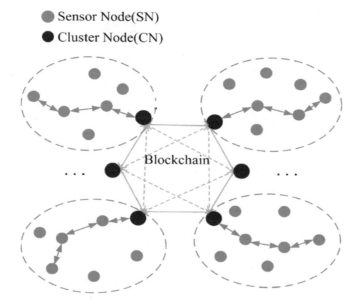

Fig. 1 Overview of the proposed system

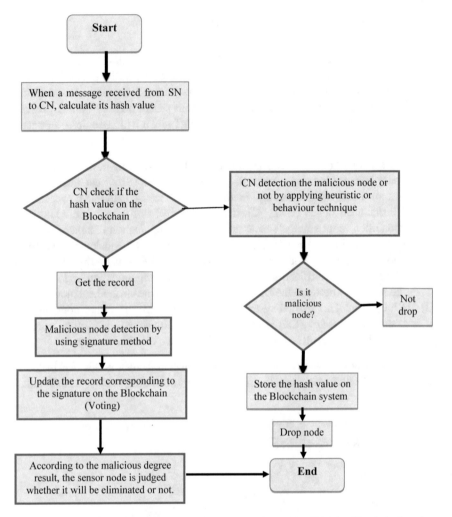

Fig. 2 Flowchart of malicious sensor node detection performed by CN using Heuristic detection system and signature-based system

location. In our scheme, the authentication of all sensor nodes is distributed to each cluster-head node because of the large number of sensor nodes and limited resources. Then, each cluster-head node applies LAC with Blockchain system. The Blockchain system comprises sensor nodes and cluster head nodes who want to eliminate fake information and detect an intruder and malicious nodes. It is assumed that each cluster head node has a heuristic intruder detection system and a signature-based system. The Blockchain system records all signatures (hash values) of benign nodes and all signatures and information of suspected intruder and malicious nodes. In our scheme, before the detection process starts, we have the following assumptions:

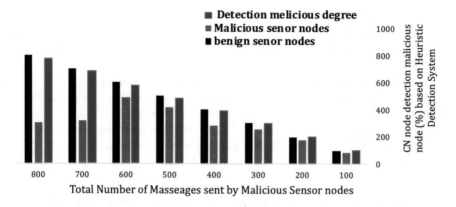

Fig. 3 CN node results in analysis using Heuristic Detection to detect malicious sensor nodes in WSNs

(1) All the sensor nodes (SNs) should be registered in the Blockchain system.
(2) All cluster head nodes (CNs) should be registered in the Blockchain system.
(3) Each cluster head node has all public keys and signatures of all sensor nodes that belong to the cluster node's exact location.
(4) Each cluster head node has a shared secret key with Blockchain.
(5) The Blockchain system has a list of all sensor nodes' public keys and signatures in the network.
(6) The Blockchain system has two lists: (a) benign identity list and (b) suspected malicious identity list.
(7) Sensitive data and information stored on the Blockchain can be represented as a record, and the following five elements represent the record:

- Suspected sensor node hash value.
- Number of votes for "malicious".
- Number of votes for "benign".
- Addresses of sensor nodes who voted "malicious".
- Addresses of sensor nodes who voted "malicious".
- Addresses of sensor nodes who voted "benign".

Where the numbers of votes for "malicious" and "benign" will be used to calculate the degree of malicious in the elimination decision formula (Sect. 4.4) to decide whether to eliminate the node. The recording of sensor nodes addresses prevents the same sensor node from illegally voting more than once. Here, the sensor node address is not an IP address but the address used on the Blockchain, such as "0yac46tu874458ef540ade6068dfe2f44e8fc6543".

In the current scheme, we assume that each cluster head node (CN) belonging to the Blockchain system installs the following two intruder detection methods:

(1) Malicious node detection using the heuristic-based method: This system is running when the cluster head node (CN) receives a message from the sensor node

(SN). In the current proposal, it is assumed that each CN detects the malicious or benign nodes by checking whether the hash value of the message is already registered as benign node identity or as suspected malicious node identity on the Blockchain system.

(2) **Malicious node detection using the signature-based method:** This technique ensures the authentication of the signatures of all nodes by investigating whether signatures already exist on the Blockchain system or not.

A flowchart shows the process of malicious sensor node detection performed by CN using a Heuristic detection system and a signature-based system (Fig. 2).

3.2 Detection of Suspected Malicious Nodes Using Heuristic Detection System on CN with Blockchain

When a CN receives a message from a sensor node, CN calculates the hash value and checks it if this value is on the Blockchain system. If the hash value does not exist on the Blockchain, a heuristic malicious detection system is performed to validate the sensor node's hash value. Suppose the heuristic or behaviour detection system determines that the received message is malicious. In that case, the CN sends the node hash value to the Blockchain system to share it and then eliminates the node and store it as a suspected malicious node identity. When other nodes receive the same message, the first validate whether the hash value of the message is already registered as benign node identity or as suspected malicious node identity on the Blockchain system by CN. If the same hash value of the message already exists on the suspected malicious identity list on the Blockchain system, the heuristic detection system on the CN judges that it is a malicious node and remove it.

3.3 Detection of Suspected Malicious Nodes Using Signature-Based System on CN with Blockchain

In contrast, if the same hash value exits on the benign identity list, a signature-based system through asymmetric encryption is executed. This technique ensures the authentication of the signatures of all nodes. The Blockchain system has a list of all node signatures in the network and shares it with all CNs. It is assumed that each sensor node should sign each message received and send it securely to all neighbors' nodes to avoid cyber-attacks through cluster head nodes. Thus, all nodes can sign and forward all messages over the network by using asymmetric encryption. In this way, each cluster head node can investigate whether signatures already exist on the Blockchain system or not. Thus, based on the result on the Blockchain, the signature detection system in the cluster head node can decide whether to delete the suspect node or not. Besides, Blockchain system shares this information among all cluster nodes, which includes all suspect malicious nodes. In this way, each cluster node can know the list of malicious nodes that have been detected through a result of voting (malicious or benign) by the other nodes and send the result to the Blockchain system. Finally, according to the voting results on the Blockchain and detection systems on the CN, the detection system decides whether to eliminate the network's suspect node.

3.4 Applying the Elimination Decision Formula

In this section, we apply the elimination decision formula to calculate the maliciousness degree. Table 1 show definitions of some symbols used in the elimination decision formula.

As we mentioned above, when the CN's detection system validates the hash value of the sensor node and finds the hash value exists on the Blockchain records. Then, the CN detection system could decide whether to remove the suspected node based on the result of maliciousness degree, calculated by the elimination decision formula using the following equations below.

In Eq. (1), when the degree of malicious D_m is smaller or equal to a threshold of malicious degree T_m, the result is satisfied; therefore, the sensor node is not removed from the Blockchain network.

$$D_m \leq T_m \qquad (1)$$

In Eq. (2), when the degree of malicious D_m is greater than the threshold of malicious degree T_m, the result is not satisfied; therefore, the sensor node is removed from the Blockchain network.

$$D_m > T_m \qquad (2)$$

Table 1 Definitions of some symbols used in the elimination decision formula

Symbols	Definitions
D_m	Degree of malicious
T_m	Threshold of malicious degree
M_v	Total number of votes for malicious
b_v	Total number of votes for benign
T_v	Threshold of total votes
V_r	Rate of voting confidence
S_r	Rate of self-confidence
R_d	Result of the malicious detection system

When $M_v + b_v \geq T_v$: **The** detection system uses only the voting result on the Blockchain and calculates the maliciousness degree using Eq. (3).

$$Dm = \frac{M_v}{M_v + b_v} \tag{3}$$

The CN detection system calculates the maliciousness degree using Eq. (4), the results of voting on the Blockchain, and its malicious detection results by heuristic or signature-based methods. Here, it is assumed that the malicious detection system outputs 1 when the sensor node is malicious and 0 when the sensor node is benign. That is, $R_d \in \{0, 1\}$.

$$Dm = \frac{Mv}{M_v + b_v} \times V_r + R_d \times S_r \tag{4}$$

where V_r and S_r are computed by the following Eqs. (2) and (3):

$$\mathbf{V_r} = \frac{\mathbf{M_{v+b_v}}}{\mathbf{T_v}} \tag{5}$$

$$\mathbf{S_r} = \mathbf{1 - V_r} \tag{6}$$

Example: Assume that a message is sent from the sensor node to cluster head node, and voting for the sensor node hash value on the Blockchain is 10 "malicious" votes ($M_v = 20$) and 5 "benign" votes ($b_v = 5$). Also, the malicious detection system in the CN judges the sensor node to be malicious ($R_d = 1$), the threshold for total votes is set to 20 ($T_v = 20$), and the threshold for maliciousness degree is set to 0.5 ($T_m = 0.5$). The malicious degree D_m in this example is calculated as follows:

$$Dm = \frac{10}{10 + 5} \times \frac{10}{10 + 5} + 1 \times \left(1 - \frac{10 + 5}{20}\right) = \frac{3}{4}$$

Based on the D_m result, the sensor node will be deleted from the Blockchain network because of Dm's value greater than the threshold of malicious degree ($\frac{3}{4} > 0.5$).

4   Experimentation Analysis and Results

The proposed scheme was implemented in the simulation environment using OMNeT++ software. OMNeT++ is considered a common tool to develop wireless sensor networks in the simulation environment as observed in wireless networks literature. The proposed scheme was performed by specifying network areas with the distribution of sensor nodes (SNs) based on cluster head nodes (CNs) in the network

Table 2 Wireless sensor network simulation setup

Simulation parameters	Value
Simulation tool	OMNeT++
Simulation environment	800 × 800
Number of sensor nodes *SNi*	50, 100, 200, 400, 600, 800
Number of cluster head nodes	6
Numberofmalicious nodes	100, 200, 300, 400, 500
Numberofbenign nodes	100, 200, 300
Transmission range	150 M
Packet size	256 Kbps
The transmission interval of CN	30 s
Transmission of benign nodes time interval	10 s

topological order. Also, we created a private virtual Blockchain with communication connectivity with cluster head nodes (CNs). To achieve that, we installed Geth, an Ethereum client, and interacted with Geth through a Python script. The simulation parameters were set in the cluster head nodes (CNs) with connectivity with sensor nodes. Also, the function assigned to CN nodes to detect malicious sensor nodes in the network such as hijacked nodes by the assessment of hash values of each sensor node, the signature of the sensor node, degree of malicious (D_m) and a total number of votes for malicious (Mv). The simulation parameters used in the proposed scheme is presented in Table 2.

4.1 Result Analysis of CN Function Based on Heuristic Detection System

The simulation results for the function of CN node were observed to verify the proposed scheme's performance reliability in terms of detecting the malicious sensor nodes through computing the hash value and validate it through the Blockchain system. The results statistic seen for malicious sensor node detection and identification through CN and Blockchain using Heuristic Detection System was found quite consistent and remarkable. The CN nodes were determined for the detection and identification, where a malicious node send a fake message to the CN node in the network. Similarly, this message was also received by the CN node. The CN node performed the necessary security verification process to match the node hash value with hash values in the Blockchain system. The results indicated that this sensor node did not verify the security condition of the hash value by matching its value. After that, the CN node sends the node hash value to the Blockchain system to share it, and then eliminates it and stores it as a suspected malicious node identity. Also, the Blockchain generates an alarm message to acknowledge a malicious node's existence in the network. The simulation results verify that the CN node with Blockchain successfully identified a malicious node in the network. This verifies that the CN node detection rate of malicious node based on hash value assessment was entirely accurate in the wireless sensor network against the fake message. Subsequently, the number of malicious sensor nodes was increased in the deployed WSN infrastructure to verify performance reliability with many fake news, which was also found quite exceptional for the CN node. The CN node detects the maximum number of fake messages in their location, whose statistics are shown in Fig. 5.

4.2 Result Analysis of CN Function Based on Signature-Based System

Our proposed model's results have also been evaluated the CN function based on Signature-Based System, where this method ensures the authentication of the signatures of all nodes correctly by CN with Blockchain system. In the simulation, we used malicious nodes to send messages with fake signatures. Here in this scenario, the hash values and sensor nodes ID in fake messages were kept similar to benign nodes, but the signatures were different for all introduced malicious nodes. During the simulation, the CN nodes have checked for assessment of signatures fake of all malicious nodes, which was found quite remarkable by assessing sensor nodes signatures in the network with the Blockchain system. Moreover, the statistical analysis observed during the simulation for a CN node based on Signature-Based System is shown in Fig. 4, where the malicious node signature detection is presented in graphical form as captured during the simulation.

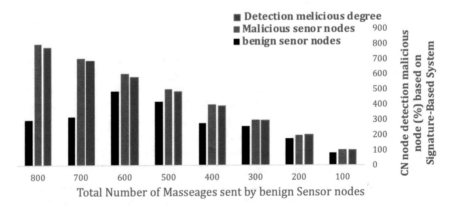

Fig. 4 CN node results in analysis using signature-based system to detect malicious sensor nodes in WSNs

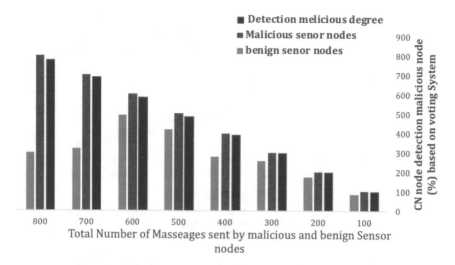

Fig. 5 CN node results analysis using voting system to detect malicious and benign sensor nodes in WSNs

4.3 Result Analysis of CN Function Based on Voting-System for Malicious or Benign Sensor Nodes

The CN results were also seen to detect malicious or benign sensor nodes during the operational network. In case one of the nodes is hijacked, the node will start to send fake information, but signed by its correct identifier, and then the voting method will detect its up-normal activity as shown in the current experiment of our approach in this study. In this way, each cluster node can know the list of malicious nodes that have been detected through a result of voting (malicious or benign) by the other nodes and send the result to the Blockchain system. Finally, according to the voting results on the Blockchain and detection systems on the CN, the detection system decides whether to eliminate the network's suspect node. The statistical analysis extracted from the simulation tool is shown in Fig. 5, where both malicious and benign nodes broadcast messages in the network. However, those correct messages received by CN directly from sensor nodes are also assessed by a voting system based on the value of the degree of malicious (D_m). The statistical results analysis for fake messages of malicious nodes by using a voting system, which was captured during the simulation, is shown in Fig. 5 (Fig. 3).

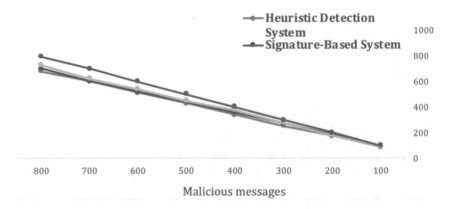

Fig. 6 Overall detection rate of malicious messages in WSNs

4.4 Result Analysis of CN Functions Based on Heuristic Detection System, Signature-Based System and Voting-System

In the simulation, the overall proposed scheme was also evaluated after combining the three detection systems to assess malicious nodes' overall detection rate. The overall results statistic showed that 94.9% of malicious messages were detected and identified successfully during the simulation for our scheme, as shown in Fig. 6.

5 Conclusion

In this research, a novel approach of Blockchain-based Heuristic, Signature and Voting methods for detecting malicious attacks for Wireless Sensor Networks was proposed. The proposed scheme will counter malicious sensor attacks in deployed WSNs with minimal network resource utilization. The proposed scheme uses three functions with Blockchain technology to ensure the security of the network and maintain a secure communication infrastructure for WSNs. The three functions of the proposed scheme back up each other to identify malicious attacks in a more precise way in the network. Similarly, the functions work independently in the network, but the authentication mechanism backs up each other to identify malicious attacks at a high rate. In the first function, the CN node performs a heuristic malicious detection system to validate the hash value of the sensor node. Suppose the heuristic detection system determines that the received message is malicious. In that case, the CN sends the node hash value to the Blockchain system to share it and then eliminates the node and store it as a suspected malicious node identity. The second function is the signature-based method. If the same hash value exits on the benign identity list, a signature-based system through asymmetric encryption is executed. This technique ensures the authentication of the signatures of all nodes. The Blockchain system has a list of all node signatures in the network and shares it with all CNs. It is assumed that each sensor node should sign each message received and send it securely to all neighbors' nodes to avoid cyber-attacks through cluster head nodes. Thus, all nodes can sign and forward all messages over the network by using asymmetric encryption. In this way, each cluster head node can investigate whether signatures already exist on the Blockchain system or not. Thus, based on the result on the Blockchain, the signature detection system in the cluster head node can decide whether to delete the suspect node or not. The voting method is the third function, in case one of the nodes is hijacked, the node will start to send fake information, but signed by its correct identifier, and then the voting method will detect its up-normal activity. In this way, each cluster node can know the list of malicious nodes that have been detected through a result of voting (malicious or benign) by the other nodes and send the result to the Blockchain system. According to the voting results on the Blockchain and detection systems on the CN, the detection system decides whether to eliminate the network's suspect node. Finally, we carried out simulation experiments; the overall experimental results show that our scheme can effectively suppress sensor nodes' malicious attacks.

References

1. Adil, M., Almaiah, M.A., Omar , A., Almomani, O.: An anonymous channel categorization scheme of edge nodes to detect jamming attacks in wireless sensor networks. Sensors. **20**(8), 2311 (2020)
2. Adil, M., Khan, R., Almaiah, M.A., Al-Zahrani, M., Zakarya, M., Amjad, M.S., Ahmed, R.: MAC-AODV based mutual authentication scheme for constraint oriented networks. IEEE Access. **4**(8), 44459–44469 (2020)
3. Al , A.K., Almaiah, M.A., Almomani, O., Al-Zahrani, M., Al-Sayed, R.M., Asaifi, R.M., Adhim, K.K., Althunibat, A., Alsaaidah, A.: Improved Security Particle Swarm Optimization (PSO) algorithm to detect radio jamming attacks in mobile networks. Quintana **11**(4), 614–624 (2020)
4. Prasad, R., Rohokale, V.: Cyber threats and attack overview. In: Cyber Security: The Lifeline of Information and Communication Technology 2020, pp. 15–31. Springer, Cham (2020)
5. Vasilyev, V., Shamsutdinov, R.: Security analysis of wireless sensor networks using SIEM and multi-agent approach. In: 2020 Global Smart Industry Conference (GloSIC), pp. 291–296, 17 November 2020. IEEE (2020)
6. Ammar, M., Russello, G., Crispo, B.: Internet of Things: a survey on the security of IoT frameworks. J. Inf. Secur. Appl. **1**(38), 8–27 (2018)
7. Rathee, G., Sandhu, R., Saini, H., Sivaram, M., Dhasarathan, V.: A trust computed framework for IoT devices and fog computing environment. Wirel. Netw. **26**(4), 2339–2351 (2020)
8. Khan, M.A., Salah, K.: IoT security: review, blockchain solutions, and open challenges. Future Gener. Comput. Syst.. **1**(82), 395–411 (2018)
9. Merkow, M.S., Breithaupt, J.: Information Security: Principles and Practices. Pearson Education, Indianapolis (2014)
10. Zou, Y., Zhu, J., Wang, X., Hanzo, L.: A survey on wireless security: technical challenges, recent advances, and future trends. Proc. IEEE. **104**(9), 1727–1765 (2016)
11. Tomić, I., McCann, J.A.: A survey of potential security issues in existing wireless sensor network protocols. IEEE Internet of Things J. **4**(6), 1910–1923 (2017)

12. Kumar, J.S., Patel, D.R.: A survey on internet of things: Security and privacy issues. Int. J. Comput. Appl. **90**(11), 20–26 (2014)
13. Sicari, S., Rizzardia, A., Griecob, L.A., Coen-Porisini, A.: Security, privacy and trust in Internet of Things: the road ahead. Comput. Netw. **76**, 146–164 (2015)
14. Lin, J., Yu, W., Zhang, N., Yang, X., Zhang, H., Zhao, W.: A Survey on Internet of Things: architecture, enabling technologies, security and privacy, and applications. IEEE IoT J. **4**(5), 1125–1142 (2017)
15. Moinet, A., Darties, B. and Baril, J.L.: Blockchain based trust and authentication for decentralized sensor networks (2017). arXiv preprint arXiv:1706.01730
16. Adil, M., Khan, R., Almaiah, M.A., Binsawad, M., Ali, J., Al , A., Ta, Q.T.: An efficient load balancing scheme of energy gauge nodes to maximize the lifespan of constraint oriented networks. IEEE Access. **11**(8), 148510–148527 (2020)
17. Khan, M.N., Rahman, H.U., Almaiah, M.A., Khan, M.Z., Khan, A., Raza, M., Al-Zahrani, M., Almomani, O., Khan, R.: Improving energy efficiency with content-based adaptive and dynamic scheduling in wireless sensor networks. IEEE Access. **25**(8), 176495–176520 (2020)
18. Adil, M., Khan, R., Ali, J., Roh, B.H., Ta, Q.T., Almaiah, M.A.: An energy proficient load balancing routing scheme for wireless sensor networks to maximize their lifespan in an operational environment. IEEE Access. **31**(8), 163209–163224 (2020)
19. Marchang, J., Ibbotson, G., Wheway, P.: Will blockchain technology become a reality in sensor networks? In: 2019 Wireless Days (WD), 24 April 2019, pp. 1–4. IEEE (2019)
20. Almaiah, M.A., Dawahdeh, Z., Almomani, O., Alsaaidah, A., Al-khasawneh, A., Khawatreh, S.: A new hybrid text encryption approach over mobile ad hoc network. Int. J. Electr. Comput. Eng. (IJECE). **10**(6), 6461–6471 (2020)

Artificial Intelligence and Blockchain Applications for Smart Cyber Ecosystems

A Framework Using Artificial Intelligence for Vision-Based Automated Firearm Detection and Reporting in Smart Cities

Muhammad Hunain, Talha Iqbal, Muhammad Assad Siyal,
Muhammad Azmi Umer, and Muhammad Taha Jilani

Abstract For a few decades, mega-cities are facing some huge challenges. Among them, the prevention of crime seems to be more challenging than others. The safety of citizens in the dense urban population with conventional practices are unable to control the increasing crime rate. This work is aimed to develop a framework for the autonomous surveillance of public places, with visual-based handheld arms detection in a near real-time. It scans all the objects that come in front of the camera and when any type of weapon comes in contact with a lens it gives an alert, locks that object and the person holding it and identifies the person using facial recognition. If the alert does not get responded in a few minutes, the system will automatically notify the 3rd person or agency about the incident. It can also manually highlight any object in a frame to keep track of its movement for security purposes. Machine and Deep Learning techniques were used to train models for object detection and facial recognition. The model achieved an accuracy of 97.33% in object detection and 90% in facial recognition.

1 Introduction

Over the past few decades, various urban areas within developing countries have experienced a growing population and rural-to-urban migration rate. It is estimated that nearly half the population of the world is now living in the cities [1], now making them mega-cities which can be seen by World Economic Forum (WEF) reports'

M. Hunain · T. Iqbal · M. A. Siyal
DHA Suffa University, Karachi, Pakistan
e-mail: Hunain@techonventures.com

T. Iqbal
e-mail: Talha@techonventures.com

M. A. Umer (✉)
DHA Suffa University, and KIET Karachi, Karachi, Pakistan
e-mail: azmi.umer@dsu.edu.pk

M. T. Jilani
Karachi Institute of Economics and Technology, Karachi, Pakistan
e-mail: m.taha@pafkiet.edu.pk

© The Author(s), under exclusive license to Springer Nature Switzerland AG 2021
Y. Maleh et al. (eds.), *Artificial Intelligence and Blockchain for Future Cybersecurity Applications*, Studies in Big Data 90,
https://doi.org/10.1007/978-3-030-74575-2_13

statistics in Fig. 1. This rapid transition has presented many challenges, including risks to the immediate and surrounding environment, to natural resources, to health conditions, to social cohesion, and individual rights [2]. The later has introduced the safety and security concerns for the citizens living in a megacity. Similarly, for governments and administrative agencies, one of the most important consideration is to monitor and control the criminal activities. Table 1 has described the number of incidences and crime rates in major cities of India.

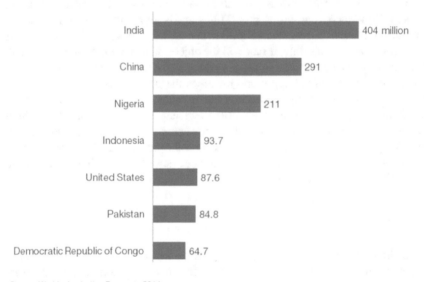

Source: World urbanization Prospects 2014

Fig. 1 Urban population growth [3]

Table 1 Incidences and crime rates in mega cities [4]

Year	No. of incidence	Crime rate
2009	8,91,576	826.5
2010	11,19,621	1037.8
2011	11,49,059	713.2
2012	11,03,858	685.2
2013	12,03,514	748.8

In conventional practice, such issues are addressed by using CCTV based surveillance and monitoring only. However, current developments in ICT have opened new opportunities to develop some intelligent methods for effective control and monitoring of crime. Over the past few years, some topics have been top in research areas in computer technological era. Those are detection, tracking, and understanding the moving objects to prevent crime. Similarly, Intelligent visual surveillance system (IVSS) are one of the surveillance system that refers to automate visual monitoring process involving interpretation and analysis of object detection and behavior, also the tracking of that object to understand the current scene of that visual events. Two main tasks that are highly focused are discussed in [5] i.e. scene anomaly and large area surveillance control. All detection and tracking of moving objects in a sequence and behavior analysis are in scene interpretation. The control task multiple cameras are to tackle captured or fixed objects which are in motion in a wide-area surveillance.

Detection of moving objects is a hectic task as well as it is an important task for any video surveillance system. Secondly, tracking is required in upper-level applications after detection because it requires the location and shape of objects in every camera region or frame via detection algorithm [6]. A video surveillance might embody a minimum of one sensing unit capable of being operated in a very scanning mode and a video process unit coupled to the sensing unit, the video process unit to receive and method image information from the sensing unit and to find scene events and target activity [7]. Similarly, a system proposed in [8], which was mostly based on hardware devices like motion sensors, light sensors, alarms, etc. It detects the anomaly and reports the user through push notification on any handheld device like a mobile or laptop.

The campus security system was proposed in [9]. This system is consist of a school gate state monitor, an entrance guide terminal and a base station, in this system entrance gate terminal monitors the presence of entrance guard or check whether the guard is on duty or not in the campus. Second is the school gate state monitor which monitors the state of the gate whether it is open or close. Third is a base station which receives information from the entrance guide terminal and school gate state monitor and generate alarm signals when the entrance guard is missing and the school gate is opened. This campus security system can monitor in real-time and can alert when detects an anomaly, which helps to improve the security system of the campus.

As the violent criminals, burglars and intruders have become so dangerous for the properties and lives of people. Protection and security for households become a necessity. Anti-Intruder Monitoring and Alarm [10] with the purpose to help homeowners and make them informed about criminals and alarm triggering decisions. The alarm system uses images and locations of sensed motion and offers the option of allowing multiple key holders to receive security alerts via cellular network short message service (SMS). The alarm system also gives the option of sending distress messages to the police or trusted neighbors. The security system can be easily controlled by using a mobile device or remote control. The algorithm of this system has been designed simply and made the probability of false alarms almost non-existence.

According to the research carried out, there is no such application/software which is capable of doing surveillance and as well as identifying objects and people in real-time. Some products have some similarities in terms of facial recognition, data extraction, object detection, notifying 3rd party or security agency and generate an alarm system. But no one is completely satisfied by implementing all functionalities mentioned above in a single program as in the proposed system. These are the aspect that lead to this system an upend over previously launched products. Rest of the chapter is organized as follows: Sect. 2 is an overview of the related work. Section 3 has described the methodology. Section 4 has discussed the experimental evaluations and results, while Sect. 5 has discussed the conclusion and possible future work.

2 Related Work

In today's modern life there is an increasing interest in the precautionary and protective measures in the world and private space of social welfare. Therefore, there is a need to look for the surveillance arrangement to provide a safe and sound environment for the citizens. Currently, technologies like cameras, sensors, microphones, and detectors are being used. Trespasser detection is the new increasing demand in the commercial and private sectors. However, it is difficult to eradicate the concept of using these technologies without being detected. Hence looking at this flaw [11] proposed the idea of a multi-sensor intelligent system that can operate on the principle of entropy from several sources to find the danger or any internet breach. Therefore they developed a generic ontology that allowed the integration of all the input heterogeneous knowledge in a homogeneous way.

Handheld gun detection was performed in [12]. They used Convolutional Neural Network (CNN) to detect guns from cluttered scenes. They particularly used Deep Convolution Network (DCN) through transfer learning. The model was evaluated on a benchmark Internet Movie Firearms Database (IMFDB). Similarly, CNN has been used in [13] for gun detection. They got training accuracy of 93% and testing accuracy of 89%. Gun detection was also performed in [14] using color-based segmentation. They used k-means clustering to omit objects other than the weapons from the images. Harris interest point detector and Fast Retina Keypoint (FREAK) were used to locate the weapons in the segmented images.

Nowadays home security and its safety become one of the biggest concerns for homeowners. Leveraging audio/video recording and communication devices provides methods for information about crime. An approach was proposed in [15], which includes a method of comprising, a method of receiving from an audio/video recording, and a communication device. It has a first alert signal and a first video signal, the first video signal including images captured by a camera of the A/V recording and communication device, transmitting to a client device, in response to receiving the first alert signal and the first video signal, a second alert signal and a second video signal, the second video signal including the images captured by the camera of the A/V recording and communication device, receiving a report signal from the client device; It work on the images captured by the camera of the A/V recording and communication device, that a crime may have been committed, posting an offer of a reward for information about the crime.

An intelligent visual surveillance system has been proposed in [16] with the help of cameras attached in the network to observe the people and vehicles. The system modules are proposed to perform critical works like the management of cameras, tracking objects, recognition of people via biometric technology, monitoring the crowd to catch anomaly. Similarly, [17] is also based on the video surveillance system in which the system uses metadata rule for analyzing and exchange of information between intelligent video surveillance system that analyzes the required data through streaming on camera. The metadata rule is just to enhance the indexing method by indexing a large database and collaboratively searches and manages the integrated security environment more accurately and efficiently. The system focused on both high-level and low-level context to utilize metadata as a raw back source for security system services. Physical sensors (metal detector, cameras, scanners) in public areas are for the low-level context of the system. The situation is being captured in the high-level context-aware system by analyzing the context data coming through sensors in the low-level system. The system also provides the tracking system by moving an object in the field of view called FOVs. The system also supports real-time tracking of moving objects by tilting, panning and zooming in FOVs.

The digital surveillance system is pre-install by the ubiquitous approach and generates a huge amount of video streaming and other data as well. The development of the cloud environment has empowered to deploy intelligent video surveillance technologies through Web Services to enhance public security. The introduction of the novel system and the combination of cloud computing techniques with the automatized license plate recognition engines have been discussed in [18]. Its approach was to analyze big data to detect as well as to keep track of a target vehicle in a city with a license plate number issued to vehicles. Likewise, [19] has discussed the reviews about the recent development techniques of relevant technologies like pattern recognition and computer vision. They have discussed the multi-camera tracking, topologies of computing with integrated cameras, multi-level frames object detection and tracking, identification and some sort of re-identification, and both static and active cameras' cooperative video security. The detailed explanation of the technical aspect used by these terminologies and comparison of pros and cons between different approaches for solution has been provided. It mainly focuses on the connection and integration of different modules within the application. They have also focused on improving the efficiency, accuracy, and complexity. An intelligent video surveillance system (IVSS) has also been proposed in [20] by having a functionality detection and identification of anomaly and alarming situations by sensing the moving objects. The main motive of this system design was to reduce video processing and transmission, therefore, allowing a huge number of cameras deploying on the system to satisfy its usage as a security solution with safety integration in smart cities. Here alarming and detection were performed based on moving objects using the feature parameters of performed detection results and also using ontologies and semantic reasoning.

Threat-detection in a distributed multi-camera surveillance system was proposed in [21]. They observed the threats by analyzing the motion of an object in software installed at the first camera then detection of a suspicious object at the camera when motion of the object does not match to a motion flow model at the first camera. Then the tracking process is being entertained from the first frame to the second camera frame based upon the suspicious detection of objects. Just like the first camera, the second camera processing for detection is being done via the same software installed in it. As the first camera and assigned threat scores aside when motion of the object does not match to a motion flow model at the second camera, like the initial one and finally generating an alarm based on part of the threat scores detected at these frames of cameras and notifying the authorities.

The security system has been used for safety for homes and other areas greatly. The security system proposed in [22] consists of a main automatic circuit which have motions detector for activating an audible alarm and provides further detections to identify criminals and crime. It has an emergency light flasher which is manually activated by the user. It provides an inside home control panel. Inside the home control panel also responds to remote manually. This system has been used more effectively that easily terminate possible home invasion or robbery. This system also enhances safety and security.

An intelligent image processing method for the video surveillance systems was proposed in [23]. It includes a technology of tracking and detecting multiple moving objects, which can be easily applied to business and home surveillance systems consisting of a network video recorder (NVR) and internet protocol (IP) camera. It also provides the easiest way for detection and tracking, in which it uses the red-green-blue (RGB) color background modeling with a sensitivity parameter to extract moving regions, the blob-labeling to group moving objects and the morphology to eliminate noises. If it comes to the tracking of the fast-moving object then this method can define the direction as well as the velocity of the group formed by the objects which are in motion.

An intelligent video/sound analysis and ID database framework was proposed in [24]. It may define a security zone or gathering of zones. The framework may distinguish vehicles and people entering or leaving the zone through picture acknowledgement of the vehicle or individual when compared with prerecorded data available in a database. The framework may alarm the security workforce as to warrants or other data found relating to the perceived vehicle or individual coming out because of a database seek. The framework may analyze pictures of a presume vehicle, for example, an undercarriage picture to standard vehicle pictures recorded in the database. The framework may additionally take in the standard occasions and areas of vehicles or people followed by the framework and to make security workforce ready upon deviation from standard movement.

Parallel execution of an ongoing canny video surveillance system on Illustrations Preparing Unit (GPU) was portrayed in [25]. The system depends on foundation subtraction and made out of movement detection, camera attack detection (moved camera, out-of-center camera, and secured camera discovery), surrendered object detection, and object tracking algorithms. As the calculation algorithms have diverse qualities, their GPU executions have distinctive acceleration rates. Test results demonstrate that when all the available algorithms run simultaneously, parallelization in GPU influences the system to up to 21.88 times quicker than the central processing unit partner, empowering real-time analysis of a higher number of cameras.

3   Methodology

Machine and Deep Learning techniques were used to train models for object detection and facial recognition. Further Transfer Learning was performed on the inception R-CNN V2 dataset. Extra layers were added to identify the weapon. Facial Identification was performed using the inception Haar Cascade Frontal Face dataset while Recognition was done using the MTCNN method. Object Locking was done using OCF-CRS algorithm. Data Extraction from social media was done using jsoup while the 3rd party notification was implemented using Twilio SMS. The complete GUI built in python using PyQt v4.11. The comparison of the proposed system with existing work is described in Table 2.

High Level Architecture, Software Architecture, Sequence diagram, and State diagram of the system are shown in Fig. 2, 3, 4, and 5 respectively.

Table 2 Comparison with existing work

Existing work	Features					
	Object detection	Object tracking	Specification of unethical object	Database maintenance	Alarming system	Extraction of culprit's information
Proposed system	Yes	Yes	Yes	Yes	Yes	Yes
[26]	Yes	Yes	No	Yes	Yes	No
[27]	Yes	No	No	No	No	No
[28]	Yes	No	No	Yes	No	No
[29]	Yes	Yes	No	Yes	No	Yes
[30]	Yes	Yes	No	No	Yes	No
[31]	Yes	Yes	No	Yes	No	No

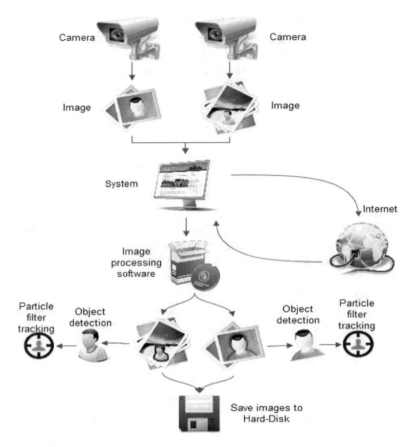

Fig. 2 High level architecture

3.1 Transfer Learning

Innovation plays an important role in the utilization of a pre-trained model. For instance, a model can be used without making any changes into it, for example, it can be used in an application to categorize new photos. The pre-trained model can be used in coordination with other neural network model. In this case, the load of the pre-trained model can be frozen considering the fact that they are not updated based on the newly trained model. Similarly, the load can be refreshed based on the training of new model. However, there could be a lower learning rate. This allows pre-trained model to behave like a weight initialization program during the training of the new model. Some of its common usages are as classifier and standalone feature extractor. The pre-trained model can be directly used as a classifier to classify new photos. The pre-trained model or some segment of the model can also be used to pre-process new photos and to extract useful attributes.

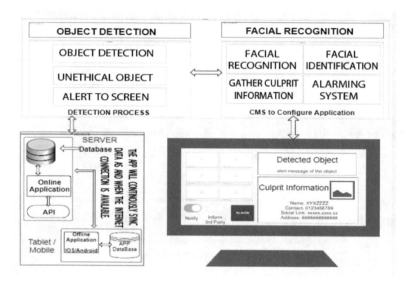

Fig. 3 Software architecture

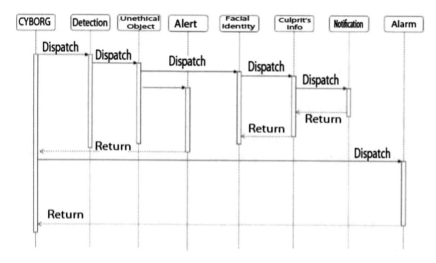

Fig. 4 Sequence diagram

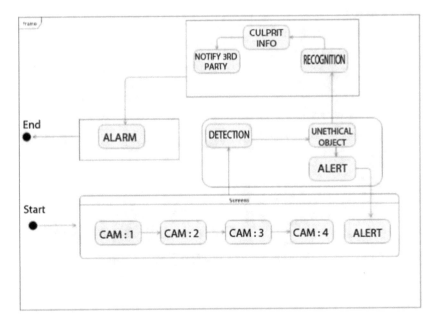

Fig. 5 State diagram

3.2 Model Selection

There are many models for image identification. Some of them includes VGG (e.g. VGG16 or VGG19), GoogLeNet (e.g. InceptionV2), and Residual Network (e.g. ResNet50). They are usually used for transfer learning becasue of their architectural innovations. The process of loading the InceptionV3 pre-trained model is described below.

3.2.1 InceptionV3 Pre-trained Model

The InceptionV3 is commonly known as the third iteration of the inception architecture. Initially, it was developed for GoogLeNet. The model requires the colour images should be 299 × 299. The script described in this section was taken from [32]. Run the following script to load the model.

```
1 # example of loading the inception v3 model
2 from keras.applications.inception_v3 import InceptionV3
3 # load model
4 model = InceptionV3()
5 # summarize the model
6 model.summary()
```

Listing 1.1 Loading the inception v3 model [32]

After executing the above code, the model will be loaded. It will also download the required weights. It will summarize the model architecture to ensure that the model got loaded correctly. Now download photos to train the model and save these photos in current working directory with some filename like "knife1.jpg". A pre-trained model can be utilized to group new photos among the 1,000 known classes. The photo needs to be stacked and reshaped to a 224 × 224 square. Moreover, the pixel esteems are scaled in the route required by the model. The model works on different tests. Along these lines the components of a stacked picture should be extended by 1 for one picture with 224 × 224 pixels and three channels.

```
# load an image from file
image = load_img('knife.jpg', target_size=(224, 224))
# convert the image pixels to a numpy array
image = img_to_array(image)
# reshape data for the model
image = image.reshape((1, image.shape[0], image.shape[1], image.shape[2]))
# prepare the image for the VGG model
image = preprocess_input(image)
```
Listing 1.2 Loading an image from file [32]

Now the model is loaded and ready to make a prediction. This means that it will calculate the probability of the image belonging to each of the thousand classes.

```
# predict the probability across all output classes
yhat = model.predict(image)
# convert the probabilities to class labels
label = decode_predictions(yhat)
# retrieve the most likely result, e.g. highest probability
label = label[0][0]
```
Listing 1.3 Prediction [32]

After combining all of this together, a new image is loaded and the prediction is made to its most likely class.

```
# load an image from file
image = load_img('knife.jpg', target_size=(224, 224))
# convert the image pixels to a numpy array
image = img_to_array(image)
# reshape data for the model
image = image.reshape((1, image.shape[0], image.shape[1], image.shape[2]))
# prepare the image for the Inception model
image = preprocess_input(image)
# load the model
model = InceptionV3()
# predict the probability across all output classes
yhat = model.predict(image)
# convert the probabilities to class labels
label = decode_predictions(yhat)
# retrieve the most likely result, e.g. highest probability
label = label[0][0]
# print the classification
print('%s (%.2f%%)' % (label[1], label[2]*100))
```
Listing 1.4 Combined Code [32]

3.3 Object Tracking

For tracking the object we have used MIL Tracker that is designed in OpenCV. The
reason for choosing the MIL tracker is its high accuracy. The script described in this
section was taken from [33]. For object tracking using OpenCV, create a new .py file
and insert the following script:

```python
import cv2
import time

if __name__ == '__main__' :

    # Start default camera
    video = cv2.VideoCapture(0);

    # Find OpenCV version
    (major_ver, minor_ver, subminor_ver) = (cv2.__version__).
    split('.')

    # With webcam get(CV_CAP_PROP_FPS) does not work.
    # Let's see for ourselves.

    if int(major_ver) < 3 :
        fps = video.get(cv2.cv.CV_CAP_PROP_FPS)
        print ("Frames per second using video.get(cv2.cv.
    CV_CAP_PROP_FPS): {0}".format(fps))
    else :
        fps = video.get(cv2.CAP_PROP_FPS)
        print ("Frames per second using video.get(cv2.
    CAP_PROP_FPS) : {0}".format(fps))


    # Number of frames to capture
    num_frames = 120;


    print ("Capturing {0} frames".format(num_frames))

    # Start time
    start = time.time()

    # Grab a few frames
```

```
33      for i in range(0, num_frames) :
34          ret, frame = video.read()
35
36
37      # End time
38      end = time.time()
39
40      # Time elapsed
41      seconds = end - start
42      print ("Time taken : {0} seconds".format(seconds))
43
44      # Calculate frames per second
45      fps  = num_frames / seconds;
46      print ("Estimated frames per second : {0}".format(fps))
47
48      # Release video
49      video.release()
50  import cv2
51  import sys
52
53
54
55  if __name__ == '__main__' :
56
57      # Set up tracker.
58      (major_ver, minor_ver, subminor_ver) = (cv2.__version__).
        split('.')
59      # Instead of MIL, you can also use
60
61      tracker_types = ['BOOSTING', 'MIL','KCF', 'TLD', 'MEDIANFLOW'
        , 'GOTURN', 'MOSSE', 'CSRT']
62      tracker_type = tracker_types[2]
63
64      if int(minor_ver) < 3:
65          tracker = cv2.Tracker_create(tracker_type)
66      else:
67          if tracker_type == 'BOOSTING':
68              tracker = cv2.TrackerBoosting_create()
69          if tracker_type == 'MIL':
70              tracker = cv2.TrackerMIL_create()
71          if tracker_type == 'KCF':
72              tracker = cv2.TrackerKCF_create()
73          if tracker_type == 'TLD':
74              tracker = cv2.TrackerTLD_create()
75          if tracker_type == 'MEDIANFLOW':
76              tracker = cv2.TrackerMedianFlow_create()
77          if tracker_type == 'GOTURN':
78              tracker = cv2.TrackerGOTURN_create()
79          if tracker_type == 'MOSSE':
80              tracker = cv2.TrackerMOSSE_create()
81          if tracker_type == "CSRT":
82              tracker = cv2.TrackerCSRT_create()
83
84      # Read video
```

```
85    video = cv2.VideoCapture("20180106_153946.mp4")
86
87    # Exit if video not opened.
88    if not video.isOpened():
89        print ("Could not open video")
90        sys.exit()
91
92    # Read first frame.
93    ok, frame = video.read()
94    if not ok:
95        print ("Cannot read video file")
96        sys.exit()
97
98    # Define an initial bounding box
99    bbox = (287, 23, 86, 320)
100
101   # Uncomment the line below to select a different bounding box
102   bbox = cv2.selectROI(frame, False)
103
104   # Initialize tracker with first frame and bounding box
105   ok = tracker.init(frame, bbox)
106
107   while True:
108       # Read a new frame
109       ok, frame = video.read()
110       if not ok:
111           break
112
113       # Start timer
114       timer = cv2.getTickCount()
115
116       # Update tracker
117       ok, bbox = tracker.update(frame)
118
119       # Calculate Frames per second (FPS)
120       fps = cv2.getTickFrequency() / (cv2.getTickCount() -
      timer);
121
122       # Draw bounding box
123       if ok:
124           # Tracking success
125           p1 = (int(bbox[0]), int(bbox[1]))
126           p2 = (int(bbox[0] + bbox[2]), int(bbox[1] + bbox[3]))
127           cv2.rectangle(frame, p1, p2, (255,0,0), 2, 1)
128       else :
129           # Tracking failure
130           cv2.putText(frame, "Tracking failure detected",
      (100,80), cv2.FONT_HERSHEY_SIMPLEX, 0.75,(0,0,255),2)
131
132       # Display tracker type on frame
133       cv2.putText(frame, tracker_type + " Tracker", (100,20),
      cv2.FONT_HERSHEY_SIMPLEX, 0.75, (50,170,50),2);
134
135       # Display FPS on frame
```

```
136      cv2.putText(frame, "FPS : " + str(int(fps)), (100,50),
      cv2.FONT_HERSHEY_SIMPLEX, 0.75, (50,170,50), 2);
137
138      # Display result
139      cv2.imshow("Tracking", frame)
140
141      # Exit if ESC pressed
142      k = cv2.waitKey(1) & 0xff
143      if k == 27 : break
```

Listing 1.5 Object Tracking [34]

3.4 SMS Alert on Weapon Detection

3.4.1 SMS API

Twilio's Programmable SMS API [34] was used to send the SMS alert on weapon detection. Using this REST API, messages can be send and received. It can also keep track of the send messages. Moreover, it can recover and change the messages history.

3.4.2 SMS API Authentication

Hyper-text Transfer Protocol appeal to the API are ensured with HTTP Basic confirmation. Use the following script to do the basic authentication.

```
1 # Your account Sid and Auth Token from twilio.com/console
2 # and set the environment variables. See http://twil.io/secure
3 account_sid = os.environ['TWILIO_ACCOUNT _SID']
4 auth_token =os.environ['TIWILIO_AUTH_TOKEN' ]
```
Listing 1.6 Authentication [34]

3.4.3 Send an SMS with Twilio's API

For sending a new message from a Twilio's phone number to an outside number, execute the following script:

```
1 # Download the helper library from https://www.twilio.com/docs/
      python/install
2 import OS
3 from twilio.rest import client
4 # Your account Sid and Auth Token from twilio.com/console
5 # and set the environment variables. See http://twil.io/secure
6 account_sid=os.environ['TWILIO_ACCOUNT_SID']
7 auth_token =os.environ['TWILIO_AUTH_TOKEN']
8
```

```
 9 client = Client(account_sid, auth_token)
10
11 message = client.messages.create(
12 body='Weapon detacted!',
13 from ='+15017122661',
14 to='+15558675310'
15 )
16 print(message.sid)
```
Listing 1.7 Sending a new message [34]

4   Experimental Evaluations and Results

4.1   Evaluation Testbed

This system is based on the python model for the object, face feature and their response time. Therefore, the only professional way of testing the system is through checking the model usability and the response time for the particular model used in this system. Further system can capture the video via camera and makes its frame too slow to compare the object with the system's database and gives a comparative result with a percentage of accuracy about the object. System can capture multiple frames means that multiple object can be detected at the same time as shown in Fig. 6, while Fig. 7 is showing the Facial Recognition using the machine learning model.

Fig. 6 Object detection

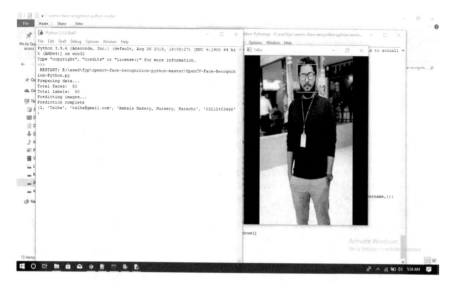

Fig. 7 Facial recognition

4.2 Results and Discussion

Data Model training was done on about 1700+ pictures for object detection. Results were tested and found an accuracy of about 97.3% with the performance rate a bit low due to camera resolution and dataset unavailability. For facial recognition, training was done for at most 25 images per person with an accuracy rate of about 90%. The notification test was done using the Twilio API and got the response time performance rate of about 5 s for a single message. The accuracy of existing work was found to be around 83 to 93% except for the two products i.e. the faceter which has an accuracy of 98.33% but limited to face only. Further, they use blockchain technology to gather and retrieve data. While the second one is Chinese surveillance which has claimed the accuracy of 98%. So far the proposed system has meanwhile improved and collectively highly functional by achieving the accuracy level up to 97.33%.

5 Conclusion and Future Work

An Intelligent Surveillance System proposed in this chapter. This system can be used to fulfill the concept of Smart Cities where surveillance is one of the fundamental building blocks. Yet there are things which can be added to make the system more appropriate like this system can be enhanced by adding sensors based detection and recognition, weapon integration with camera, enhancing accuracy up to maximum level, third-party control panel, and client's database configuration.

References

1. Penelope Hawkins, S.B., Kozul-Wright, R.: Current challenges to developing country debt sustainability (2019). https://unctad.org/en/PublicationsLibrary/gds2018d2_en.pdf
2. Cohen, B.: Urbanization in developing countries: current trends, future projections, and key challenges for sustainability. Technol. Soc. **28**(1), 63–80 (2006). http://www.sciencedirect.com/science/article/pii/S0160791X05000588
3. This is how megacities are being held back by violence (2015). https://www.weforum.org/agenda/2015/06/this-is-how-megacities-are-being-held-back-by-violence/
4. Crimes in mega cities (2013). http://ncrb.gov.in/StatPublications/CII/CII2013/Chapters/2-Crimes%20in%20Mega%20Cities.pdf
5. Kim, I.S., Choi, H.S., Yi, K.M., Choi, J.Y., Kong, S.G.: Intelligent visual surveillance–a survey. Int. J. Control Autom. Syst. **8**(5), 926–939 (2010)
6. Joshi, K.A., Thakore, D.G.: A survey on moving object detection and tracking in video surveillance system. Int. J. Soft Comput. Eng. **2**(3), 44–48 (2012)
7. Alsmirat, M.A., Jararweh, Y., Obaidat, I., Gupta, B.B.: Internet of surveillance: a cloud supported large-scale wireless surveillance system. J. Supercomput. **73**(3), 973–992 (2017)
8. Hsu, T.-J.: Home security system, US Patent 8,421,624, 16 April 2013
9. Luo, H.: Intelligent home security system, US Patent App. 11/307,417, 9 August 2007
10. Matthews, V.O., Noma-Osaghae, E., Uzairue, S.I.: An analytics enabled wireless anti-intruder monitoring and alarm system. Int. J. Sci. Res. Sci. Eng. Technol. **4**(9), 5–11 (2018)
11. Castro, J., Delgado, M., Medina, J., Ruiz-Lozano, M.: Intelligent surveillance system with integration of heterogeneous information for intrusion detection. Expert Syst. Appl. **38**(9), 11:182–11:192 (2011)
12. Verma, G.K., Dhillon, A.: A handheld gun detection using faster R-CNN deep learning. In: Proceedings of the 7th International Conference on Computer and Communication Technology, pp. 84–88. ACM (2017)
13. Lai, J., Maples, S.: Developing a real-time gun detection classifier (2017)
14. Tiwari, R.K., Verma, G.K.: A computer vision based framework for visual gun detection using Harris interest point detector. Procedia Comput. Sci. **54**, 703–712 (2015)
15. Siminoff, J., Cziment, A.: Leveraging audio/video recording and communication devices to reduce crime and enhance public safety, US Patent App. 15/675,726, 22 February 2018
16. Fookes, C., Denman, S., Lakemond, R., Ryan, D., Sridharan, S., Piccardi, M.: Semi-supervised intelligent surveillance system for secure environments. In: 2010 IEEE International Symposium on Industrial Electronics, pp. 2815–2820. IEEE (2010)
17. Nam, Y., Rho, S., Park, J.H.: Intelligent video surveillance system: 3-tier context-aware surveillance system with metadata. Multimed. Tools Appl. **57**(2), 315–334 (2012)
18. Chen, Y.-L., Chen, T.-S., Huang, T.-W., Yin, L.-C., Wang, S.-Y., Chiueh, T.-C.: Intelligent urban video surveillance system for automatic vehicle detection and tracking in clouds. In: 2013 IEEE 27th International Conference on Advanced Information Networking and Applications (AINA), pp. 814–821. IEEE (2013)
19. Wang, X.: Intelligent multi-camera video surveillance: a review. Pattern Recogn. Lett. **34**(1), 3–19 (2013)
20. Calavia, L., Baladrón, C., Aguiar, J.M., Carro, B., Sánchez-Esguevillas, A.: A semantic autonomous video surveillance system for dense camera networks in smart cities. Sensors **12**(8), 10:407–10:429 (2012)
21. Ozdemir, H.T., Lee, K.C.: Threat-detection in a distributed multi-camera surveillance system, US Patent 8,760,519, 24 June 2014

22. Nguyen, J.H.: Security system, US Patent App. 13/694,428, 5 June 2014
23. Kim, J.S., Yeom, D.H., Joo, Y.H.: Fast and robust algorithm of tracking multiple moving objects for intelligent video surveillance systems. IEEE Trans. Consum. Electron. **57**(3), 1165–1170 (2011)
24. Pederson, J.C.: Intelligent observation and identification database system, US Patent App. 15/231,132, 24 November 2016
25. Guler, P., Emeksiz, D., Temizel, A., Teke, M., Temizel, T.T.: Real-time multi-camera video analytics system on GPU. J. Real-Time Image Proc. **11**(3), 457–472 (2016)
26. Mansell, J.P., Riley, W.M.: Vehicle tracking and security system, US Patent 5,223,844, 29 June 1993
27. Lawrence, S., Giles, C.L., Tsoi, A.C., Back, A.D.: Face recognition: a convolutional neural-network approach. IEEE Trans. Neural Netw. **8**(1), 98–113 (1997)
28. Coffin, J.S., Ingram, D.: Facial recognition system for security access and identification, US Patent 5,991,429, 23 November 1999
29. China's surveillance cameras (2017). https://youtu.be/_yKga54tx6U
30. Nuuo hybrid surveillance system (2010). https://www.youtube.com/watch?v=KiZm2ZbZZj4
31. Faceter (2018). https://icodrops.com/faceter/
32. Brownlee, J.: Transfer learning in keras with computer vision models. Machine Learning Mastery. https://machinelearningmastery.com/how-to-use-transfer-learning-when-developing-convolutional-neural-network-models/. Accessed 17 Apr 2021
33. Mallick, S.: Object tracking using OpenCV (C++/Python), Learn OpenCV. https://learnopencv.com/object-tracking-using-opencv-cpp-python/. Accessed 17 Apr 2021
34. Twilio SMS API. https://www.twilio.com/docs/sms/api. Accessed 17 Apr 2021

Automated Methods for Detection and Classification Pneumonia Based on X-Ray Images Using Deep Learning

Khalid El Asnaoui, Youness Chawki, and Ali Idri

Abstract Recently, researchers, specialists, and companies around the world are rolling out deep learning and image processing-based systems that can fastly process hundreds of X-Ray and Computed Tomography (CT) images to accelerate the diagnosis of pneumonia such as SARS, covid-19, etc., and aid in its containment. Medical image analysis is one of the most promising research areas; it provides facilities for diagnosis and making decisions of several diseases such as MERS, covid-19, etc. In this paper, we present a comparison of recent deep convolutional neural network (CNN) architectures for automatic binary classification of pneumonia images based on fined tuned versions of (VGG16, VGG19, DenseNet201, Inception_ResNet_V2, Inception_V3, Resnet50, MobileNet_V2 and Xception) and a retraining of a baseline CNN. The proposed work has been tested using chest X-Ray & CT dataset, which contains 6087 images (4504 pneumonia and 1583 normal). As a result, we can conclude that the fine-tuned version of Resnet50 shows highly satisfactory performance with rate of increase in training and testing accuracy (more than 96% of accuracy).

Keywords Computer-aided diagnosis · Pneumonia automatic detection · CT and X-Ray images · Pneumonia · Coronavirus · Covid-19 · Deep learning

K. El Asnaoui (✉)
National School of Applied Sciences (ENSAO), Department of Electronics, Computer Sciences, and Telecommunications, Laboratory Smart Information, Communication and Technologies "SmarICT Lab",
Mohammed First University, BP: 669, 60000 Oujda, Morocco

Y. Chawki
Faculty of Sciences and Techniques, Moulay Ismail University, Errachidia, Morocco

A. Idri
Software Project Management Research Team, ENSIAS, Mohammed V University, Rabat, Morocco

© The Author(s), under exclusive license to Springer Nature Switzerland AG 2021   257
Y. Maleh et al. (eds.), *Artificial Intelligence and Blockchain for Future Cybersecurity Applications*, Studies in Big Data 90,
https://doi.org/10.1007/978-3-030-74575-2_14

1   Introduction

Epidemics and chronic diseases have killed numerous individuals throughout history and caused significant emergencies that have set aside a long effort to survive. Two words are utilised epidemic and outbreak to portray a malady inside populaces that emerge over a timeframe [1, 2]. Indeed, we can define epidemic as the occurrence of more cases of illnesses, injury or other health condition than expected in a given area or among a specific group of persons during a specific period. For the most part, the cases are pretending to have a common cause [2]. The outbreak is distinguished from an epidemic as more localized or less likely to evoke public panic.

Past epidemics include pneumonia. The pneumonia is an infection of the lungs, most often caused by a virus or bacteria. The infection affects the pulmonary alveoli, the tiny balloon-shaped sacs at the end of the bronchioles (Fig. 1). It usually affects only one of the lung's 5 lobes (3 lobes in the right lung and 2 in the left), hence the term lobar pneumonia. When pneumonia also reaches the bronchial tubes, it is called "Bronchopneumonia". It is the most important cause of death in the world for children younger than 5 years (about 12.8% of annual deaths) [3, 4]. It is also a leading cause of morbidity and mortality in adults worldwide and in particular in China [5–7]. Pneumonia is the third leading cause of death in Japan with a higher mortality rate for the elderly, particularly among individuals ≥80 years old [8]. Excluding lung cancer, in Portugal, Pneumonia is the huge cause of respiratory death [9].

Several Coronavirus have passed over the species barrier to cause deadly pneumonia in humans since the beginning of the 21st century. To know the pathogenesis of these deadly epidemics, the specialists need to inspect the structure of the infections and its component. This permits them to explain and provide information for the development of effective treatment and possibly vaccines [10]. Based on Table 1, that shows the major pandemics that have occurred over time. We will summarise the epidemiology and history of the type of Coronavirus in particular: SARS, MERS and Covid-19. SARS-Cov (Severe Acute Respiratory Syndrome Coronavirus) [11,

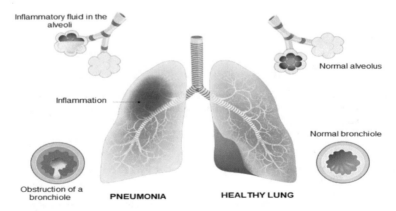

Fig. 1   Pneumonia diagram

Table 1 Major pandemics that have occurred over time

Name	Time period	Type/Pre-human host	Death toll
Spanish Flu	1918–1919	H1N1 virus/Pigs	40–50M
Asian Flu	1957–1958	H2N2 virus	1.1M
Hong Kong Flu	1968–1970	H3N2 virus	1M
HIV/AIDS	1981-Present	Virus/Chimpanzees	25–35M
Swine Flu	2009–2010	H1N1 virus/Pigs	200 000
SARS	2002–2003	Coronavirus/Bats, Civets	774
Ebola	2014–2016	Ebolavirus/Wild animals	11 000
MERS	2015-Present	Coronavirus/Bats, Camels	850
Covid-19	2019-Present	Coronavirus-Unknown (possibly Bats or pangolins)	Coronavirus Cases: 89.711.341 Deaths: 1,936,554 Recovered: 64,572,624 January 10, 2021, 11:36 GMT

12] is an acute respiratory illness caused by a coronavirus, characterized by fever, coughing, breathing difficulty, and usually pneumonia. SARS appeared first time in China exactly in the province of Guangdong in 2002 and spread to the world through air travel routes. Approximately 8098 people were affected, causing 774 deaths [13, 14] with a lethal rate of about 10% [15]. It is suggested to originate from bats [13, 16]. SARS symptoms are usually the same as flu symptoms: fever, chills, muscle aches, headache and occasionally diarrhea. After about one-week, other symptoms appear like fever of 38 °C or higher, dry cough, breath shortness [15].

MERS-Cov (Middle East Respiratory Syndrome Coronavirus) is a viral respiratory illness caused by a virus [17]. It appeared first in the Middle East and exactly in Saudi Arabia in 2012 [18, 19]. Other cases were identified in Jordan [20], Qatar [21] then spread to the world. MERS is a zoonotic virus that can be transmitted between animals and humans. Indeed, the World Health Organization has confirmed that humans are affected by contact with affected dromedary camels [22–24]. Studies have shown that the way the virus is transmitted from animals to humans is not yet understood, and the human-to-human transmission is very limited unless there is close contact [17, 25, 26]. The different MERS symptoms are as follows: Fever, Cough (Dry, Productive), Shortness of breath, Diarrhea, Myalgia, Headache, Nausea, Vomiting, Abdominal pain, Chest pain, Sore throat, Hemoptysis [17, 21, 27–30].

The world is currently experiencing a dangerous viral epidemic caused by a virus that has killed tens of thousands of people. This new virus called Covid-19 was identified in Wuhan [5, 31–41], China, in December 2019. It belongs to the Corona family of viruses, but it is more deadly and dangerous than the rest of the coronaviruses [42, 43]. First cases of the disease have been related to a live animal seafood market in Wuhan, denoting to a zoonotic origin of the epidemic [36, 41, 44–47]. The routes of transmission, treatments, and results of Covid-19 continually receiving much research attention in the world [31]. Indeed, researchers have identified three main modes of virus transmission: close person-to-person contact, aerosol transmission and transmission by touch [10, 42, 48, 49]. The Coronavirus is very dangerous because it can have up to two weeks of incubation without symptoms. We can cite the symptoms of Covid-19: high fever, dry cough, tiredness, shortness of breath, aches and pains, sore throat and very few people will report diarrhea, nausea or a runny nose [10, 43, 50]. As the number of patients infected by this disease increases, it becomes increasingly complex for radiologists to finish the diagnostic process in the constrained accessible time [51]. Medical images analysis is one of the most promising research areas; it provides facilities for diagnosis and making decisions of a number of diseases such as MERS, COVID-19. Recently, many efforts and more attention are paid to imaging modalities and Deep Learning (DL) in pneumonia. Therefore, interpretation of these images requires expertise and necessitates several algorithms in order to enhance, accelerate and make an accurate diagnosis. Following this context, DL algorithms [52] have obtained better performance in detecting pneumonia and demonstrated high accuracy compared with the previous state of the art methods. Motivated by the fastest and accurate detection rate of pneumonia using DL, our work will present a comparison of recent deep convolutional neural network architectures for automatic binary classification of X-Ray and CT images between normal and pneumonia in order to answer the following research questions: (**1**). Are there any DL techniques which distinctly outperforms other DL techniques? (**2**). Can DL used to early screen pneumonia from CT and X-Ray images? (**3**). What is the diagnostic accuracy that DL can be attained based on CT and X-Ray images?

Our paper's contributions are as follows: (**1**) We design fined tuned versions of (VGG16, VGG19, DenseNet201, Inception_ResNet_V2, Inception_V3, Xception, Resnet50, and MobileNet_V2) and retraining of a baseline CNN. (**2**) To avoid over-fitting in different models, we used weight decay and L2-regularizers. (**3**) The various models have been tested on chest X-Ray & CT datasets [53, 54] for binary classification and outperform state-of-the-art algorithms.

The remainder of this paper is organized as follows. Section 2 deals with some related work. In Sect. 3, we describe our proposed method. Section 4 presents some results obtained and interpreting the results. The conclusions are given in the last section.

2   Related Works

Up to this point, there is no compelling method to prevent the occurrence of lung abnormalities such as cancer and pneumonia. Therefore, early detection and accurate screening methods the most punctual indications of lung abnormalities are the initial steps to limit the risk of suffering. In this section, a brief review of some important contributions from the existing literature is presented. Pneumonia remains one of the diseases that is increasingly becoming research hotspots in recent years. Indeed, Toğaçar et al. [55] employed Convolutional Neural Network (CNN) as feature extractor based on lung X-Ray images and used some existing CNN models like AlexNet, VGG16 and VGG19 for classification between normal and pneumonia. Using the algorithm of minimum redundancy maximum relevance, the authors were able to reduce the number of deep features. A step of classification was then done using a decision tree, k-NN, linear discriminant analysis, linear regression, and SVM. Liang and Zeng [56] proposed a new deep learning framework to classify child pneumonia image by combining residual thought and dilated convolution. Thereby, to overcome the over-fitting and the model's degradation problems, the proposed method used a residual structure. The authors used also dilated convolution to resolve the issue of loss of feature space information breed by the increment in depth of the model. A deep learning method to identify and localize the pneumonia in Chest X-Rays images has been suggested by [57]. The identification model is based on Mask-RCNN that can incorporate global and local features for pixel-wise segmentation. The investigation of post-stroke pneumonia prediction models using advanced machine learning algorithms, specifically deep learning approaches has been presented in [58]. Indeed, the authors have used the classical classification methods (logistic regression, support vector machines, and extreme gradient boosting). They also implemented methods based on multiple layer perceptron neural networks and recurrent neural networks to use the temporal sequence information in electronic health record systems. The obtained results showed that the deep learning-based predictive model achieved the optimal performance compared to many classical machine learning methods. In [59], the authors proposed an automated detection and localization method of pneumonia on chest X-Ray images using machine learning solutions. They presented two CNN (RetinaNet and Mask R-CNN). The proposed method was validated on a dataset of 26,684 images from Kaggle Pneumonia Detection Challenge. Bhandary et al. [52] have reported a deep learning framework for examining lung pneumonia and cancer. Thus, they proposed two different deep learning techniques: the first one was a Modified AlexNet (MAN). It was intended to classify chest X-Ray images into normal, and pneumonia class using Support Vector Machine and its performance was validated with pre-trained deep learning (AlexNet, VGG16, VGG19 and ResNet50). Simultaneously, the second method implemented a fusion of handcrafted and learned features in the MAN to improve classification accuracy during lung cancer assessment. To assist radiologists for better diagnosis, [60] suggested a method for detection consolidations in chest X-Ray images using deep learning. Authors have used a deep convolutional neural network pre-trained with ImageNet data to improve the models' accuracy. Then, to enhance the models' generalisation, they proposed a three-step pre-processing approach: removing the confounding variables, histogram matching and improving the contrast of colorful image.

3 Proposed Contribution

Deep learning methods have recently demonstrated huge potential with state-of-the-art performance on image processing and computer vision [61]. These techniques have been applied in various medical imaging modalities with high performance [62] in segmentation, detection, and classification. Some DL methods incorporate skin cancer detection, breast cancer detection, and classification, lung cancer detection [62], etc. Even though these methods have shown huge achievement in medical imaging success, they require a large amount of data, which is yet not available in this field of applications. Following the context of no availability of medical imaging dataset and motivated by the success of deep learning and medical image processing, our work is going to deeply compare different fine-tuned [52] architectures: (VGG16, VGG19, DenseNet201, Inception_ResNet_V2, Inception_V3, Xception, Resnet50, and MobileNet_V2). The following sections detail the proposed models.

3.1 Proposed Baseline CNN Architecture

Generally, a CNN model consists of five layers: input layer, convolutional layers, pooling layers, full-connection layers, and output layer (Fig. 2). Moreover, it is known that a CNN model can be trained end-to-end to allow the feature extraction and selection, and finally classification or prediction. Understanding how the network interprets an image and processes it is difficult. However, it has been shown that features extracted by the layers of a network work better than human-built features [63].

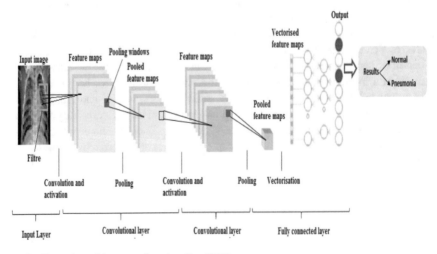

Fig. 2 The main architecture of our baseline CNN

The proposed baseline CNN for our experiment has the following architecture:

- Input layer: In our experiment, the inputs are X-Ray and CT images. The parameters are defining the image dimension (244 × 244).
- Convolutional layers: a convolution is a linear operation consisting of a set of weights with the input. It is designed for two-dimensional input; the multiplication is performed between a two-dimensional array of weights (filters) and an array of input data. In the proposed architecture we have 3 layers with a filter of size 3 × 3 and zero padding.
- Pooling Layers: represent a technique to down-sample feature maps by summarizing the presence of features in patches of the feature map. There are two types of pooling methods that are average pooling and max pooling. In the proposed architecture, we used max-pooling in order to calculate the maximum value in each patch for every feature map. The max-pooling is set to 2 × 2 with a stride of 2.
- Rectified Linear Unit (ReLU) layers: we have used 4 ReLU layers for each convolutional layer.
- Fully connected layers: They treat the input data as a simple vector and produce an output as a single vector.

3.2 Deep Learning Architectures

Deep learning architectures are highly used to diagnose pneumonia since 2016 [52, 53], the most investigated DL techniques are VGG16, VGG19, Inception_V3, DenseNet201, Xception, Resnet50, Inception_ResnetV_2, and MobileNet_V2. We have chosen these 8 techniques due to the accuracies they offer.

- **VGG16 and VGG19**

Proposed in 2014 by Simonyan and Zisserman, Visual Geometry Group (VGG) is a convolutional neural network architecture that won the ILSVR competition in 2014 [64]. The major characteristic of this architecture is that instead of having a large number of hyper-parameters, they concentrated on simple 3 × 3 size kernels in the convolutional layers and 2 × 2 size in the max-pooling layers. In the end, it has 2 Fully Connected (FC) layers trailed by a softmax for output. The most familiar VGG models are VGG16 and VGG19, which include 16 and 19 layers, respectively. The difference between VGG16 and VGG19 is that VGG19 has one more layer in each of the three convolutional blocks [65].

- **Inception_V3**

Inception models are a type of Convolutional Neural Networks developed by Szegedy in 2014 [66]. The inception models differ from the ordinary CNN in the structure where the inception models are inception blocks that mean lapping the same input tensor with multiple filters and concatenating their results. Inception_V3 is a new version of the inception model presented for the first time in 2015 [67]. It is an improved version of inception_V1 and inception_V2 with more parameters. Indeed, it has a block of parallel convolutional layers with 3 different sizes of filters (1×1, $3 \times 3, 5 \times 5$).

Additionally, 3×3 max pooling is also performed. The outputs are concatenated and sent to the next inception module. This model accepts an input image size of 299 $\times$ 299 pixels.

- **Resnet50**

Resnet50 is a deep residual network developed by [68] and is a subclass of convolutional neural networks used for image classification. It is the winner of ILSVRC 2015. The principal innovation is the introducing of the new architecture network-in-network using residual layers. The Resnet50 consists of five steps, each with a convolution and identity block, each convolution block and each identity block have 3 convolution layers. Resnet50 has 50 residual networks and accepts images size of 224×224 pixels.

- **Inception_ResNet_V2**

Inception_ResNet_V2 is a convolutional neural network trained on more than a million images from the ImageNet database [69]. It is a hybrid technique combining the inception structure and the residual connection. The model accepts images of 299 $\times$ 299 image, and its output is a list of estimated class probabilities. The advantages of Inception_Resnet_V2 are converting inception modules to Residual Inception blocks, adding more Inception modules and adding a new type of Inception module (Inception-A) after the Stem module.

- **DenseNet201**

Dense Convolutional Network (DenseNet201) is a convolutional neural network with 201 layers deep and accepts an input image size of 224×224 [70]. DenseNet201 is an improvement of ResNet that includes dense connections among layers. It connects each layer to every other layer in a feed-forward fashion. Unlike traditional convolutional networks with L layers that have L connections, DensNet201 has $L(L + 1)/2$ direct connections. Indeed, compared to traditional networks, DenseNet can improve the performance by increasing the computation requirement, reducing the number of parameters, encouraging feature reuse and reinforcing feature propagation.

- **MobileNet_V2**

MobileNet_V2 [71] is a convolutional neural network being an improved version of MobileNet_V1. It is made of only 54 layers and has an input image size of $224 \times$ 224. Its main characteristic is instead of performing a 2D convolution with a single kernel, instead of performing a 2D convolution with a single kernel. It uses depthwise separable convolutions that consist of applying two 1D convolutions with two kernels. That means, less memory and parameters are required for training leading to a small and efficient model. We can distinguish two types of blocks: the first one is a residual block with a stride of 1, the second one is block with a stride of 2 for downsizing. For each block, there are three layers: the first layer is 1×1 convolution with ReLU6, the second layer is the depthwise convolution, and the third layer is another 1×1 convolution but without any non-linearity.

- **Xception**

Xception, presented by Chollet [72], is a convolutional neural network that is 71 layers deep. It is an improved version of Inception architecture and involves depthwise separable convolutions. More precisely, Xception replaces the standard Inception modules with depthwise separable convolutions. It showed good results compared to VGG16, Resnet and Inception in classical classification problems. Xception has an input image size of 299×299.

4 Experimental Results and Analysis

We compared all mentioned above models once they have been fine-tuned for the automatic binary classification on two new publicly available image datasets (chest X-Ray & CT dataset [53, 54]). As it can be observed in Fig. 3, which shows the diagram of the main steps necessary to compare the different models: data acquisition, data pre-processing, training and classification. The following sections give out in detail the steps of this comparison.

4.1 Dataset

This present work introduces two publicly available image datasets which contain X-Ray and computed tomography images. The first dataset [53] is a chest X-Ray & CT dataset composed of 5856 images with two categories (4273 pneumonia and 1583

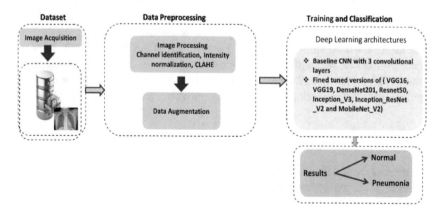

Fig. 3 Block diagram of the process of X-Ray and CT classification

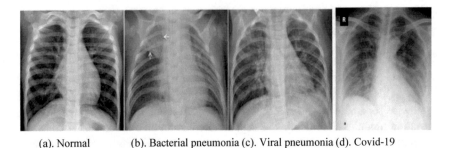

(a). Normal   (b). Bacterial pneumonia (c). Viral pneumonia (d). Covid-19

Fig. 4 Examples of Chest X-Rays in patients with pneumonia

normal) while the second one is named Covid Chest X-ray Dataset [54] containing 231 Covid-19 Chest X-Ray images. We joined the second dataset to the first one to form a joint dataset which finally composed of 6087 images (jpeg format) and has two classes (4504 pneumonia and 1583 normal). The pneumonia class contains images of bacterial pneumonia, viral pneumonia and Covid19. As can be seen from Fig. 4 that illustrates an example of chest X-Rays in patients with pneumonia, the normal chest X-Ray (Fig. 4(a)) shows clear lungs with no zones of abnormal opacification. Moreover, Fig. 4(b) shows a focal lobar consolidation (white arrows). Also, Fig. 4(c) shows a more diffuse "interstitial" pattern in both lungs [53] while Fig. 4(d) presents an image of a patient infected by covid19 [54].

4.2 Data Pre-processing and Splitting

The next stage is to pre-process input images using different pre-processing techniques. The motivation behind image pre-processing is to improve each input image's quality of visual information (to eliminate or decrease noise present in the original input image, enhance the quality of image through increased contrast, and delete the low or high freqes etc.). In this study, we used intensity normalization [73] and Contrast Limited Adaptive Histogram Equalization (CLAHE) [74, 75]. Intensity normalization is a straightforward pre-processing step in image processing applications [73]. In our contribution, we normalize the input image (Fig. 5(b)) to the standard normal distribution using min-max normalization (Eq. 1).

$$X_{norm} = \frac{x - x_{\min}}{x_{\max} - x_{\min}} \tag{1}$$

Furthermore, before feeding input image into the proposed models, CLAHE is necessary to improve the contrast in images [74, 75]. Figure 5 illustrates an example of using these techniques.

For data splitting, we used in this experiment 60% of the images for training and 40% of the images for testing. We ensure that the images chosen for testing are not

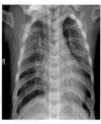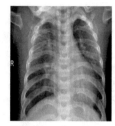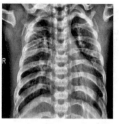

| (a). Original image | (b). Normalized image | (c). CLAHE |

Fig. 5 X-ray image pre-processing

used during training to perform the binary classification task successfully. Moreover, we observed that the dataset is imbalanced. Thereby 75% of the images represent the pneumonia class. To overcome this issue, we resampled the dataset by using data augmentation.

4.3 Data Augmentation

Data augmentation is used for the training process after dataset pre-processing and splitting and aims to avoid the risk of over-fitting. Moreover, the strategies we used include geometric transforms such as rescaling, rotations, shifts, shears, zooms and flips (Table 2). We generated from each single input image 2 new images with different augmentation techniques. Therefore, the total number of images in the normal class was increased by 2 times.

4.4 Training and Classification Dataset

After data pre-processing, splitting and data augmentation techniques, our training dataset size is increased and ready to be passed to the feature extraction step with the proposed models to extract the appropriate and pertinent features. The extracted features from each proposed model are flattened together to create the vectorized feature maps. The generated feature vector is passed to a multilayer perceptron to classify each image into corresponding classes. Finally, the performance of the proposed method is evaluated on test images using the trained model. We repeated each experiment three times and reported their average results.

Table 2 Data augmentation used

Argument	Parameter value	Description
Rescale	1/255.0	Scale images from integers 0–255 to floats 0–1
Rotation range	90	Degree range of the random rotations
Horizontal and Vertical shift range	0.2	The parameter value of horizontal and vertical shifts (20%) is a fraction of the given dimension
Shear range	0.2	Controls the angle in counterclockwise direction as radians in which our image will allow to be sheared
Zoom range	0.2	Allows the image to be "zoomed out" or "zoomed in"
Horizontal flip	True	Controls when a given input is allowed to be flipped horizontally during the training process
Fill mode	Nearest	This is the default option where the closest pixel value is chosen and repeated for all the empty values

4.5 Experimental Setup

Towards an automatic binary classification based on a publicly available image dataset (Chest X-Ray dataset [53, 54]), our experimentations were carried out based on following experimental parameters: All the images of the dataset were resized to 224×224 pixels except those of Inception_V3, Inception_Resnet_V2 and Xception models that were resized to 299×299. To train the models, we set the batch size to 32 with the number of epochs set to 300. The training and testing samples are initiated to 159 and 109, respectively. Adam with $\beta 1 = 0.9$, $\beta 2 = 0.999$ is used for optimization, and learning rate initiated to 0.00001 and decreased it to 0.000001. Moreover, we used weight decay and L2-regularizers to reduce over-fitting for the different models. A fully connected layer was trained with the ReLU, followed by a dropout layer with a probability of 0.5. We updated the last dense layer in all models to output two classes corresponding to normal and pneumonia instead of 1000 classes as was utilized for ImageNet. The implementation of the proposed models is done using a computer with Processor: Intel (R) core (TM) i7-7700 CPU @ 3.60 GHz and 8 GB in RAM running on a Microsoft Windows 10 Professional (64-bit). For implementation, Keras/Tensorflow is used as deep learning backend. Our training and testing steps run using NVIDIA Tesla P40 with 24 GB RAM.

4.6 Evaluation Criteria

After extracting the appropriate feature, the last step is to classify the attained data and assign it to a specific class [76]. Among the different classification performance properties, and since the dataset is now balanced, our study uses the following benchmark metrics: accuracy (ACC), sensitivity (SEN), specificity (SPE), precision (PRE) and F1 score (F1) [52, 76]. These metrics are defined as follows:

$$ACC = \frac{TP + TN}{TP + TN + FP + FN} \times 100 \qquad PRE = \frac{TP}{TP + FP} \times 100$$

$$SPE = \frac{TN}{TN + FP} \times 100 \qquad SEN = \frac{TP}{TP + FN} \times 100$$

$$F1 = 2 \times \frac{Recall \times Precision}{Recall + Precision} \times 100 \tag{2}$$

where: TP stands for: True Positive. FP: False Positive. TN: True Negative, and FN: False Negative.

4.7 Results and Discussion

This section presents the results for the binary classification for the chest X-Ray & CT images [53, 54] with the following architectures (Baseline CNN, Fine-tuning the top layers of VGG16, VGG19, Inception_V3, Xception, Resnet50, Inception_Resnet_V2, DenseNet201, and MobileNet_V2). Also, to check each proposed model's performance and robustness, several experiments are conducted on chest X-Ray dataset [53, 54]. The results are presented separately using training and testing curves of accuracy and loss and confusion matrix.

4.7.1 Classification Results of the Different Architectures

This subsection presents and discusses the classification results of chest X-Ray & CT images [53, 54]. Before discussing these results, let us define some parameters related to the deep learning process: the training curve is calculated from the training dataset that provides an idea of how well the model is learning. In contrast, the testing curve is calculated from a hold-out testing dataset that explains how well the model is generalizing. Simultaneously, the training and testing loss are defined as a summation of the errors made for each example in testing or training sets. Note that in contrast to accuracy, loss is not a percentage. Furthermore, the confusion matrix shows a detailed representation of images after classification [52]. To summarize, a model that generalizes well is a model that is neither over-fit nor under-fit.

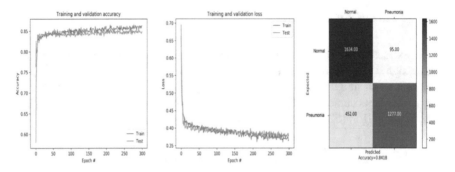

Fig. 6 Accuracy and loss curve and confusion matrix of Baseline CNN

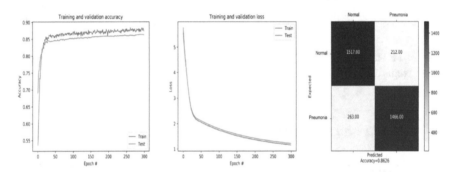

Fig. 7 Accuracy and loss curve and confusion matrix of VGG16

• **Baseline CNN**

According to the Fig. 6, it is observed that the accuracy curve of training data is rapidly increasing from epoch 0 to epoch 6 where the accuracy is equal to 83.17%, after that, it begins to increase slightly until epoch 300 where the accuracy is equal to 86.28%. The same applies to the accuracy curve of testing data with an accuracy of 84.84 for epoch 300. Concerning the loss curve of training data, it is rapidly decreasing from epoch 0 to epoch 6 where the loss is 43.77. At that point, it begins to decrease slightly until the end of training (epoch 300) where the loss is equivalent to 36.56. Same for loss curve of testing data with a loss of 37.85 for epoch 300. From the confusion matrix, it is noted that the first images (Normal class), the model recognizes 1634 images correctly, but 95 were marked as pneumonia. Likewise, for the second image's class (Pneumonia), the model was capable to identify 1277 images correctly, unlike 252 images were marked as Normal.

• **VGG16**

Figure 7 presents the accuracy, loss curve and confusion matrix of VGG16. Indeed, from the epoch 0 to epoch 11, the accuracy curve of training data is quickly increasing where it is equal to 81.05%, and then it converges to a value of 87.51%. The same applies to the accuracy curve of testing data with an accuracy of 86.32% for epoch 300. A rapid decreasing of loss curve can be noted for training data from epoch 0 to epoch 25, where the loss is equivalent to 1.43. At this epoch, a kind of stability can be observed up to the value of 1.15. The same goes for the loss curve of testing data where the loss is equal to 2.21 for epoch 300. The model can predict 1517 images correctly in the normal class from the confusion matrix, yet 212 were named pneumonia. For the Pneumonia class, the model was capable to identify 1466 images correctly, and 263 images were marked as Normal.

- **VGG19**

As it is shown in Fig. 8, the curve of training data (testing data) can be divided into two intervals: the first one starts from epoch 0 to epoch 13 (from epoch 0 to epoch 10). We can observe a quick increase in accuracy where the accuracy is equal to 81.01% (83.05%). In the second interval, the accuracy becomes stable and converges toward 87.42% (86.89%). For the loss curve of training and testing data, we see a good fit. Indeed, from epoch 0 to 18, the loss is rapidly decreasing, where it is equal to 1.27. Afterwards, it begins to increase slightly until the end of the training, equivalent to 1.31. As observed (see confusion matrix) in the normal class, the VGG19 model had the option to predict 1390 images correctly and 339 images as pneumonia. The model also was capable to classify 1582 images as pneumonia and 147 images as Normal for the Pneumonia class.

- **Inception_V3**

Figure 9 shows the accuracy, loss curve and confusion matrix of Inception_V3. Thereby, for the training and testing accuracy and from epoch 0 to epoch 7, we can see that the accuracy is increasing until the value of 91.03%. After epoch 7, the accuracy gets started to be stable where it is equal to 97.01% and 95.94% for training and testing data respectively. A good fit can be noticed for the loss curve of training data in either the quick increasing interval from epoch 0 to epoch 32 where the loss is 3.98 or in the other interval where the decreasing is slow and converges to 1.76. As shown in the confusion matrix, for the Pneumonia class, the model was

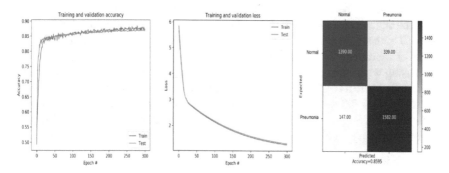

Fig. 8 Accuracy and loss curve and confusion matrix of VGG19

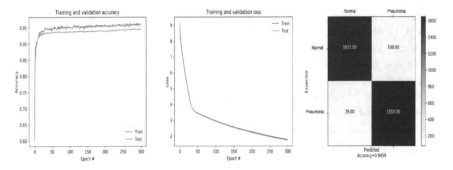

Fig. 9 Accuracy and loss curve and confusion matrix of Fine-tuned Inception_V3

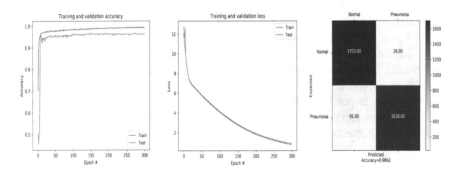

Fig. 10 Accuracy and loss curve and confusion matrix of Fine-tuned ResNet50

able to identify 1650 images as pneumonia and 79 images as Normal. Concerning class Normal, Inception_V3 model can predict 1621 images as Normal and 108 as pneumonia.

- **ResNet50**

Figure 10 illustrates the results obtained by Resnet50. In fact, from epoch 0 to 24, the accuracy values are increasing expeditiously either for training or testing data where the maximum value is 97.36%. After that, the values start to be stables (99.23% and 96.23% for training and testing data respectively). We observe a good fit for the loss curve of training and testing data, indeed, from epoch 0 to 21, the loss is briskly decreasing where it is equal to 6.89, afterwards, it gets started to be stable until the epoch 300 where it is equal to 0.85. The confusion matrix indicates that, for images of Normal class, 1703 images were predicted correctly as Normal and 26 were marked as pneumonia. Same for the second images of Pneumonia class, the model had the option to identify 1638 images correctly, unlike 91 images were labeled as Normal.

- **Inception_ResNet_V2**

Figure 11 shows that from epoch 0 to 18, the training and testing accuracy curve are increasing until the value of 95.51%. After epoch 18, the accuracy begins to be

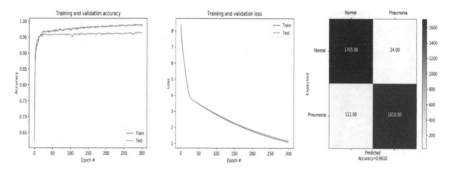

Fig. 11 Accuracy and loss curve and confusion matrix of Fine-tuned Inception_Res-Net_V2

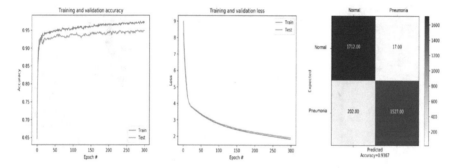

Fig. 12 Accuracy and loss curve and confusion matrix of Fine-tuned DensNet201

stable, equivalent to 99.11% and 96.41% for training and testing data. We can see an excellent fit for the loss curve for training and testing data where the values are 3.99 (epoch 24) and 1.17 (epoch 300). For the Pneumonia class, as depicted by the confusion matrix, the Inception_ResNet_V2 model was able to identify 1618 images correctly as pneumonia and 111 images as Normal. On other hand, for Normal class, 1705 were correctly classified as Normal and 24 images as pneumonia.

- **DensNet201**

The obtained accuracy curve of training data is speedily increasing until the value of 93.49% (Fig. 12). After epoch 16, the accuracy enters the stability stage where it is equivalent to 97.16% and 94.91% for training and testing data respectively. A good fit can be seen for the loss curve of training and testing data. In fact, from epoch 0 to epoch 17, the loss is quickly decreasing where it is equal to 3.99, and then it becomes stable until the epoch 300 where it is equal to 1.91. The confusion matrix depicts that for the first images (Normal class) the model can recognize 1712 images correctly in the normal class, yet 17 were named as pneumonia. The model also was able to identify 1527 images correctly, and 202 images were marked as Normal for the second images (Pneumonia class).

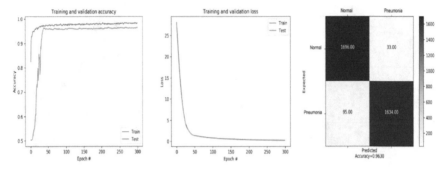

Fig. 13 Accuracy and loss curve and confusion matrix of Fine-tuned MobileNet_V2

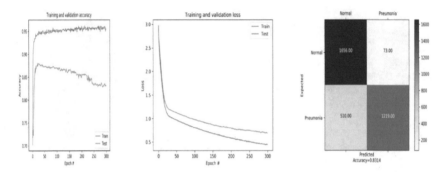

Fig. 14 Accuracy and loss curve and confusion matrix of Fine-tuned Xception

- **MobileNet_V2**

As illustrated by Fig. 13, we can observe that from epoch 0 to 16, the training and testing accuracy are increasing until the value where the accuracy is equal to 96.38%. After epoch 16, the accuracy becomes stable and it is equal to 98.27% and 96.64% for training and testing data respectively. For the loss curve of training data, an excellent fit is noticed. Until the epoch 44, the value of loss is expeditiously decreasing where the value is 1.49. Then it converges towards 0.24. Regarding the confusion matrix, for the first images (Normal class), the model was able to identify 1696 images correctly in the normal class, but 33 were classified as pneumonia. Likewise, in the Pneumonia class, 1634 images were labeled correctly as pneumonia and 95 were identified as Normal.

- **Xception**

It is noted that the accuracy of training data is fastly increasing from epoch 0 to 10 where the accuracy is equal to 93.10% (see Fig. 14). Then it gets stable until the end of training where the accuracy is equal to 95.45%. For the testing data, a quick increasing can be seen from epoch 0 to 12 where the value is 86.87%, after that, it begins to decrease till 69.03% for epoch 300. For the loss curve of training and testing data, the values are rapidly decreasing from epoch 0 to epoch 26, where the value is 1.10. After epoch 26, the value converges to 0.44 and 0.69 for training and testing data respectively. When we see this confusion matrix, we can say that for the first images (Normal class), the model has the option to recognize 1656 images correctly. Moreover, 73 were selected as pneumonia. The model also can recognize 1219 images correctly (Pneumonia class). Thus 510 images were marked as Normal for the second images (Pneumonia class).

4.7.2 Discussion

In this study, we investigated the binary classification (Normal and pneumonia) based on X-Ray images using transfer learning of recent deep learning architectures to identify the best performing architecture based on the several parameters defined in Eq. (2). First, we individually compare the deep learning architectures by measuring their accuracies. After that, we compare each deep learning architecture's accuracy and loss results to discern the outperforming architecture (Figs. 15, 16). Moreover, Table 3 illustrates a comparison between the various deep learning models used in our experiments in terms of parameters defined in Eq. (2).

For each model in Fig. 15(a) and (b), which summarize the previous training and testing accuracy figures, the plots of training and testing accuracy increase to the point of stability. It is observed that fine-tuned version of Inception_Resnet_V2, Inception_V3, Resnet50, Densnet201 and Mobilenet_V2 show highly satisfactory performance with a rate of increase in training and testing accuracy with each epoch. They outperform the baseline CNN, Xception, VGG16 and VGG19 that demonstrate low performance. From epoch 20, they start to be stable until the end of training where the training and testing accuracy of baseline CNN, VGG16 and VGG16 are equal to 85%. However, Xception reaches 83% in testing accuracy and 95% in training accuracy. In this case, the predictive model produced by Xception algorithm does not adapt well to the training set (Over-fitting). Besides, the plots of training and testing loss (Fig. 16(a) and (b)) decrease to the point of stability for each proposed model. As can be seen, the fine-tuned version of the models shows highly satisfactory performance with the rate of decrease in training and testing loss with each epoch.

Results for our multi-experiment classification are tabulated in Table 3 based on different fine-tuned versions of recent deep learning architectures. The table depicts in detail classification performances across each experiment. From the results, it is noted that the accuracy when we use Xception, baseline CNN, VGG19 and VGG16 are low compared with other DL architectures, since these last models help to obtain respectively 83.14%, 84.18%, 85.94% and 86.26% of accuracy. Unlike, the highest accuracies are reported by DensNet201 (93.66%), Inception_V3 (94.59%), Inception_Resnet_V2 (96.09%), MobileNet_V2 (96.27%) and Resnet50 (96.61%). In addition, MobileNet_V2 has been proven to obtain remarkable results in related tasks [77] whereas ResNet50 [68, 78] provides a good combination of performance and number of parameters and has proved faster training. Therefore, we recommend the MobileNet_V2 (96.27% of accuracy) and Resnet50 (96.61% of accuracy) models

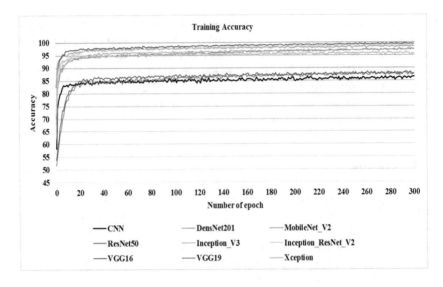

(a)

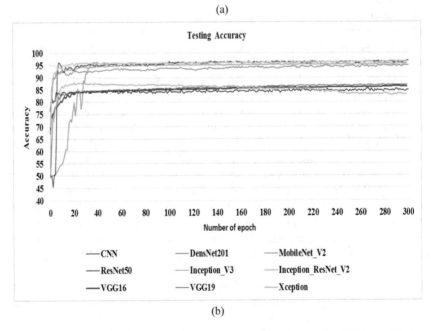

(b)

Fig. 15 Summarization of the previous figures in term of accuracy curve for different architectures

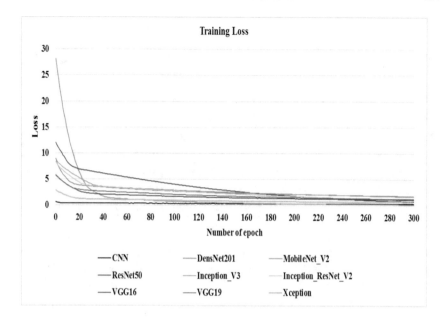

(a)

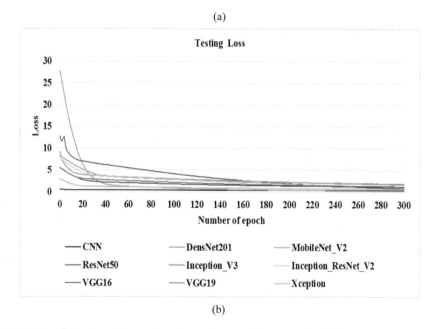

(b)

Fig. 16 Summarization of the previous figures in term of Loss curve for different architectures

Table 3 Evaluations metrics in (%)

	TP	TN	FN	FP	ACC	SEN	SPE	PRE	F1
Baseline CNN	1634	1277	452	95	84.18	78.33	93.07	94.05	85.66
VGG16	1517	1466	263	212	86.26	85.22	87.36	87.73	86.46
VGG19	1390	1582	147	339	85.94	90.43	82.35	80.39	85.11
Xception	1656	1219	510	73	83.14	76.45	94.34	95.77	85.03
DensNet201	1712	1527	202	17	93.66	89.44	98.89	99.01	93.98
Inception_V3	1621	1650	79	108	94.59	95.35	93.85	93.75	94.54
Inception_Resnet_V2	1705	1618	111	24	96.09	93.88	98.53	98.61	96.19
MobileNet_V2	1696	1634	95	33	96.27	94.61	98.02	98.06	96.30
Resnet50	1703	1638	91	26	96.61	94.92	98.43	98.49	96.67

to be used for the Computer-Aided Diagnosis systems to identify the health status of patients against pneumonia in X-ray and CT images, since the best scores of training and testing accuracy were obtained. Clinical examinations are the following step of this research work.

5 Conclusions and Future Works

In this work, we presented automated methods used to classify the chest X-Ray & CT images into pneumonia and the normal class using eight deep learning architectures (VGG16, VGG19, DenseNet201, Inception_ResNet_V2, Inception_V3, Xception, Resnet50, and MobileNet_V2) and a baseline CNN. The main goal is to answer the following research questions: (1). Are there any DL techniques which distinctly outperforms other DL techniques? (2). Can DL use to early screen pneumonia from CT and X-Ray images? (3). What is the diagnostic accuracy that DL can be attained based on CT and X-Ray images?. The experiments were conducted using chest X-Ray & CT dataset, which contains 6087 images (4504 pneumonia and 1583 normal). The pneumonia class contains images of bacterial pneumonia, viral pneumonia and Covid19. Moreover, the performances of these experiments were evaluated using various performance metrics. Furthermore, the obtained results show that the Resnet50 gave high performance (accuracy is more than 96%) against other architectures cited in this work (accuracy is lower than 96%). Due to these models' high performance, we believe that these results help doctors make decisions in clinical practice.

Ongoing work intends to develop a full system for pneumonia via deep learning detection, segmentation, and classification. In addition, the performance may be improved using more datasets, more sophisticated feature extraction techniques such as color [79], texture [80], shape [81, 82]. In addition, the performance may be improved using more datasets, more sophisticated feature extraction techniques. Also other fusion approaches would be interesting [83].

Authors' Contributions: The experiments and the programming stage were carried out by Khalid El Asnaoui. All authors wrote the paper, and all approve this submission.

References

1. Orbann, C., Sattenspiel, L., Miller, E., Dimka, J.: Defining epidemics in computer simulation models: how do definitions influence conclusions? Epidemics **19**, 24–32 (2017)
2. Centers for Disease Control and Prevention: Principles of Epidemiology in Public Health Practice (DHHS Publication SS1978), 3rd edn. U.S. Department of Health and Human Services, Atlanta (2012)
3. O'Brien, K.L., Baggett, H.C., Abdullah Brooks, W., Feikin, D.R., Hammitt, L.L., Higdon, M.M., Howie, S.R.C., Knoll, M.D., Kotloff, K.L., Levine, O.S., Madhi, S.A., Murdoch, D.R., Christine Prosperi, J., Scott, A.G., Shi, Q., Thea, D.M., Zhenke, W., Zeger, S.L., Adrian, P.V., Akarasewi, P., Anderson, T.P., Antonio, M., Awori, J.O., Baillie, V.L., Bunthi, C., Chipeta, J., Chisti, M.J., Crawley, J., DeLuca, A.N., Driscoll, A.J., Ebruke, B.E., Endtz, H.P., Fancourt, N., Wei, Fu., Goswami, D., Groome, M.J., Haddix, M., Hossain, L., Yasmin Jahan, E., Kagucia, W., Kamau, A., Karron, R.A., Kazungu, S., Kourouma, N., Kuwanda, L., Kwenda, G., Li, M., Machuka, E.M., Mackenzie, G., Mahomed, N., Maloney, S.A., McLellan, J.L., Mitchell, J.L., Moore, D.P., Morpeth, S.C., Mudau, A., Mwananyanda, L., Mwansa, J., Ominde, M.S., Onwuchekwa, U., Park, D.E., Rhodes, J., Sawatwong, P., Seidenberg, P., Shamsul, A., Simões, E.A.F., Sissoko, S., Somwe, S.W., Sow, S.O., Sylla, M., Tamboura, B., Tapia, M.D., Thamthitiwat, S., Toure, A., Watson, N.L., Zaman, K., Zaman, S.M.A.: Causes of severe pneumonia requiring hospital admission in children without HIV infection from Africa and Asia: the PERCH multi-country case-control study. The Lancet **394**(10200), 757–779 (2019)
4. Liu, L., Oza, S., Hogan, D., Chu, Y., Perin, J., Zhu, J., Lawn, J.E., Cousens, S., Mathers, C., Black, R.E.: Global, regional, and national causes of under-5 mortality in 2000–15: an updated systematic analysis with implications for the Sustainable Development Goals. The Lancet **388**(10063), 3027–3035 (2016)
5. Tian, Y., Yiqun, W., Liu, H., Si, Y., Yao, W., Wang, X., Wang, M., Junhui, W., Chen, L., Wei, C., Tao, W., Gao, P., Hu, Y.: The impact of ambient ozone pollution on pneumonia: a nationwide time-series analysis. Environ. Int. **136**, 105498 (2020)
6. Prina, E., Ranzani, O.T., Torres, A.: Community-acquired pneumonia. The Lancet **386**(9998), 1097–1108 (2015)
7. Welte, T., Torres, A., Nathwani, D.: Clinical and economic burden of community-acquired pneumonia among adults in Europe. Thorax **67**(1), 71–79 (2012)
8. Kondo, K., Suzuki, K., Washio, M., Ohfuji, S., Fukushima, W., Maeda, A., Hirota, Y.: Effectiveness of 23-valent pneumococcal polysaccharide vaccine and seasonal influenza vaccine for pneumonia among the elderly–selection of controls in a case-control study. Vaccine **35**(36), 4806–4810 (2017)
9. Hespanhol, V., Bárbara, C.: Pneumonia mortality, comorbidities matter? Pulmonology **26**(3), 123–129 (2020)
10. Yang, J., Zheng, Y., Gou, X., Pu, K., Chen, Z., Guo, Q., Ji, R., Wang, H., Wang, Y., Zhou, Y.: Prevalence of comorbidities in the novel Wuhan coronavirus (COVID-19) infection: a systematic review and meta-analysis. Int. J. Infect. Dis. 10 (2020)

11. Drosten, C., Günther, S., Preiser, W., van der Werf, S., Brodt, H.-R., Becker, S., Rabenau, H., Panning, M., Kolesnikova, L., Fouchier, R.A.M., Berger, A., Burguière, A.-M., Cinatl, J., Eickmann, M., Escriou, N., Grywna, K., Kramme, S., Manuguerra, J.-C., Müller, S., Rickerts, V., Stürmer, M., Vieth, S., Klenk, H.-D., Osterhaus, A.D.M.E., Schmitz, H., Doerr, H.W.: Identification of a novel coronavirus in patients with severe acute respiratory syndrome. N. Engl. J. Med. **348**(20), 1967–1976 (2003)

12. Ksiazek, T.G., Erdman, D., Goldsmith, C.S., Zaki, S.R., Peret, T., Emery, S., Tong, S., Urbani, C., Comer, J.A., Lim, W., Rollin, P.E., Dowell, S.F., Ling, A.-E., Humphrey, C.D., Shieh, W.-J., Guarner, J., Paddock, C.D., Rota, P., Fields, B., DeRisi, J., Yang, J.-Y., Cox, N., Hughes, J.M., LeDuc, J.W., Bellini, W.J., Anderson, L.J.: A novel coronavirus associated with severe acute respiratory syndrome. N. Engl. J. Med. **348**(20), 1953–1966 (2003)

13. Walls, A.C., Park, Y.J., Tortorici, M.A., Wall, A., McGuire, A.T., Veesler, D.: Structure, function, and antigenicity of the SARS-CoV-2 spike glycoprotein. Cell **181**(2), 281–292 (2020)

14. Roussel, Y., Giraud-Gatineau, A., Jimeno, M.T., Rolain, J.M., Zandotti, C., Colson, P., Raoult, D.: SARS-CoV-2: fear versus data. Int. J. Antimicrob. Agents **55**(5), 105947 (2020)

15. Wu, C., Liu, Y., Yang, Y., Zhang, P., Zhong, W., Wang, Y., Wang, Q., Xu, Y., Li, M., Li, X., Zheng, M., Chen, L., Li, H.: Analysis of therapeutic targets for SARS-CoV-2 and discovery of potential drugs by computational methods. Acta Pharm. Sin. B **10**(5), 766–788 (2020)

16. Hu, B., Zeng, L.-P., Yang, X.-L., Ge, X.-Y., Zhang, W., Li, B., Xie, J.-Z., Shen, X.-R., Zhang, Y.-Z., Wang, N., Luo, D.-S., Zheng, X.-S., Wang, M.-N., Daszak, P., Wang, L.-F., Cui, J., Shi, Z.-L.: Discovery of a rich gene pool of bat SARS-related coronaviruses provides new insights into the origin of SARS coronavirus. PLOS Pathog. **13**(11), e1006698 (2017)

17. World Health Organization: Middle East respiratory syndrome coronavirus (MERS-CoV). https://www.who.int/emergencies/mers-cov/en/. Accessed Mar 2020

18. van Boheemen, S., de Graaf, M., Lauber, C., Bestebroer, T.M., Stalin Raj, V., Zaki, A.M., Osterhaus, A.D.M.E., Haagmans, B.L., Gorbalenya, A.E., Snijder, E.J., Fouchier, R.A.M.: Genomic characterization of a newly discovered coronavirus associated with acute respiratory distress syndrome in humans. mBio **3**(6), e00473 (2012)

19. Zaki, A.M., Van Boheemen, S., Bestebroer, T.M., Osterhaus, A.D., Fouchier, R.A.: Isolation of a novel coronavirus from a man with pneumonia in Saudi Arabia. N. Engl. J. Med. **367**(19), 1814–1820 (2012)

20. Hijawi, B., Abdallat, M., Sayaydeh, A., Alqasrawi, S., Haddadin, A., Jaarour, N., El Sheikh, S., Alsanouri, T.: Novel coronavirus infections in Jordan, April 2012: epidemiological findings from a retrospective investigation. Eastern Mediterr. Health J. **19**(Supp. 1), S12–S18 (2013)

21. Farooq, H.Z., Davies, E., Ahmad, S., Machin, N., Hesketh, L., Guiver, M., Turner, A.J.: Middle East respiratory syndrome coronavirus (MERS-CoV)—surveillance and testing in North England from 2012 to 2019. Int. J. Infect. Dis. **93**, 237–244 (2020)

22. World Health Organization: Middle East respiratory syndrome coronavirus (MERS-CoV). https://www.who.int/news-room/fact-sheets/detail/middle-east-respiratory-syndrome-coronavirus-(mers-cov). Accessed Mar 2019

23. Mohd, H.A., Al-Tawfiq, J.A., Memish, Z.A.: Middle East respiratory syndrome coronavirus (MERS-CoV) origin and animal reservoir. Virol. J. **13**(1), 87 (2016)

24. Azhar, E.I., El-Kafrawy, S.A., Farraj, S.A., Hassan, A.M., Al-Saeed, M.S., Hashem, A.M., Madani, T.A.: Evidence for camel-to-human transmission of MERS coronavirus. N. Engl. J. Med. **370**(26), 2499–2505 (2014)

25. Ki, M.: 2015 MERS outbreak in Korea: hospital-to-hospital transmission. Epidemiol. Health **37**, e2015033 (2015)

26. Oboho, I.K., Tomczyk, S.M., Al-Asmari, A.M., Banjar, A.A., Al-Mugti, H., Aloraini, M.S., Alkhaldi, K.Z., Almohammadi, E.L., Alraddadi, B.M., Gerber, S.I., Swerdlow, D.L., Watson, J.T., Madani, T.A.: 2014 MERS-CoV outbreak in Jeddah—a link to health care facilities. N. Engl. J. Med. **372**(9), 846–854 (2015)

27. Wang, N., Rosen, O., Wang, L., Turner, H.L., Stevens, L.J., Corbett, K.S., Bowman, C.A., Pallesen, J., Shi, W., Zhang, Y., Leung, K., Kirchdoerfer, R.N., Becker, M.M., Denison, M.R., Chappell, J.D., Ward, A.B., Graham, B.S., McLellan, J.S.: Structural definition of a neutralization-sensitive epitope on the MERS-CoV S1-NTD. Cell Reports **28**(13), 3395–3405 (2019)

28. Baharoon, S., Memish, Z.A.: MERS-CoV as an emerging respiratory illness: a review of prevention methods. Travel Med. Infect. Dis. **32**, 101520 (2019)

29. Al-Omari, A., Rabaan, A.A., Salih, S., Al-Tawfiq, J.A., Memish, Z.A.: MERS coronavirus outbreak: implications for emerging viral infections. Diagn. Microbiol. Infect. Dis. **93**(3), 265–285 (2019)

30. Aguanno, R., ElIdrissi, A., Elkholy, A.A., Embarek, P.B., Gardner, E., Grant, R., Mahrous, H., Malik, M.R., Pavade, G., VonDobschuetz, S., Wiersma, L., Van Kerkhove, M.D.: MERS: progress on the global response, remaining challenges and the way forward. Antiviral Res. **159**, 35–44 (2018)

31. Guo, H., Zhou, Y., Liu, X., Tan, J.: The impact of the COVID-19 epidemic on the utilization of emergency dental services. J. Dent. Sci. **15**(4), 564–567 (2020)

32. Lippi, G., Plebani, M., Henry, B.M.: Thrombocytopenia is associated with severe coronavirus disease 2019 (COVID-19) infections: a meta-analysis. Clin. Chim. Acta **506**, 145–148 (2020)

33. Amrane, S., Tissot-Dupont, H., Doudier, B., Eldin, C., Hocquart, M., Mailhe, M., Dudouet, P., Ormières, E., Ailhaud, L., Parola, P., Lagier, J.-C., Brouqui, P., Zandotti, C., Ninove, L., Luciani, L., Boschi, C., La Scola, B., Raoult, D., Million, M., Colson, P., Gautret, P.: Rapid viral diagnosis and ambulatory management of suspected COVID-19 cases presenting at the infectious diseases referral hospital in Marseille, France, - January 31st to March 1st, 2020: a respiratory virus snapshot. Travel Med. Infect. Dis. **36**, 101632 (2020)

34. Cortegiani, A., Ingoglia, G., Ippolito, M., Giarratano, A., Einav, S.: A systematic review on the efficacy and safety of chloroquine for the treatment of COVID-19. J. Crit. Care **57**, 279–283 (2020)

35. Tang, B., Li, S., Xiong, Y., Tian, M., Yu, J., Xu, L., Zhang, L., Li, Z., Ma, J., Wen, F., Feng, Z., Liang, X., Shi, W., Liu, S.: COVID-19 pneumonia in a hemodialysis patient. Kidney Med. **2**(3), 354–358 (2020)

36. Wilder-Smith, A., Chiew, C.J., Lee, V.J.: Can we contain the COVID-19 outbreak with the same measures as for SARS? Lancet Infect. Dis. **20**(5), e102–e107 (2020)

37. Driggin, E., Madhavan, M.V., Bikdeli, B., Chuich, T., Laracy, J., Biondi-Zoccai, G., Brown, T.S., Der Nigoghossian, C., Zidar, D.A., Haythe, J., Brodie, D., Beckman, J.A., Kirtane, A.J., Stone, G.W., Krumholz, H.M., Parikh, S.A.: Cardiovascular considerations for patients, health care workers, and health systems during the COVID-19 pandemic. J. Am. Coll. Cardiol. **75**(18), 2352–2371 (2020)

38. Cheng, Y., Luo, R., Wang, K., Zhang, M., Wang, Z., Dong, L., Li, J., Yao, Y., Ge, S., Xu, G.: Kidney disease is associated with in-hospital death of patients with COVID-19. Kidney Int. **97**(5), 829–838 (2020)

39. Pung, R., Chiew, C.J., Young, B.E., Chin, S., Chen, M.-C., Clapham, H.E., Cook, A.R., Maurer-Stroh, S., Toh, M.P.H.S., Poh, C., Low, M., Lum, J., Koh, V.T.J., Mak, T.M., Cui, L., Lin, R.V.T.P., Heng, D., Leo, Y.-S., Lye, D.C., Lee, V.J.M., Kam, K.-Q., Kalimuddin, S., Tan, S.Y., Loh, J., Thoon, K.C., Vasoo, S., Khong, W.X., Suhaimi, N.-A., Chan, S.J.H., Zhang, E., Olivia, Oh., Ty, A., Tow, C., Chua, Y.X., Chaw, W.L., Ng, Y., Abdul-Rahman, F., Sahib, S., Zhao, Z., Tang, C., Low, C., Goh, E.H., Lim, G., Hou, Y., Roshan, I., Tan, J., Foo, K., Nandar, K., Kurupatham, L., Chan, P.P., Raj, P., Lin, Y., Said, Z., Lee, A., See, C., Markose, J., Tan, J., Chan, G., See, W., Peh, X., Cai, V., Chen, W.K., Li, Z., Soo, R., Chow, A.L.P., Wei, W., Farwin, A., Ang, L.W.: Investigation of three clusters of COVID-19 in Singapore: implications for surveillance and response measures. The Lancet **395**(10229), 1039–1046 (2020)

40. Zhang, S., Diao, M., Yu, W., Pei, L., Lin, Z., Chen, D.: Estimation of the reproductive number of novel coronavirus (COVID-19) and the probable outbreak size on the Diamond Princess cruise ship: a data-driven analysis. Int. J. Infect. Dis. **93**, 201–204 (2020)

41. El Zowalaty, M.E., Järhult, J.D.: From SARS to COVID-19: a previously unknown SARS-related coronavirus (SARS-CoV-2) of pandemic potential infecting humans–Call for a One Health approach. One Health **9**, 100124 (2020)
42. Li, Q., Guan, X., Peng, W., Wang, X., Zhou, L., Tong, Y., Ren, R., Leung, K.S.M., Lau, E.H.Y., Wong, J.Y., Xing, X., Xiang, N., Yang, W., Li, C., Chen, Q., Li, D., Liu, T., Zhao, J., Liu, M., Wenxiao, T., Chen, C., Jin, L., Yang, R., Wang, Q., Zhou, S., Wang, R., Liu, H., Luo, Y., Liu, Y., Shao, G., Li, H., Tao, Z., Yang, Y., Deng, Z., Liu, B., Ma, Z., Zhang, Y., Shi, G., Lam, T.T.Y., Wu, J.T., Gao, G.F., Cowling, B.J., Yang, B., Leung, G.M., Feng, Z.: Early transmission dynamics in Wuhan, China, of novel coronavirus–infected pneumonia. N. Engl. J. Med. **382**(13), 1199–1207 (2020)
43. World Health Organization: Novel Coronavirus (2019-nCoV) Situation Report-28. https://www.who.int/docs/default-source/coronaviruse/situation-reports/20200217-sitrep-28-covid-19.pdf?sfvrsn=a19cf2ad_2. Accessed Mar 2020
44. Kandel, N., Chungong, S., Omaar, A., Xing, J.: Health security capacities in the context of COVID-19 outbreak: an analysis of International Health Regulations annual report data from 182 countries. The Lancet **395**(10229), 1047–1053 (2020)
45. Sun, J., He, W.-T., Wang, L., Lai, A., Ji, X., Zhai, X., Li, G., Suchard, M.A., Tian, J., Zhou, J., Veit, M., Su, S.: COVID-19: epidemiology, evolution, and cross-disciplinary perspectives. Trends Mol. Med. **26**(5), 483–495 (2020)
46. Roosa, K., Lee, Y., Luo, R., Kirpich, A., Rothenberg, R., Hyman, J.M., Yan, P., Chowell, G.: Real-time forecasts of the COVID-19 epidemic in China from February 5th to February 24th, 2020. Infect. Dis. Model. **5**, 256–263 (2020)
47. Chinese National Health Commission: Reported cases of 2019-nCoV. https://ncov.dxy.cn/ncovh5/view/pneumonia?from=groupmessage&isappinstalled=0. Accessed Mar 2020
48. Lu, C.W., Liu, X.F., Jia, Z.F.: 2019-nCoV transmission through the ocular surface must not be ignored. Lancet **395**(10224), e39 (2020)
49. Liu, R., Han, H., Liu, F., Lv, Z., Kailang, W., Liu, Y., Feng, Y., Zhu, C.: Positive rate of RT-PCR detection of SARS-CoV-2 infection in 4880 cases from one hospital in Wuhan, China, from Jan to Feb 2020. Clin. Chim. Acta **505**, 172–175 (2020)
50. Simcock, R., Thomas, T.V., Estes, C., Filippi, A.R., Katz, M.S., Pereira, I.J., Saeed, H.: COVID-19: global radiation oncology's targeted response for pandemic preparedness. Clin. Transl. Radiat. Oncol. **22**, 55–68 (2020)
51. El Asnaoui, K., Chawki, Y.: Using X-ray images and deep learning for automated detection of coronavirus disease. J. Biomol. Struct. Dyn. 1–12 (2020). https://doi.org/10.1080/07391102.2020.1767212
52. Abhir Bhandary, G., Ananth Prabhu, V., Rajinikanth, K.P., Thanaraj, S.C., Satapathy, D.E., Robbins, C.S., Zhang, Y.-D., João Manuel, R.S., Tavares, N.S., Raja, M.: Deep-learning framework to detect lung abnormality – a study with chest X-Ray and lung CT scan images. Pattern Recogn. Lett. **129**, 271–278 (2020)
53. Kermany, D., Zhang, K., Goldbaum, M.: Labeled optical coherence tomography (OCT) and Chest X-Ray images for classification. Mendeley Data **2**(2) (2018)
54. Cohen, J.P., Morrison, P., Dao, L.: COVID-19 image data collection. arXiv:2003.11597 (2020). https://github.com/ieee8023/covid-chestxray-dataset
55. Toğaçar, M., Ergen, B., Cömert, Z., Özyurt, F.: A deep feature learning model for pneumonia detection applying a combination of mRMR feature selection and machine learning models. IRBM **41**(4), 212–222 (2020)
56. Liang, G., Zheng, L.: A transfer learning method with deep residual network for pediatric pneumonia diagnosis. Comput. Methods Programs Biomed. **187**, 104964 (2020)
57. Jaiswal, A.K., Tiwari, P., Kumar, S., Gupta, D., Khanna, A., Rodrigues, J.J.: Identifying pneumonia in chest X-rays: a deep learning approach. Measurement **145**, 511–518 (2019)
58. Ge, Y., Wang, Q., Wang, L., Wu, H., Peng, C., Wang, J., Xu, Y., Xiong, G., Zhang, Y., Yi, Y.: Predicting post-stroke pneumonia using deep neural network approaches. Int. J. Med. Inform. **132**, 103986 (2019)
59. Sirazitdinov, I., Kholiavchenko, M., Mustafaev, T., Yixuan, Y., Kuleev, R., Ibragimov, B.: Deep neural network ensemble for pneumonia localization from a large-scale chest x-ray database. Comput. Electr. Eng. **78**, 388–399 (2019)

60. Behzadi-khormouji, H., Rostami, H., Salehi, S., Derakhshande-Rishehri, T., Masoumi, M., Salemi, S., Keshavarz, A., Gholamrezanezhad, A., Assadi, M., Batouli, A.: Deep learning, reusable and problem-based architectures for detection of consolidation on chest X-ray images. Comput. Methods Programs Biomed. **185**, 105162 (2020)
61. Alom, M.Z., Taha, T.M., Yakopcic, C., Westberg, S., Sidike, P., Nasrin, M.S., Van Esesn, B.C., Awwal, A.A.S., Asari, V.K.: The history began from alexnet: a comprehensive survey on deep learning approaches. arXiv preprint arXiv:1803.01164 (2018)
62. Litjens, G., Kooi, T., Bejnordi, B.E., Setio, A.A.A., Ciompi, F., Ghafoorian, M., van der Jeroen, A.W.M., Ginneken, B.L., Sánchez, C.I.: A survey on deep learning in medical image analysis. Med. Image Anal. **42**, 60–88 (2017)
63. Ouhda, M., El Ansaoui, K., Ouanan, M, Aksasse, B.: Content-based image retrieval using convolutional neural networks. In: First International Conference on Real Time Intelligent Systems, pp. 463–476. Springer, Cham (2017)
64. Simonyan, K., Zisserman, A.: Very deep convolutional networks for large-scale image recognition. arXiv preprint arXiv:1409.1556 (2014)
65. Zhang, Q., Wang, H., Yoon, S.W., Won, D., Srihari, K.: Lung nodule diagnosis on 3D computed tomography images using deep convolutional neural networks. Procedia Manuf. **39**, 363–370 (2019)
66. Szegedy, C., Liu, W., Jia, Y., Sermanet, P., Reed, S., Anguelov, D., Erhan, D., Vanhoucke, V., Rabinovich, A.: Going deeper with convolutions. In: Proceedings of the IEEE Conference on Computer Vision and Pattern Recognition, pp. 1–9 (2015)
67. Szegedy, C., Vanhoucke, V., Ioffe, S., Shlens, J., Wojna, Z.: Rethinking the inception architecture for computer vision. In: Proceedings of the IEEE Conference on Computer Vision and Pattern Recognition, pp. 2818–2826 (2016)
68. He, K., Zhang, X., Ren, S., Sun, J.: Deep residual learning for image recognition. In: Proceedings of the IEEE Conference on Computer Vision and Pattern Recognition, pp. 770–778 (2016)
69. Szegedy, C., Ioffe, S., Vanhoucke, V., Alemi, A.: Inception-v4, inception-ResNet and the impact of residual connections on learning. In: Proceedings of the AAAI Conference on Artificial Intelligence, vol. 31, no. 1, February 2017
70. Huang, G., Liu, Z., Van Der Maaten, L., Weinberger, K.Q.: Densely connected convolutional networks. In: Proceedings of the IEEE Conference on Computer Vision and Pattern Recognition, pp. 4700–4708 (2017)
71. Sandler, M., Howard, A., Zhu, M., Zhmoginov, A., Chen, L.C.: MobileNetV2: inverted residuals and linear bottlenecks. In: Proceedings of the IEEE Conference on Computer Vision and Pattern Recognition, pp. 4510–4520 (2018)
72. Chollet, F.: Xception: deep learning with depthwise separable convolutions. In: Proceedings of the IEEE Conference on Computer Vision and Pattern Recognition, pp. 1251–1258 (2017)
73. Kassani, S.H., Kassani, P.H., Wesolowski, M.J., Schneider, K.A., Deters, R.: Classification of histopathological biopsy images using ensemble of deep learning networks. arXiv preprint arXiv:1909.11870 (2019)
74. Kharel, N., Alsadoon, A., Prasad, P.W.C., Elchouemi, A.: Early diagnosis of breast cancer using contrast limited adaptive histogram equalization (CLAHE) and Morphology methods. In: 2017 8th International Conference on Information and Communication Systems (ICICS), pp. 120–124. IEEE, April 2017
75. Makandar, A., Halalli, B.: Breast cancer image enhancement using median filter and CLAHE. Int. J. Sci. Eng. Res. **6**(4), 462–465 (2015)
76. Blum, A., Chawla, S.: Learning from labeled and unlabeled data using graph mincuts (2001)
77. Apostolopoulos, I.D., Aznaouridis, S.I., Tzani, M.A.: Extracting possibly representative COVID-19 biomarkers from X-ray images with deep learning approach and image data related to pulmonary diseases. J. Med. Biol. Eng. **40**, 462–469 (2020)
78. Farooq, M., Hafeez, A.: COVID-ResNet: a deep learning framework for screening of Covid 19 from radiographs. arXiv preprint arXiv:2003.14395 (2020)

79. El Asnaoui, K., Chawki, Y., Aksasse, B., Ouanan, M.: A new color descriptor for content-based image retrieval: application to coil-100. J. Digit. Inf. Manag. **13**(6), 473 (2015)
80. El Asnaoui, K., Chawki, Y., Aksasse, B., Ouanan, M.: Efficient use of texture and color features in content-based image retrieval (CBIR). Int. J. Appl. Math. Stat. **54**(2), 54–65 (2016)
81. Chawki, Y., El Asnaoui, K., Ouanan, M., Aksasse, B.: Content frequency and shape features based on CBIR: application to color images. Int. J. Dyn. Syst. Differ. Eqn. **8**(1–2), 123–135 (2018)
82. Ouhda, M., El Asnaoui, K., Ouanan, M., Aksasse, B.: Using image segmentation in content-based image retrieval method. In: International Conference on Advanced Information Technology, Services and Systems, pp. 179–195. Springer, Cham (2017)
83. El Asnaoui, K.: Design ensemble deep learning model for pneumonia disease classification. Int. J. Multimed. Inf. Retr. **10**, 55–68 (2021). https://doi.org/10.1007/s13735-021-00204-7

Using Blockchain in Autonomous Vehicles

Nidhee Kamble, Ritu Gala, Revathi Vijayaraghavan, Eshita Shukla, and Dhiren Patel

Abstract Autonomous vehicles have the potential to revolutionize the automotive industry and are gaining immense attention from academia as well as industry. However, facets of autonomous vehicle systems related to the interconnection of independent components pose vulnerabilities to the system. These vulnerabilities aren't guaranteed to be solved by traditional security methods. Blockchain technology is a powerful tool that can aid in improving trust and reliability in such systems. This paper provides a survey on how blockchain can help improve not only security but also other aspects of the AV systems, focussing on the two major blockchain ecosystems as of this writing - Ethereum and Bitcoin. Our survey found that blockchain technology can assist in different use cases related to AVs, such as providing shared storage, enhancing security, optimizing vehicular functionalities, and enhancing related industries. This paper suggests directions for improvement in the sectors of Autonomous Vehicles (AV), which can be achieved with the incorporation of blockchain into Intelligent Transport Systems (ITS) or individual vehicular units.

Keywords Blockchain · Distributed Ledger Technology (DLT) · Autonomous Vehicle (AV) · Connected Vehicles (CV) · Intelligent Transportation System (ITS)

N. Kamble (✉) · R. Gala · R. Vijayaraghavan · E. Shukla · D. Patel
VJTI, Mumbai, India
e-mail: ndkamble_b17@ce.vjti.ac.in

R. Gala
e-mail: rsgala_b17@ce.vjti.ac.in

R. Vijayaraghavan
e-mail: rvijayaraghavan_b17@ce.vjti.ac.in

E. Shukla
e-mail: epshukla_b17@ce.vjti.ac.in

© The Author(s), under exclusive license to Springer Nature Switzerland AG 2021
Y. Maleh et al. (eds.), *Artificial Intelligence and Blockchain for Future Cybersecurity Applications*, Studies in Big Data 90,
https://doi.org/10.1007/978-3-030-74575-2_15

1 Introduction

Transport systems have evolved from being a status symbol to being necessary for the current day and age. We cannot imagine a world without the means of transport that we have at our disposal today. With the advancement of associated technologies, we see a shift to the usage of electric vehicles and autonomous vehicles, which are expected to reduce the strict operating requirements (e.g. personal driving license), energy usage, and environmental impact. Autonomous Vehicles (AVs) are intended to be eco-friendly and energy-conscious and provide a comfortable user experience, cause an increase in consumer savings, and reduce the number of traffic deaths. With reduced private ownership of vehicles, the value of the AV's service will not be based on the brand, but on the quality of service and experience provided.

However, there are specific issues that need to be addressed before AVs can become ubiquitous. AVs rely on trust in the sharing and communication of information within components of a single vehicular unit, or multiple vehicles interacting with each other in a Vehicular Ad-hoc Network (VANET). AVs use a multitude of technologies to make this communication possible. The state information consists of combinations of location and time references of objects for precise and continuous position tracking related to other objects or vehicles around the AV. The working of the AV happens in stages - sight (sensors), communication (Vehicle-to-Everything (V2X) technology), and movement (actuators). Asmaa Berdigh and Khalid El Yassini (2017) [4] give an overview of these technologies. V2V focuses on wireless communication of relevant information between vehicles to provide a more efficient driving experience, like better safety. Vehicular utilisation of multimedia services using V2I uses cellular network infrastructures. Intelligent Transportation Systems are better managed Vehicle-to-Roadside or V2R connectivity, using real-time updates on road statuses.

The two main tasks of AVs include perception and prediction. The shared information and data, the signals from LiDAR, GPS, etc., are susceptible to multiple security threats and attacks. Apart from the data security issues, there arises the concern of liability management in accidents caused by AVs. Blockchain is most known for being extremely secure for storing data, in the sense that modifying previously entered data is impossible without affecting any other blocks (in the blockchain). Blockchain technology can offer a seamless decentralized platform where information about insurance, proof of ownership, patents, repairs, maintenance and tangible/intangible assets can be securely recorded, tracked and managed. In this paper, we survey the use of blockchain technology to help tackle these issues and concerns in AVs. We also suggest room for improvement in the current vehicular functionalities, and how blockchain technology can be leveraged to improve related industries.

The rest of the paper is organized as follows: Sect. 2 provides an overview of blockchain and autonomous vehicles and discusses autonomous vehicles' issues. In Sect. 3, current use cases which use blockchain to solve these problems in AVs are discussed. Section 4 discusses the analysis of these recent use cases, with suggested directions to address them. Conclusion and future directions are presented in Sect. 5 with references at the end.

2 Background

We define some of the terminologies that are common across different papers that we have reviewed.

2.1 Autonomous Vehicles

The terms 'self-driving vehicles' or 'autonomous vehicles' refer to vehicles that navigate without human intervention by the integration of hardware sensors and software algorithms of intelligence.

As per the NHTSA [29] guidelines, autonomous vehicles have the following levels:

- Level 0: No Automation
  This level consists of completely manual driving.
- Level 1: Driving Assistance
  The vehicle can assist with steering or accelerating/braking but not both simultaneously. A driver is required to drive the vehicle.
- Level 2: Partial Automation
  At this level, steering and accelerating/braking can be performed simultaneously, but the driver must monitor the driving environment and perform the remaining driving operations.
- Level 3: Conditional Automation
  At this level, the car can perform all aspects of driving, but a driver must be present if the system requests so.
- Level 4: High-Driving Automation
  This a fully functional driving system that requires no assistance and does not need the driver to pay much attention
- Level 5: Fully Autonomous (Unconditional)
  In this system, human occupants are just passengers and not drivers. This is the highest level of automation.

For this paper, we will consider autonomous vehicles to be those of Level 3 and higher. AVs use a multitude of technologies integrated and thus have various components. These components all have different uses but should, as explained by Alberto Broggi et al. (2008) [7], contribute to giving five major functionalities to the AV:

1. Vehicular state estimation (static/dynamic);
2. Information retrieval about the surrounding (static/moving objects);
3. Information collection on driver/occupant state (to prevent casualties or report them);

4. Communication with other vehicles and other infrastructure (traffic lights or stop signs);
5. Enabling access to a Positioning System (perhaps GPS).

2.2 Technologies Used in AV Systems

Perception in AVs happens through raw information inputted through Vehicle-to-Vehicle (V2V) components or sensors. The critical process of obstacle detection (to detect static and moving objects) is done in the perception task. Based on perception, AVs act by maps, weather, traffic data, topological conditions, and surrounding vehicles positions. Ultrasonic, LiDAR (light detection and ranging), RADAR (radio detection and ranging), and cameras aid in perception. Ultrasonic sensors are mainly used in parking sensors, and radar is only used for extremely long-distance tracking used for Adaptive Cruise Control (ACC). Cameras are generally only used to find lane markings and display signs such as speed limits on a vehicle's dashboard. The combination of RADAR and LiDAR can capture images and transfer them through electrical interfaces. The in-vehicle micro-computer will process the information acquired and analyse the data to make driving decisions by making an almost instantaneous 3D map of the area around the vehicle. The use of the created 3D map, combined with GPS, is used for tackling the problem of identifying an ego vehicle's position, a critical piece of information required for autonomous vehicles.

Accurate perception is the key to ensuring safety in an AV. Perception aids AVs make decisions spontaneously, using quantifiable variables that estimate environmental factors (surrounding vehicle's location/condition, pedestrians locations/conditions, vehicle occupant's conditions, maps, weather and traffic data). It uses many sensors like GPS (Global Positioning System) LiDAR (Light Detection and Ranging for accurate reliable and cost-effective mapping), RADAR (Radio Detection and Ranging used for Adaptive Cruise Control [ACC]) and ultrasonic sensors (used for parking). Obstacle Detection (a crucial task) is accomplished using Computer Vision (using a camera that transmits captured information to in-vehicle microprocessors). Some suggested techniques include KITTI for pedestrian and cyclist detection, PSPnet by Zhao et al. (2012) [28].

Technologies used for AVs build upon the native functions of traditional, level 0 vehicles to optimize them specifically. Correa, et al. (2017) [9] propose a design for a parking system for AVs, implemented on a Vehicular Sensor Networks with minimal infrastructural overhead. The research simulates a parking layout using mathematical models, defining its accessibility rate in parking place availability for AVs. Geng and Cassandras (2012) [11] propose methods used by traditional AV systems for navigation in geographical scenarios, like VANETs, ultrasounds, in addition to oft-used GPS and LoS (Line of Sight) with their analyses. Received Signal Strength (RSS), the Time of Arrival (ToA) and the Time Difference of Arrival (TDoA) both in anchor-based solutions and in cooperative approaches, are used in GPS-denied environments. One of the notable mentions in enlisting previous research is Roadside

Units (RSUs), used to utilise AVs' unused resources—like a rechargeable battery and storage capacity—using IPARK [28], a system for guided parking over infrastructure-less VANETs.

2.3   Vehicular ad-hoc Networks (VANETs), Intelligent Transport Systems (ITS) and Connected Vehicles (CVs)

A 'Vehicular Ad-hoc Network (VANET)' is a group of stationary and moving vehicles connected via a wireless network. An 'Intelligent Transport System (ITS)' is an infrastructure where vehicles are connected using smart devices [12]. The term 'Connected Autonomous Vehicles (CAVs)' refers to a group of autonomous vehicles that may connect to the internet and provide improved data sharing in the form of risk data, sensory and localization data and environmental perception.

Figure 1 is the depiction of how network users can access and utilize deployed applications. Each participant is registered on the blockchain (Ethereum blockchain) and has an address (Ethereum address). Benjamin Leiding et al. [5], made possible by Ethereum Blockchain, applications for enforcement of provision rules regarding services are available to all network users. The cost of running the chain is self-regulating, which happens because of a price being paid for each transaction in the form of Ethereum - gas. Consequently, each automobile pays a fee for each transaction made. This concept of making cars pay for the infrastructure and computing has a limitation: The most loyal customers (who use the charging station more frequently), incur more penalty. Although this means that there is a big incentive for providing RSUs and other essential tools, the fee goes to miners and computational services for mining transactions and mining-pools.

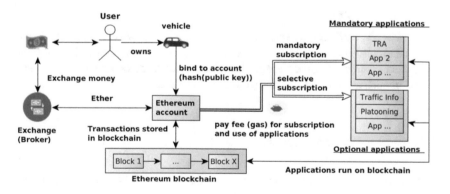

Fig. 1 Ethereum-based service provision and rule enforcement in self-managed VANETs

2.4   Blockchain

A blockchain [6] is a growing list of records, called blocks linked using cryptography. Each block contains a cryptographic hash of the previous block, a timestamp, and transaction data (Fig. 2) (generally represented as a Merkle tree) [21]. Blockchains can be either public (everyone can view and verify the data), private (governed by a single entity), consortium (semi-private, shared across various organizations with restricted access) or hybrid (features of both private and public blockchains). The most popular blockchains are the Bitcoin network and the Ethereum blockchain.

The key properties of blockchain are:

1. Decentralized: There is no centralized authority, as the blockchain is not owned by a single entity.
2. Secure: The data stored is in encrypted form using hash functions, making it secure.
3. Immutable: Data once inserted into the blockchain, cannot be changed due to the blockchain structure itself, thus making it tamper-resistant.
4. Transparent: Since it is a distributed ledger, anyone can access the data on the blockchain.

Components of a blockchain are:

1. Node: User or computer within a blockchain network
2. Transaction: the smallest building block of a blockchain system
3. Block: a data structure used for keeping a set of transactions which is distributed to all nodes in the network
4. Chain: a sequence of blocks in a particular order
5. Miners: specific nodes which perform block verification and add nodes to the chain
6. Consensus: a set of rules and regulations mutually agreed upon by all the nodes in the blockchain

A *'state channel'* is an off-chain channel through which two or several blockchain users can atomically exchange blockchain-compliant information to be added on-chain later when closing the channel. The channel is closed on either completion or failure of such atomic transactions (transfer or exchange).

2.5   Scalability with Blockchains

Another issue with current IoT networks is that of scalability. As the number of devices connected through an IoT network grows, current centralised systems to authenticate, authorise and connect different nodes in a network will turn into a bottleneck. This would necessitate huge investments into servers that can handle a large amount of information exchange, and the entire network can go down if the server becomes unavailable.

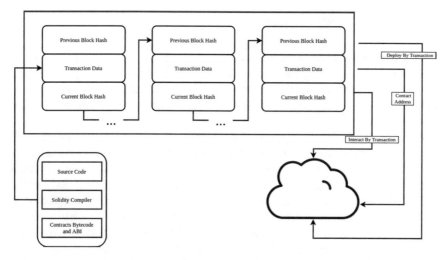

Fig. 2 Structure of blockchain

In public blockchains such as Bitcoin, very time and computationally intensive mining-based consensus mechanisms are often used to establish trust between entirely anonymous parties. Thus significant transaction times result which results not only in poor performance but also poor scalability. This leads to the creation of side chains to offload transaction processing from the main chain.

In business cases to business (B2B) and business to consumer (B2C) interactions, the use of private and permissioned blockchain is preferred. Private blockchains have a reduced number of nodes, which results in a far faster consensus mechanism and in general, improves scalability and performance.

There are now new blockchains coming up, termed blockchain 3.0, which are based on the principles of DLT. These blockchains improve scalability and performance by the use of DAG (Directed Acyclic Graph) and novel validation and voting mechanism [14].

Another mechanism proposed by Lyubomir Stoykov et al. [18] is the VIBES architecture. VIBES uses configurable input parameters like network-information and number of miners, to provide a flexible solution. The simulator includes information about throughput and cost-per-transaction. To bypass most heavy computations for large scale applications, the paper suggests improving scalability via fast-forward computing. This helps complete simulations before time. Nodes try to estimate computational costs and ask for permission to fast forward. After greenlighting the operation, the orchestrator declares the operation complete and skips forward.

2.6   Consensus Mechanism

A consensus mechanism is used to tackle fault-tolerance. It is used to arrive at a group consensus regarding the data to be added to the network or the network's state. The famous consensus mechanisms are Proof of Work (used by Bitcoin, Litecoin, and Monero), Proof of Stake (used by Ethereum 2.0 and Dash), Proof of Vote and Proof of Burn.

Ole Meyer et al. (2018) [23] propose a consensus algorithm for autonomous vehicles that do not rely on a central authority for control and monitoring. An autonomous entity in the system called an agent, can predict dangerous situations and trigger a protocol to resolve or avoid it. Parallel solutions generated simultaneously might lead to suboptimal solutions. In such cases, priority is allotted based on a parameter specific to the situation. For example, to avoid an impending collision between two autonomous cars, the car about to reach a place first should slow down and the other halt, instead of both the cars halting. The parameter for priority ordering can also be a characteristic that can be independently determined by both the parties; the information regarding which can be easily obtained by them. The protocol must yield invariant solutions. This ensures that the consensus can be achieved even without communication between the participants, reducing overhead for exchange over a network and hardware prerequisites for facilitating it.

2.7   Use of Blockchain to Ensure Security

An essential worry within autonomous vehicles is the high dependence on IoT devices. These IoT devices are often vulnerable to Distributed Denial of Service (DDoS) attacks. Blockchain technology can prove to be extremely useful in this aspect. Blockchain eliminates a single point of failure-based attacks and provides a medium for auditable and traceable changes. Further, blockchains provide help with authentication and identification of devices over a distributed database.

2.8   Problems and Improvements Associated with AVs

With the expectation of AVs becoming a norm, the number of AVs on the road will go on increasing. As self-driving vehicles are equipped with more sensors and network connectivity than non-autonomous ones, the number of security vulnerabilities and thus, the attack surface of an AV is undoubtedly increased. Adversaries today are becoming increasingly skillful [27]. These skills coupled with feasible low-cost offensive devices, can enable them to break into car security systems easily and in the worst case, allow complete unauthorized control of the vehicle or data tampering. Further, with autonomy, comes a lack of accountability. When autonomous vehicles are involved in accidents (collisions between themselves, or collisions with conventional vehicles, pedestrians or other objects), how should such events be recorded for forensic purposes to determine liability? Also, how could such recorded events be verified, trusted, and not tampered with? Such issues become critical when there exist incentives for various parties involved to tamper with the recorded events to avoid punitive penalties [8].

The expected functionalities of autonomous vehicles could be enhanced due to the integration of vehicle sensors and blockchain. The revolution of autonomous vehicles and the aid of blockchain technology could affect closely related industries. For instance, blockchain in these AVs could negate middle parties' need, be it brokers in fleet management systems or ride-sharing companies like Uber.

3   Use of Blockchain in AVs

3.1   Decentralised Storage and Security Mechanism

On surveying, we noticed that blockchain could serve as shared storage to facilitate accident management and can be used to tackle AVs' security attacks. Below is the detailed summary of the two cases.

Accident Reporting and Verification. Hao Guo, Ehsan Meamari and Chien-Chung Shenis (2018) [13] focus on event recording mainly for accident forensics. They propose Proof of Event as a consensus mechanism, a recording and broadcasting mechanism for the events. The collections of records are accepted as new nodes to the blockchain depending on the verifier and participant nodes (vehicles). The credit score is a measure of how 'trusted' a vehicle is. This includes being a witness or a verifier to an accident. Since the Proof of Event protocol provides no tangible award, credit scores are an attempt at incentivisation. Higher credit scores may reflect lower insurance premiums on the vehicle.

Further, they adumbrate the protocols necessary for implementing the system. Proposal for a reward-based smart vehicle data-sharing framework is proposed by Singh (2017) [20] for intelligent vehicle communication using blockchain. The concept is abstract and introduces a blockchain network model for communication over a VCC (Vehicular Cloud Communication) for reporting safety–critical incidents and (the possibility or occurrences of) hazards to drivers. It uses Proof of Driving as the consensus mechanism where crypto tokens provide the incentivisation in IVTP (Intelligent Vehicle Trust Points).

Narbayeva, Saltanat et al. (2020) [22] present a mathematical foundation to use blockchain technology to increase information integrity by sending parameters of each vehicle's current state, verified by the signals of neighbouring vehicles. The authors have developed a tracking system for car actions using the blockchain system based on the Exonum platform.

Security in Connected Autonomous Vehicles. AVs are more susceptible to malicious cyber-attacks due to increased Vehicle-to-Vehicle (V2V) communication via VANETs [15]. Vrizlynn L.L. Thing et al. (2016) [26] classify attacks possible on autonomous vehicles. The two classes of attacks are physical access and remote access attacks. Physical access attacks include invasive attacks like code modification, code injection, packet sniffing, packet fuzzing and in-vehicle spoofing. Remote access attacks include external signal spoofing and jamming. The security issues concerning CAVs are addressed in the paper by Rathee, Geetanjali, et al. (2019) [25]. They have proposed a blockchain-based solution where each IoT device (sensor/actuator) and the vehicle are registered to the network before acquiring any of the services. Initially, the vehicular number, along with IoT device data, will be stored on the blockchain. Given the high amount of computation power and time needed for the large amount of data generated further, they propose that only the IoT devices store relevant information to the blockchain, which can also be analyzed. Any alteration on information can then easily be detected as it will alter previous records as well.

First, there is no reliable mechanism to keep track of compromised sensors which are a crucial part of the ecosystem of CAVs. Additionally, in a scenario where CAVs are used for a cab-booking service, technical experts may hack into the system and change important information like accidents the car has been associated with, for personal gains. Data falsification attacks are a primary security issue where vehicles in a network rely on other vehicles' information.

The standard encryption schemes like AES will not be feasible for CVs since they produce many data as mentioned by Jolfaei, A., & Kant, K (2019) [16]. Key management could become an issue for each of the devices and they can cause a potential weakness in the system.

Anil Saini et al. (2019) [3] propose a new blockchain network to accommodate priority vehicles. The regular blockchain networks have many drawbacks. Some limitations include dealing via crypto-currencies (instead of trust messages/events) and higher latency (reduced by using 2-levels in the proposed network). The proposed network would use 2-levels and the first level will consist of authorised nodes (placed in different areas). If an RSU node wants to become a participant of the network, it must first get verified by the authorised nodes. The second level consists of registered RSU nodes. The vehicles register with its nearby RSU, after which the RSU verifies the vehicle's identity and stores it on the blockchain. The RSU also receives information generated by the vehicles, like traffic congestion, accident-related information, etc. This information is distributed to the neighboring roadside nodes to be validated in the blockchain network of RSUs. There exists no central authority in this entire process, thus enabling decentralization.

3.2 Blockchain to Improve AV Functionalities

While surveying, we noticed that blockchain could improve an autonomous vehicle's functionality in the ways mentioned below.

Verifying Vehicle Lifecycle. The automotive supply chain industry can be quite complex, ranging from government regulatory parties, manufacturers, suppliers, and vendors to spare parts suppliers. *P. K. Sharma, N. Kumar and J. H. Park (2019)* [24] delineate into each phase of the automotive industry (regulator, manufacturer, dealer, leasing company, user, maintenance, scrap) and explained the benefits of using smart contracts for the digitization of this process. They give a complete overview of the process. They propose a blockchain and smart contract-based scalable distributed framework model for the lifecycle tracking of vehicles. A miner node selection algorithm based on the Fruit Fly Optimization algorithm (FOA) has been suggested to avoid the mining process during the block generation carried out by a unique miner pool and limited by miners.

Insurance and Payments. M. Demir, O. Turetken and A. Ferworn (2019) [19] propose a tamper-free ledger of events as an insurance record of motor vehicles for the provision of evidence in the event of a dispute. This can include all aspects of insurance transactions. The system uses a permissioned blockchain (Hyperledger based) for obtaining, sharing and verifying insurance records will help stakeholders as a reliable sharing platform and a ledger of events. Alejandro Ranchal Pedrosa and Giovanni Pau (2018) [2] provide a detailed algorithm for the payment of a refuelling scenario in autonomous vehicles using Ethereum State Channels. The use of these state channels aims to support instant and reliable trading of information, goods and currency.

Charging Stations and Power Requirements. Alejandro Ranchal Pedrosa and Giovanni Pau (2018) [2] suggest using Ethereum State Channels as an unforgeable recording, flexibility and scalability for Machine to Machine (M2M) transactions in charging stations. A detailed algorithmic approach has been developed to cover all pertinent use cases during the interactions between the AV and the charging station. The use of these state channels aims to support instant and reliable trading of information, goods and currency.

Fabian Knirsch et al. (2017) [17] provides a protocol for allowing the driver of an electric vehicle to find the cheapest charging station in a given location radius. The bids sent by different charging stations are stored on a blockchain to provide transparency and verifiability. The phases of requesting and serving corresponding charging locations have been elaborated upon (Fig. 3).

Instead of the Vehicle Engine Control Unit, Raspberry Pi is proposed by Felix Kohlbrenner et al. (2019) [10]. This can be used to collect the required data and utilize the real vehicle bus data. This provides a similar environment with similar restrictions. Once the charger is plugged in, the event of "start charging is triggered". Various vehicle information is recorded and saved in a hash (updated when required),

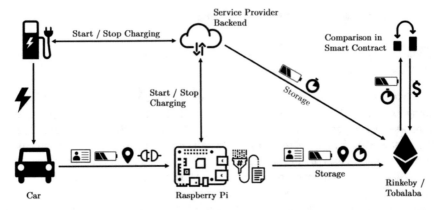

Fig. 3 Details of how a charging event would occur and the entities/tasks involved in this process

this upholds the trustworthiness of the charging system. Still, user-data is saved off-chain due to privacy concerns. With the help of a signed transaction, the charging status is stored on the Ethereum Blockchain. A payment request is triggered when a certain charging threshold is reached (any payment method can be used).

Parking for AVs. There are intelligent parking management architectures that are suited specifically for the system heterogeneity of AVs. Jennath, H. S. et al. (2019) [15] propose a blockchain-based solution for the creation of parking pools using a non-fungible token system for rentals of users' unused land for a stipulated amount of time with little or no legal hassles. Additionally, this method leverages income from unused property, which is an added advantage. Smart contracts over blockchain enforce the contractual agreement between the participants ensuring financial transparency in the proposed system. This system can be implemented in the present scenario for traditional cars and extended to AVs in the future. With an increase in automation levels, some decision-making tasks—such as the inclusion of vehicles in parking pools—may also be taken by AVs instead of humans.

3.3 Optimizing Related Industries

The AV industry is not standalone and affects other related industries, like transport and freight, human involvement, inter-industry dependency, and consumer experience. Advancements in AV sectors by the integration of blockchain will, by extension, affect these industries. Additionally, it can also be used to address corresponding improvements.

Vehicle Sharing. Using Proof of Work consensus algorithm for the validation of Demand Response, Abubaker Zain et al. (2019) [1] present in this paper a blockchain based mechanism to provide users with real-time availability of on-network intelligent vehicles. In the system, vehicles can provide services, as part of a fleet on a single Intelligent Transport System (ITS) network. This paper uses the Proof of Work consensus algorithm to validate Demand Response (DR) events.

Freight Industries. *Dogar, Ghulam and Javaid, Nadeem (2019)* [9] proposed a system in which vehicles that belong to a fleet can be part of a single Intelligent Transport System (ITS) network, providing services to all the autonomous vehicles and carrying out their jobs normally. Special vehicles that are part of a fleet will be registered with their respective organization only by registering with the Intelligent Vehicle Trust Point (IVTP). To facilitate the assignment of tasks and task-completion, an incentive-based blockchain-based Fleet Management System (BFMS) is proposed. Such a system can prove extremely useful in parking and charging cases where queries (for bids) can be used to provide the intelligent-vehicle options, from which the best can be chosen.

4 Analysis

The analysis of the previous section is divided into the following categories.

4.1 Relevance of Blockchain

DLTs vs Blockchain. While many use cases of AVs rightly require blockchain, there has been a trend to misuse blockchain as a technology, which means using them without a proper consensus mechanism. Many use cases simply require storage immutability, which can easily be provided by permissioned Distributed Ledger Technologies (DLT), and using a blockchain in such cases is not exclusively required.

Tamper Resistance. Specific research papers focus on 'tamper-free ledgers to ensure data integrity over AV communication. Distinguishing the terms tamper-free, tamper-tolerant, and tamper-resistant, has implications on understanding what the technology provides. A blockchain is tamper-resistant: It resists (the possibility of) being modified, by design. In the possibility of a modification, its protocols are resilient enough for it to resist the effects of tampering. Based on our study, the terms tamper-free and tamper-tolerant point at something possible to be tampered with, the results of which can be rectified later - by rollback, late control, or implementational modifications.

Lack of Appropriate Consensus Mechanisms. The prevalent consensus mechanisms for blockchain—Proof of Work, Stake, and Authority—are criticized in a few research papers for their inability to maintain the decentralization of control in the blockchain, eventually resulting in the concentration of power in the regions with higher computational power and resources, respectively. However, proposed alternatives to these, as stated in the papers, lack incentivization. For shared records of AV lifecycle and logs for vehicle sharing, each participant on the chain should verify the on-chain information by its existence alone. Since the verification results from the consensus mechanism, which operates only on the on-chain data, it follows that the data source must also be on-chain. These data sources must be intrinsic to the blockchain for verification to happen as a part of the working. Unless it is made possible to embed some kind of metadata in the AV records that make its source on-chain, the verification remains external in all systems currently proposed, rendering the consensus mechanism of little use by itself. Looking at the potential of blockchains as an ecosystem, we opine that it remains underutilised in such use cases.

4.2   Issues with the Use of Blockchain in AV Systems

Scalability. The concept of transparency in the blockchain is based on the fact that each node in the blockchain stores a separate copy of the entire data present on the blockchain. This isn't feasible for AVs due to rapid generation of large amounts of data. An increase in the number of vehicles (nodes) will add to this data, decreasing the system's efficiency. A possible solution would be to store only the bare minimum information on the blockchain and store the rest of the data on a shared file system like IPFS.

Feasibility of Computation. Blockchain consensus mechanism requires a large amount of computational power. These computations may not be feasible on AVs, which might, in turn, result in low throughput of the system, by causing an increase in latency.

4.3   Future of Related Industries

Exploring the current proposals and analyzed possibilities, advancements in the AV sector with blockchain or DLTs would improve the experience around providing insurance, with extended services around providing a clean driving record, or for vehicle lending or sharing. DLTs will facilitate mainstream adoption of car sharing by scheduling and matching rides without a middleman's need. Distributed ledger technologies can allow information on vehicle availability to be made publicly accessible so that users and car owners can match journeys easily.

Blockchain could also aid in effective supply chain management in the freight industry. However, simply using blockchain technology does not ensure the effective transport and delivery of goods. Tampering with RFID tags attached to goods and cases of smuggling can lead to incorrect information stored on the blockchain, which voids the use of blockchain in the first place.

4.4 Using Cryptocurrency

With vehicles becoming driverless, payment can be tackled by providing a payment method that is intrinsic or facilitated by the blockchain infrastructure itself. This would mean that payments for parking and toll, payment can be made using cryptocurrencies.

However, the use of cryptocurrencies will be unfavourable in case of a 51% miner attack. However, this kind of attack requires massive computation on popular blockchain platforms like Bitcoin and Ethereum. In the case of smaller blockchains, it is not difficult to amass the computational power for these attacks, and such an attack could be possible. Therefore, autonomous vehicles must be very careful before selecting their desired blockchain for payments.

Further, the volatility of crypto currencies is a significant limitation for adopting blockchain-based payments-especially if it is to be integrated as a long term solution with autonomous vehicles. This volatility is a consequence of state-specific fiscal policies and standards, and not an intrinsic property of cryptocurrencies itself. An optimistic approach might predict that this stability increases; an overly optimistic approach might say that fiat currencies shall be measured in terms of cryptocurrencies in the future (converse of the present scenario). A practical approach is to gauge the market behaviours due to fiat-crypto exchange interactions and adoptions and see how one system can address the weakness (es) of another.

4.5 Resolution of Security Issues

It is impossible to address all security attacks mentioned in Sect. 3A, but blockchain-based solutions can be implemented to prevent specific security attacks. The issues of code modification and code injection can be reduced by incorporation of a permissioned blockchain. This will prevent unauthorized access to the AVs and thus minimise the possibility of such attacks. External signals like GPS and LiDAR signals can be verified using blockchain to prevent external signal spoofing attacks.

Table 1 summarizes the advantages and disadvantages of the proposed methodologies in the use of blockchain in AVs.

Table 1 Advantages and disadvantages of methodologies in the use of AVs

Reference No	Use case in AVs	Purpose	Advantage	Disadvantage
[13]	Accident reporting	Decentralized storage	Using the data of events from various sources and the generated Hash digest obtained using the "Proof of Event" mechanism (with Dynamic Federation Consensus)	The proposed mechanism will not work optimally in areas that are sparsely populated due to which there may not be verifiers or witnesses
[20]	Accident reporting and verification	Decentralised storage and security mechanism	The proposed intelligent vehicle trust point methodology provides fast and secure communication between smart vehicles and stores details about the communication history, which can be beneficial during accidents	The current proposed methodology does not cover multiple vehicle communication as of yet
[22]	Security in connected autonomous vehicles	Decentralized storage and security mechanism	The proposed solution uses the standard ECDSA for confirming transactions and micropayments, which makes it a secure approach. The proposed solution facilitates micropayments in emergencies, wherein one car needs to be prioritized	The solution states that each vehicle's current state will be shared with its neighbours in its vicinity, spanning over a 100–150 m radius. However, there is no mention of what the current state of each vehicle would include, and what the messages to neighbouring vehicles would encompass either, to be shared over the blockchain

(continued)

Table 1 (continued)

Reference No	Use case in AVs	Purpose	Advantage	Disadvantage
[15]	Parking for AVs	Blockchain to improve AV Functionalities	The proposed solution uses non-fungible parking tokens for unused land and provides transparency and trust through the use of a blockchain system	The proposed solution does not mention how the blockchain, combined with an IoT system, will be scaled. An increase in the number of blockchain nodes will most probably decrease the system's efficacy due to increased computation
[25]	Security in Connected Autonomous Vehicles	Decentralized storage and security mechanism	The proposed methodology tracks the information provided by IoT devices, thus ensuring continuous monitoring of data, which provides security and transparency at each step	The proposed solution mentions storing all the data received from IoT devices onto a normal database at first, followed by permanent storage on the blockchain. This seems unnecessary, as duplicating the data, which will be generated in large amounts, will lead to redundancy
[18]	Security in connected autonomous vehicles	Decentralised storage and security mechanism	The solution proposed uses a lightweight permutation scheme suitable for encrypting real-time data generated by weak devices	The proposed solution performs well against the given test cases but needs to be tested more extensively

(continued)

Table 1 (continued)

Reference No	Use case in AVs	Purpose	Advantage	Disadvantage
[24]	Verifying vehicle lifecycle	Blockchain to Improve AV Functionalities	The proposed blockchain base distributed framework for automotive industry allows for significant time and cost savings and enabling manufacturers and suppliers to protect their brands against counterfeit products	For such a framework to exist in a smart city, there needs to be a standardized regulatory framework
[19]	Insurance and payments	Blockchain to Improve AV Functionalities	The use of blockchain for vehicle insurance ledger allows transparently sharing the vehicle insurance records and provides for the collective nature of contribution as participants' may not trust each other	Like the previous paper, this proposed solution would require some governance of the blockchain, perhaps in the form of a consortium
[2]	Insurance and payments	Blockchain to Improve AV Functionalities	The most promising advantage of this proposed architecture is the possibility of blockchain compliant, fast payments due to state channels	Although the use of state channels in the proposed solutions is beneficial, there are associated risks with state channels related to set up and obliviousness of the parties involved, such as improper time-locks, coin theft, data loss or forgetting to broadcast transactions on time

(continued)

Table 1 (continued)

Reference No	Use case in AVs	Purpose	Advantage	Disadvantage
[17]	Charging stations and power requirements	Blockchain to Improve AV Functionalities	The proposed solution is a blockchain-based protocol for finding the nearest and cheapest charging station, ensuring the consumer's privacy and confidentiality	The solution requires a specific number of properties to be met by a blockchain, for it to be used, and scalability remains the most discerning issue
[1]	Vehicle sharing	Optimizing Related Industries	The proposed system allows whole information about the route to be revealed to the customer by real-time traffic information. Further, there is a reduced transaction cost due to the mechanism of peer to peer car sharing, which removes the need for any bank or any reliable authority	The proposed system assumes a driverless environment. As such issues like accident verification and payment of tolls need to be tackled. This proposed architecture can be combined with other proposed architectures mentioned to create a robust
[9]	Freight industries	Optimizing related industries	The paper proposes how different fleets of special vehicles would carry out operations for various organizations for other purposes in blockchain-enabled Intelligent transport	Testing has been done for a small fleet size (120). Thus, the scalability of the proposed system is a matter yet to be determined
[10]	Charging stations and power requirements	Blockchain to Improve AV Functionalities	Decreased latency (30% faster) Reduced cost of operation	Approach violates the principle of the separation of concerns

5 Conclusion

With its key characteristics of decentralization, immutability and transparency, Blockchain has the true potential of being adopted in AVs due to its ability to tackle many issues that AVs are expected to have seamlessly. This paper has provided a comprehensive literature review on the current use cases of blockchain technology in autonomous vehicles. We first provided an overview of autonomous vehicles, followed by an overview of blockchain architecture. We then investigated the current use cases by partitioning them into three broad groups based on blockchain usage in Autonomous Vehicles - as decentralized storage and security mechanism, for Improving AV Functionalities optimizing Related Industries. Finally, we provided a brief analysis of these use cases, discussing their relevance and issues. As a future scope, Bitcoin's Lightning Network (LN) can be implemented for payment channels or primary payment rail coordination for freight chain activities. LN is a second-layer solution enabling Bitcoin to scale to over a million transactions per second (compared to 7 of Bitcoin) with payments routed peer-to-peer within milliseconds. As our analysis suggests, there is significant scope for the integration of blockchain technology in AVs. Our survey mainly indicates that more research needs to be conducted using blockchain for the different facets of AVs mentioned in this paper.

References

1. Abubaker, Z., et al.: Decentralized mechanism for hiring the smart autonomous vehicles using blockchain. In: Barolli, L., Hellinckx, P., Enokido, T. (eds.) BWCCA 2019. LNNS, vol. 97, pp. 733–746. Springer, Cham (2020). https://doi.org/10.1007/978-3-030-33506-9_67
2. Pedrosa, A.R., Pau, G.: ChargeltUp: on blockchain-based technologies for autonomous vehicles. In: Proceedings of the 1st Workshop on Cryptocurrencies and Blockchains for Distributed Systems (CryBlock 2018). Association for Computing Machinery, New York, NY, USA, pp. 87–93 (2018)
3. Saini, A., Sharma, S., Jain, P., Sharma, V., Khandelwal, A.K.: A secure priority vehicle movement based on blockchain technology in connected vehicles. In: Proceedings of the 12th International Conference on Security of Information and Networks (SIN 2019). Association for Computing Machinery, New York, NY, USA, Article 17, pp. 1–8 (2019)
4. Berdigh., A., Yassini, K.E.: Connected car overview: solutions, challenges and opportunities. In: Proceedings of the 1st International Conference on Internet of Things and Machine Learning (IML 2017), pp. 1–7, Article 56. Association for Computing Machinery, New York, NY, USA (2017)
5. Leiding, B., Memarmoshrefi, P., Hogrefe, D.: Self-managed and blockchain-based vehicular ad-hoc networks. In: Proceedings of the 2016 ACM International Joint Conference on Pervasive and Ubiquitous Computing: Adjunct (UbiComp 2016), pp. 137–140. Association for Computing Machinery, New York, NY, USA (2016)
6. Blockchain - Wikipedia: https://en.wikipedia.org/wiki/Blockchain. Accessed 04 Feb 2021
7. Broggi, A., Zelinsky, A., Özgüner, Ü., Laugier, C.: Intelligent Vehicles. In: Siciliano, B., Khatib, O. (eds.) Springer Handbook of Robotics, pp. 1627–1656. Springer, Cham (2016). https://doi.org/10.1007/978-3-319-32552-1_62
8. Correa, A., Boquet, G., Morell, A., Lopez Vicario, J.: Autonomous car parking system through a cooperative vehicular positioning network. Sensors 17(4), 848 (2017)

9. Dogar, G., Javaid, N.: Blockchain Based Fleet Management System for Autonomous Vehicles in an Intelligent Transport System (2019)
10. Kohlbrenner, F., Nasirifard, P., Löbel, C., Jacobsen, H.-A.: A blockchain-based payment and validity check system for vehicle services. In: Proceedings of the 20th International Middleware Conference Demos and Posters (Middleware 2019), pp. 17–18. Association for Computing Machinery, New York, NY, USA (2019)
11. Geng, Y., Cassandras, C.G.: A new smart parking system infrastructure and implementation. Procedia Soc. Behav. Sci. **54**, 1278–1287 (2012)
12. Guerrero-Ibañez, J., Zeadally, S., Castillo, C., Juan. : Sensor Technologies for Intelligent Transportation Systems. Sensors. **18**, 1212 (2018)
13. Guo, H., Meamari, E., Shen, C.-C.: Blockchain-inspired Event Recording System for Autonomous Vehicles, pp. 218–222 (2018)
14. Improving Performance and Scalability of Blockchain Networks. https://www.wipro.com/blogs/hitarshi-buch/improving-performance-and-scalability-of-blockchain-networks/. Accessed 14 Feb 2021
15. Jennath, H.S., Adarsh, S., Chandran, N., Ananthan, R., Sabir, A., Asharaf, S.: Parkchain: a blockchain powered parking solution for smart cities. Front. Blockchain **2**, 6 (2019)
16. Jolfaei, A., Kant, K.: Privacy and security of connected vehicles in intelligent transportation system. In: 2019 49th Annual IEEE/IFIP International Conference on Dependable Systems and Networks–Supplemental Volume (DSN-S), pp. 9–10. IEEE, June 2019
17. Knirsch, F., Unterweger, A., Engel, D.: Privacy-preserving blockchain-based electric vehicle charging with dynamic tariff decisions. Computer Science - Research and Development (2017)
18. Stoykov, L., Zhang, K., Jacobsen, H.-A.: VIBES: fast blockchain simulations for large-scale peer-to-peer networks: demo. In Proceedings of the 18th ACM/IFIP/USENIX Middleware Conference: Posters and Demos (Middleware 2017), pp. 19–20. Association for Computing Machinery, New York, NY, USA (2017)
19. Demir, M., Turetken, O., Ferworn, A.: Blockchain based transparent vehicle insurance management. In: 2019 Sixth International Conference on Software Defined Systems (SDS), Rome, Italy, pp. 213–220 (2019)
20. Singh, M., Kim, S.: Intelligent Vehicle-Trust Point: Reward based Intelligent Vehicle Communication using Blockchain, arXiv preprint arXiv:1707.07442 (2017)
21. Merkle Tree. https://en.wikipedia.org/wiki/Merkle_tree. Accessed 14 Feb 2021
22. Narbayeva, S., Bakibayev, T., Abeshev, K., Makarova, I., Shubenkova, K., Pashkevich, A.: Blockchain technology on the way of autonomous vehicles development. Transp. Res. Procedia. **44**, 168–175 (2020)
23. Meyer, O., Hesenius, M., Gries, S., Wessling, F., Gruhn, V.: A decentralized architecture and simple consensus algorithm for autonomous agents. In: Proceedings of the 12th European Conference on Software Architecture: Companion Proceedings (ECSA 2018), p. 1–4. Association for Computing Machinery, New York, NY, USA (2018). Article 7
24. Sharma, P.K., Kumar, N., Park, J.H.: Blockchain-based distributed framework for automotive industry in a smart city. IEEE Trans. Industr. Inf. **15**(7), 4197–4205 (2019)
25. Rathee, G., et al.: A blockchain framework for securing connected and autonomous vehicles. Sensors **19**(14), 3165 (2019)
26. Thing, V.L., Wu, J.: Autonomous vehicle security: a taxonomy of attacks and defences. In: 2016 IEEE International Conference on Internet of Things (iThings) and IEEE Green Computing and Communications (GreenCom) and IEEE Cyber, Physical and Social Computing (CPSCom) and IEEE Smart Data (SmartData), pp. 164–170. IEEE, December 2016
27. Yebes, J.J., Bergasa, L.M., García-Garrido, M.: Visual object recognition with 3D-aware features in KITTI urban scenes. Sensors **15**, 9228–9250 (2015)
28. Zhao, H., Lu, L., Song, C., Wu, Y.: IPARK: location-aware-based intelligent parking guidance over infrastructureless VANETs. Int. J. Distrib. Sens. Netw. **8**, 1–12 (2012)
29. Dot/NHTSA policy Statement Concerning Automated Vehicles (2016). https://www.nhtsa.gov/staticfiles/rulemaking/pdf/Autonomous-Vehicles-Policy-Update-2016.pdf. Accessed 14 Dec 2021

Crime Analysis and Forecasting on Spatio Temporal News Feed Data—An Indian Context

Boppuru Rudra Prathap, Addapalli V. N. Krishna, and K. Balachandran

Abstract Social media is a platform where people communicate, interact, share ideas, interest in careers, photos, videos, etc. The study says that social media provides an opportunity to observe human behavioral traits, spatial and temporal relationships. Based on study Crime analysis using social media data such as Facebook, Newsfeed articles, Twitter, etc. is becoming one of the emerging areas of research across the world. Using spatial and temporal relationships of social media data, it is possible to extract useful data to analyse criminal activities. The research focuses on implementing textual data analytics by collecting the data from different news feeds and provides visualization. This research's motivation was identified based on relevant work from different social media crime and Indian government crime statistics. This article focuses on 68 types of different crime keywords for identifying the type of crime. Naïve Bayes classification algorithm is used to classify the crime into subcategories of classes with geographical factors, and temporal factors from RSS feeds. Mallet package is used for extracting the keywords from the newsfeeds. K-means algorithm is used to identify the hotspots in the crime locations. KDE algorithm is used to identify the density of crime, and also our approach has overcome the challenges in the existing KDE algorithm. The outcome of research validated the proposed crime prediction model with that of the ARIMA model and found equivalent prediction performance.

Keywords Social media · Crime analysis · Crime prediction · Hotspot detection · Crime density

B. R. Prathap (✉) · A. V. N. Krishna · K. Balachandran
Computer Science and Engineering, CHRIST (Deemed to be University), Bengaluru, India
e-mail: boppuru.prathap@christuniveristy.in

A. V. N. Krishna
e-mail: adapalli.krishna@christuniversity.in

K. Balachandran
e-mail: balachandran.k@christuniversity.in

© The Author(s), under exclusive license to Springer Nature Switzerland AG 2021
Y. Maleh et al. (eds.), *Artificial Intelligence and Blockchain for Future Cybersecurity Applications*, Studies in Big Data 90,
https://doi.org/10.1007/978-3-030-74575-2_16

1   Introduction

Crime analysis using data is the emerging discipline in criminology. Law enforcement agencies are focusing on methods that enable them to predict future attacks. This enables them to utilize their limited resources effectively. These agencies' major challenges are the complexities involved in the processing of large volumes of data. The variety of geographical diversity and the complexity of crime data have made crime analysis difficult. Researchers are focusing their efforts on data mining algorithms that can extract meaningful information from crime data.

There are various sources of crime data. In this research, we have used news feed data about crime in the Bangalore region. Social media data has rich data about user emotions on a particular topic. The main advantage of this kind of data is that it has precise spatial and temporal coordinates. These spatial-temporal data can be used for crime prediction which can be processed with linguistic analysis and statistical topical modeling. Social media data can be used as auxiliary sources and traditional data sources to increase the accuracy of the prediction. However, there are also limitations in using social media data. Tweets have inconsistent information, misspellings, fly word invention, syntactic structures and symbol use that computational algorithms cannot handle. Even though the content has real-time, personalized content, it is difficult to process. Newsfeed has been used as the primary source of data in this paper. The data needs to be processed with automated text analysis, smart segregation and filtering methods.

The spatiotemporal analysis provides insights about the situational awareness of local events, enables understanding of the severity, consequences and the time-evolving nature of the crime. The spatiotemporal analysis has been done based on volume based importance. In this case, the messages are extracted from news feed data and then filtered and sorted by space and time. One of the significant challenges in using news feed data is that the critical information is largely obscured by large volumes of incomplete, inconsistent and inaccurate data.

2   Related Work

Crime is considered one of the biggest threats to the development of a country. The understanding of crime behaviors has been limited until the advent of big data. Crime generally occurs in clusters. This has direct implications on the crime prevention strategies of law enforcement agencies. Criminal activities are on the rise in Bangalore's major cities (Authorized data-Karnataka State Police [19]). This has been attributed to the population density, urban immigration and the existence of slums in the city area. By understanding the factors that drive criminal behavior in spatial and temporal terms, law enforcement agencies can identify crime clusters and take appropriate action.

Research conducted by Algahtany et al. [1] shows that crime happens due to conditions called crime generators. The first success of crime encourages the criminal to conduct the activity repeatedly within their surroundings. Criminals create a safety zone around their surroundings and then gradually expand the area. They do serial offending and repeat victimization. It is found that nearly 68% of the crime happens in the same area. The consistent occurrence of crime in a particular area creates a crime hotspot. This leads to multiple social and economic consequences for the crime clusters, e.g. depreciating house prices, increased fear, etc. The discovery of crime clusters requires focused and prompt police action. By observing the possible stimulants of crime, the risk can be identified.

Therefore big data from urban areas opens up numerous opportunities for conducting advanced investigations on crime [27]. This study also helps in forming and testing various criminal theories developed using criminology to understand the different varieties of criminal phenomena. For example, the methods from environmental criminologies, such as rational choice theory and routine activity theory, suggest that time and space play a significant role in understanding criminal activities [6].

Certain attractors for a crime such as a drug abuse, weather patterns, and specific land uses. Crime is seasonal. It is found that cold season triggers violent crime and hot weather triggers nonviolent crime [22]. There are also complex associations between crime and weather. According to [6], crime peaks at nights, weekends, and holidays.

Along with the research progress in environmental criminology, computerized mapping and spatial analysis of crime events have evolved. Various software systems, such as Geographical Information Systems (GIS) are developed to visualize crime patterns in different geographical regions [9]. The geographical position information is available in the FIR raised by the law enforcement authorities. This data is then utilized to create insights for a better understanding of criminal activities. The spatial analysis techniques and GIS mapping and modeling enable lawmakers and policing agencies to determine the distribution of criminal activity and the likelihood of their reoccurrence. GIS is essential in studying criminal trends and criminal activities [10].

Spatial mapping is a powerful data management tool and provides visual interactions to analyze crime. The environmental analysis can be done on three levels: Micro, meso, and macro. Microanalysis focuses on specific crime sites, meso analysis identifies crime patterns at the neighborhood level, and macro analysis compares crime distribution across countries. Crime pattern theory states that crime does not occur uniformly in space and time. There are definite crime hotspots [17]. Spatial analysis can help identify the hotspots based on the movement of people, their daily activities, places they go and the areas where they live and work. Some locations such as sports stadiums or shopping areas are crime generators that are more suitable for criminal offenses. Others are crime attractors that include clubs and bars which are more prone to victimization [11].

The primary psychological and physical need for urban dwellers is safety. For a city to have sustainable development, there is a need for urban crime prevention methods that are well planned, community-based, gender-sensitive and have extensive city coverage.

Hotspot mapping is the popular analytical technique used by law enforcement agencies. It enables them to identify crime trends and aids in decision making visually. Its applications include an operational briefing of police patrols, measurement, and analysis of crime patterns, performance analysis, intelligence development and crime reduction partnerships [8]. In essence, hotspot mapping helps determine where the next crime will happen to make use of data from the past to predict the future. It uses the principle that the retrospective patterns of crime are the best indicators of future crime patterns. Various mapping techniques are used to identify and explore patterns of crime [2]. These techniques can be simple as representing crime data as points and visualizing their geographical distribution, use of Geographical information systems (GIS) to shade areas or represent the crime distribution using volumetric densities of geographic distribution. Different hotspot techniques produce different results in terms of size, location, and shape of areas identified as hotspots.

Kernel density estimation is one method that can be used to visualize complex event data distributed in a particular region [15]. With the help of the visualizations, it is possible to generate insights and trends in the data. This has an advantage over numerical information. Crime data from sources such as news feeds can be used with kernel density estimation. Manna et al. [12] implemented kernel density estimation in R package. He compared the utility of these methods to visualize crime data over time. KDE provides a useful and effective visualization of data. It can also be used for exploratory data visualization purposes.

The other name for kernel density estimation is a "nonparametric" method. It is used for summarizing the data gathered over multiple dimensions. It can be used in various ways such as identifying underlying trends in data and estimating the density function. Kernel density estimation applies to observed data [15]. It is a descriptive measure and provides approximate predictions for future data. In this method, the function known as the kernel is applied to each data point. It averages the point location concerning that of other data points. It can be extended to multiple dimensions and is effective in estimating geographical density. Kernel density estimation is used to visualize spatial data patterns such as crime density in a particular region. Studies from Boppuru et al. [25] have explored the Geo-spatial crime analysis using news-feed data in Indian context discussed on analysis of Crime using different machine learning algorithms.

Kernel density estimation helps in detecting the crime hotspot in the city. The spatial pattern of each crime point is measurable using Kernel density estimation. The strength is estimated by counting the events in a unit area [20]. It can also be calculated using many events in the circle, slide circle to statistic and then divided by the circle area. ARIMA is used to capture even the complex relationships since it can take error terms and observations of the lagged terms. These models are based on regressing a variable on past values [16]. The ARIMA model's essence is that past time points of time series data can impact current and future time points. ARIMA models use this concept to forecast current as well as future values. ARIMA uses several lagged observations of time series to predict observations. A specific weight is applied to each of the past terms, and the weights can vary based on how recent they are.

Studies from Radcliffe have explored the constraints of spatiotemporal constraints in crime. To analyze the spatial and temporal patterns in crime data, it is necessary to have quantitative tools from physics, mathematics, and signal processing. Toole et al. [18] have identified the presence of multi-scale complex relationships of crime data with both space and time. Prathap et al. [5] have explored the different heuristic algorithms for predicting and analysing crime using social media data and discusses different sources of social media can be considered for crime analysis and forecasting. Studies from [7] have explored the different methods could be used for crime density identification using KDE and K-Means.

[3] derived a finite set of rules with the help of fuzzy association rule mining on demographic information data. [13] have extracted useful insights using kernel density estimation. [14] have used a self-exciting point process model for modelling crime data. However, the main problem with all of these methods is that they cannot use in areas for which no data is available since they rely on the historical information of a crime. [4] suggested a revolutionary method to study users' Twitter opinions about the same crime case tweets shared by active users, thereby defining improvements in public options and the distribution of emotions across various types of crimes.

A study performed by Kumar et al. [24] examines the relationship between crime and places in Saudi Arabia. It uses geographic information systems to identify and visualize the spatial distributions of regional and national crime rates in Saudi Arabia. The crimes that haied include assault, murder, theft, alcohol, and drug crimes over ten years, i.e., from 2003 to 2012. The role of "place" in crime analysis has become increasingly important in the area of environmental criminology. The spatial distribution of crime reflects the different organizational structures within the community. The focus of ecological criminologists includes spatial analysis rather than criminogenic causes such as developmental, biological and social characteristics of an offender.

The work by [29] has developed a practical method to implement ARIMA models. This method works in three iterative steps. It includes model identification as the first step, parameter estimation as the second step, and diagnostic checking as the third step. Model identification ensures that the time series generated will have auto correlational properties [30]. The data is transformed into the model identification step to make the stationary time series. Once the approximate model is developed, parameter estimation is done to reduce the overall amount of errors. The model adequacy is then checked with the help of diagnostic checking. This ensures that the model's future predictions fit with the historical data. This three-step iterative process is performed multiple times to identify the right model fit. The final selected model can then be used for prediction purposes [31].

Various machine learning algorithms can be applied to datasets—they are Functions, Bayes, Meta, Lazy and Multi-instance (MI), trees and rules. The research done by Ngai et al. made use of unnormalized crime dataset to analyze communities. The following are some of the machine learning algorithms [21]. In this algorithm, linear regression is utilized for the prediction of events. It is a simple regression method and describes the relationship between the input and output that can be interpreted easily.

[32] conducted a spatio-temporal analysis to understand urban crime using multi-source population sensed information, namely crime data, local meteorological data, POI distribution, and commuter trips. Explicitly, they present, for the first time, monthly temporal patterns and the geographical extent of crimes. They therefore investigate the spatial-temporal association using meteorological data and they also notice that overcast conditions will be more suspicious than other climatic conditions.

[33] performed the research on the effect of varying grid resolution, time resolution, and historical time frames on crime forecast results. To investigate this, they evaluate home burglary data from a large city in Belgium and forecast new crime incidents using a range of parameter values, comparing the effects of predictions.

2.1 Motivation and Objective of the Research

India is a rapidly urbanizing country in the world. United Nations predicted that about 86% of the developed world and 68% of the developing world would be urbanized by 2050. This means that the total urban population will be more than that of today's world population in the future. A large number of rural people are migrating to city centers. For example, the crime rates have increased from 2300 to 3000 for every 12,000 residents according to data from 1980 to 2000 [23]. Researchers have shown a close relationship between the sustainable development of cities and the quality of life of urban citizens. The primary psychological and physical need for urban dwellers is safety. For a city to have sustainable development, there is a need for comprehensive safety strategies and urban crime prevention methods that are well planned, community-based, gender-sensitive and have wide city coverage.

We are mainly focusing on India and Bangalore crime data. After reading the literature review about the various types of social media data being used for crime prediction, we got the motivation. Then we decided to consider the newsfeed data. We have narrowed down to Bangalore, Karnataka, one of India's most populous metro cities. It is also one of the top 10 cities with high incidents of crime. To identify the crime rates, we have used the government website statistics from 2008 to 2019 till monthly date reports published in [19]. Based on the motivation, we decided which area to consider, content to take, sources, etc.

Based on the motivation, the research identified the objective of creating a crime analysis framework using social media (Newsfeed data) concerning spatial and temporal data. Based on the main objective, the sub-objectives are:

To develop a crime data visualization tool that can portrait Crime Density in the Indian context.
To classify various crime data for effective investigations.
To predict and evaluate the crimes using forecasting techniques.

Table 1 Crime keywords classification

Category of crime	Crime keyword
Drug-related crimes	Drug Trafficking, Drug dealing, Drugs smuggling, Narcotics, drugs, and alcohol
Violent crimes	Rape, Murder, Terrorism, Kidnapping, Assault, Sexual Harassment, Sexual assault, Homicide, Gunshot, Intentional Killing peoples, Shootout, Gang-rape, Attempt to murder, Sexual abuse, Putting to death
Commercial crimes	Official Document Forgery, Currency Forgery, Official Seal Forgery, Official Stamp Forgery, Bribery, Counterfeiting, Cheating
Property crimes	Arson, Motor vehicle theft, Theft, Burglary, Robbery, Riots, Criminal breach of trust, Stealing, Barrage fire, Bombardment, Electric battery, Shelling, Looting, Embezzlement, Trespass, Incendiarism, Shoplifting, Vandalism
Traffic offences	Speeding, Signal Jump, Running a Red Light, drunk and drive
Other offences	Prostitution, Illegal Gambling, Adultery, Homosexuality, Weapons violation, Offense involving children, Public peace violation, Stalking, Cheating, Hurt, Counterfeiting, Dowry deaths, Outrage her modesty, Causing death by negligence, Suicide, Criminal damage

2.2 Identifying the Problem Based on Literature

Urban crime has become one of the vital problems for modern cities due to immigration and population growth. Law enforcement agencies collect vast amount of data to model and predict crime. However, they lack real-time information about crime. Social media is a data source that can model crime in real-time and predict it. Based on research studies, spatial and temporal data gathered from social media can be utilized for prediction & analysis of crime. Different social media sources can be used for Spatio-temporal crime analysis like News feeds, Twitter, Facebook, Sample data sets, Police data etc. [28].

The study says that the crime rate is increasing day by day, which demands Spatio-Temporal visualization techniques such as hotspots detections, Density identification and Forecasting for better Crime investigations. We have identified the different crime data characteristics such as types of crimes (68 Crime keywords-6 Subclasses), Newsfeeds, Frequency of crime, Geographic Locations, and Temporal facts. We focus on utilizing the newsfeed data effectively to predict crime.

2.3 List of Crime Keywords Considered

We have taken reference from the research done by [18]. Author has classified crime into 82 types. This research crime count is reduced to 68 classes of crime after analyzing the Indian crime data from (Authorized data-Karnataka State Police 2018), which is shown in Table 1.

3 Methodology

The paper focuses on developing an IT system that can automatically collect news feed data related to crimes in India and Bangalore, Karnataka. The feeds are then preprocessed and filtered based on the 68 types of crime and location. The news feed is then converted into XML, a machine-readable format. In this step, the raw data is converted into rich information. The data that is inconsistent, incomplete and lacking in predictable behavior or trends are processed. The data is then cleaned, and the features are extracted in the preprocessing steps. The system then classifies the data depending on the feature set. Visual hot-spots are given as output of the analysis.

Figure 1 shows the framework of proposed work. This framework is used to iden-

Fig. 1 Proposed framework

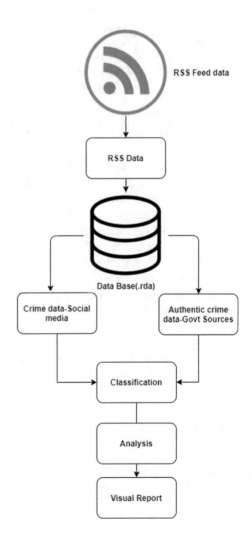

tify the pattern information and data distribution. In our research, we have attempted to combine different data analytics algorithms with statistical methods. Different statistical approaches were used in this study, such as textual data mining, factor analysis, and functional data analysis. Data mining is the technique used to process large amounts of crime dataset. It has enabled to extract of useful information about crime patterns that the police can use.

3.1 Implementation of the Process

This system architecture in which the news feed data are taken from various news websites. These news feed content are then scrapped for relevance to the 68 crime types such as theft, robbery, drunkenness, etc. The system architecture is as follows. Figure 2 explains the detailed framework of research work. The detailed framework consists of three modules namely:

1. Data mining, cleaning, and exploratory data analytics
2. Preprocessing and classification
3. Geospatial analysis and visualization detailed explanation is given below.

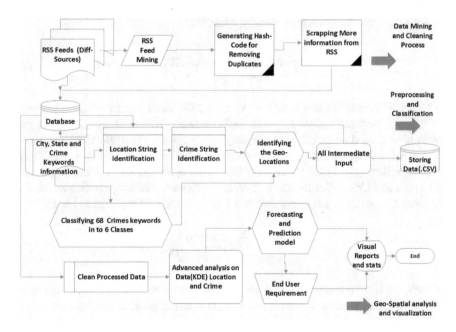

Fig. 2 Detailed framework

3.2   Proposed Analytic Approach

Detailed description of the analytical approach Explained in the following subsection (3.2.1).

3.2.1   Kernel Density Estimation

Kernel density estimation is a non-parametric way to estimate the probability density function of the random variable. KDE is used to smoothen the density of the points. The bandwidth of the kernel is a free parameter that has a strong influence on the resulting estimate. KDE can be used to visualize the shape of some data. The bandwidth affects how smooth the resulting curve is. KDE is calculated by weighing the distance of all the data points. Changing the bandwidth changes the shape of the kernel.

$$d(s) = \#S \in \pi r^2 \tag{1}$$

In Eq. 1, r is the circle radius, C(s, r) is the circle center, and #S is the event count (crimes) in a circle.

   Thus the definition of Kernel density estimation:

$$f(x) = \frac{1}{nh} \sum_{i=1}^{n} k(\frac{x - x_i}{h}) \tag{2}$$

In Eq. 2, h–Bandwidth and h > 0, x- Variable, Xi- Mean and (x − Xi) represents the distance between estimated points and events Xi. Where Xl, ... Xn is the randomly selected newsfeed data sample, Kernel function is depicted by k(), (x − Xi) gives the distance between the event Xi and the estimated points. The existing KDE algorithm (Eq. 2) Identified following two major problems 1. Identification of more potential Crime Geo locations. 2. Visualization of specific crime Geo locations. To solve this problem proposed modified a analytic method depicted in Eq. (3).

$$f(x) = \frac{1}{nh} \sum_{i=1}^{n} k\left(\frac{x - x_i}{h_i^2}\right) h_{mean} \tag{3}$$

In Eq.3, h-Maximum distance covered, n-Total number of crimes, x-Specific crime density, x_i-Crimes mean, k-Kernel function, h_i-Specific geographic distance.

   From the formula, it is found that bandwidth influences KDE. The point density change is smooth when h_i increases and the change are rough when h_i decreases. Kernel density estimation is derived from the moving window and is represented by the point process smooth intensity. In this Research, we had identified 6 classes of crimes mapped and identified the density of crimes using KDE. The h value is modified to get accurate latitude and longitude values of the crime hotspots. The proposed method result is verified and validated with public government data (Karnataka State Police 2018) and ARIMA Time series model.

4   Results and Discussion

4.1   Geo Spatial Crime Visualization (Hotspot Detection) Using Naïve Bayes and K-Means Algorithms–India

From Fig. 3 we can understand that cities such as Delhi, Bangalore, Hyderabad, etc. have a larger crime concentration. Figure 4 shows the crime statistics which shows violent crime are the most committed crime across India. The concentration of other crimes such as burglary, fraud, kill, murder etc. is more concentrated in the cities than other locations. Assault and gambling constitute 42% and 38% of the crimes committed in India. Kernel density estimation algorithm is used to identify the density of all crimes in India and Bangalore.

Figure 5 depicts the State-wise crime analysis statistics. The analysis was done based on location state-wise shows that Andhra Pradesh and Delhi are the top states for crime incidents. This shows that a high crime density is present in urban areas compared to rural areas in India.

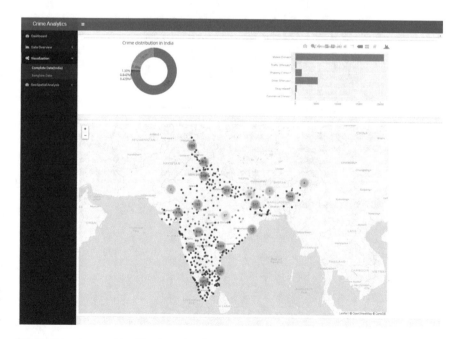

Fig. 3   Crime hotspot identification using K-Means algorithm-India

Fig. 4 Crime statistics-India

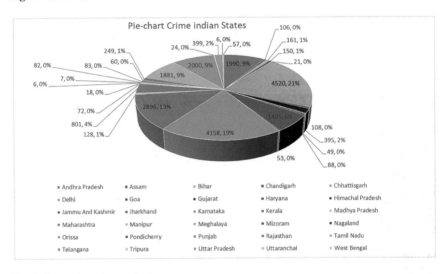

Fig. 5 State-wise crime analysis

4.2 Geo Spatial Crime Visualization (Hotspot Detection) Using Naïve Bayes and K-Means Algorithms–Bangalore

Figure 6 gives the overall picture of the crime rates in Bengaluru city with Geo-location wise. Figure 7 shows the analysis of various crimes that occur in the Geolo-cation wise. It has been found that Violent-crimes are a more reported crime in the city.

Figures 6, 7, and 8 analysis says that violent crime is the top crime in Bengaluru city with 46%. Property crimes constitute another 12% in the city. We can gain these insights with the help of Hindu news feeds. This analysis can help police officials to plan their patrolling activities.

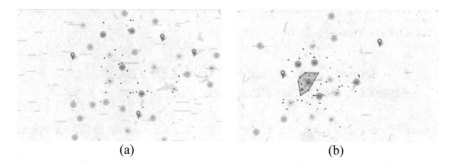

(a)   (b)

Fig. 6 **a** and **b**. Geo spatial crime hot spots in India using KNN Algorithm-Bengaluru

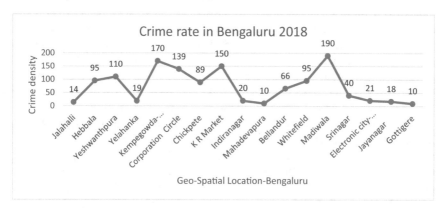

Fig. 7 Geo-spatial crime density-Bengaluru

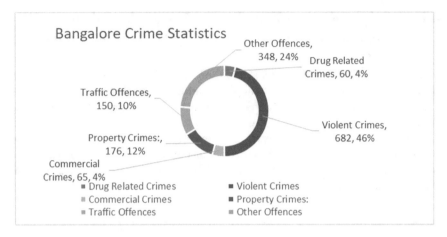

Fig. 8 Crime statistics-Bengaluru

4.3 Geo Spatial Crime Density Analysis Using KDE Algorithm–India and Bangalore

The analytics system has the facility to see the visualizations for individual crimes for both India and Bangalore. Figure 9 shows the crime density analysis of the crime data from India. It is found that Karnataka, Delhi, and Andhra Pradesh are the top states for violent crimes. There are also other crimes such as theft, burglary, etc. that are more prevalent in Kerala. The probable reason could be due to dense population in these areas. 68 types of crime are used such as drunkenness, burglary, theft, assault, fraud, gambling, harassment, killing, molestation, suicide, trespass, robbery, warrant, vandalism, murders and hurting others, etc. 68 types of crimes classified into 6 classes for ease of visualization and Comparative analysis of the shift in criminal activity over one year is calculated. It also shows the clustering of various crime types concerning commercial areas. The user can choose the crime type from the drop-down menu and get the corresponding visualization.

Figure 10 shows the crime density for all 6 crime classes such as Drug-Related Crimes, Violent Crimes, Commercial Crimes, Property Crimes, Traffic Offences, and Other Offences in the context of Bengaluru city. The total density of all 6 crimes identified as 1544. Figure 10 also shows that the central region of Bengaluru more on specific Geographic locations like Yeshwanthpura, Kempegowda Majestic, Corporation Circle, K R Market, Chickpete, etc. are most affected by crime. This is probably because the population density is higher in these Geographical areas and due to lot of moving population.

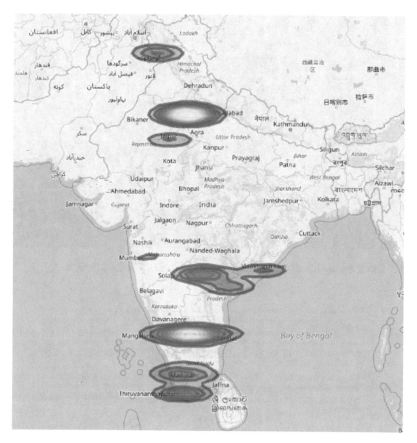

Fig. 9 Crime density identification using KDE for all 6 crime classes-India

4.4 Time Series Analysis Using ARIMA Model

4.4.1 Forecasting Analysis - 1 Day–India

Figure 11 shows the time series forecasting of crime occurrences in India for 6 h. The graph represents 2 types of lines dotted and thick line. The dotted line represents the prediction of crime with respect to time in the Indian context. Thick line represents the crimes identified in a specific period. The data can also predict individual crime level for 1 h, 1 day and 1 month, etc. It is found that the crime occurrences are more in the March time period.

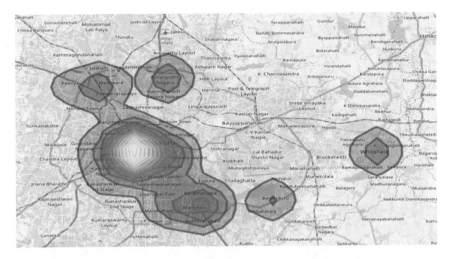

Fig. 10 Crime density identification using KDE for all 6 crime classes-Bangalore

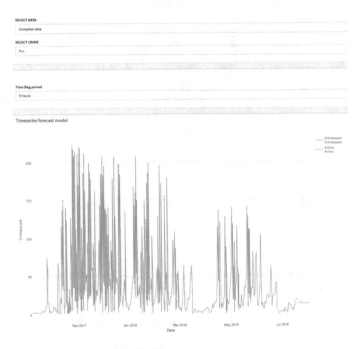

Fig. 11 Crime forecasting analysis India-6 h

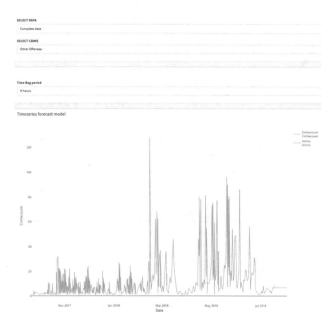

Fig. 12 Time series analysis one day (Bangalore)

4.4.2 Forecasting Analysis - 1 Day–Bangalore

Figure 12 gives the time series forecasting using Arima model of all crime in Bangalore. Forecasting of individual crimes can also be done in the system.

4.4.3 Validation of Newsfeed Crime Data with RTI (Government Authentic) Data

This Research work validated with the Government authenticated data (Karnataka State police 2018). Figure 13 types of Crime heads compared with actual crime count from news feed and Karnataka state police data, which is publicly available Crime head count. The major finding is that the densities of crimes do not match RTI (Karnataka State police 2018) data to the proposed model. Still, the sequence matching like Theft is highest in both graphs and counterfeiting is lowest in both figures. As per finding the result the sequence of crimes follows Theft, Assault, Cheating, Burglary, Kidnapping and abduction, Robbery, Molestation, Narcotic drugs, Riots, Murder, Suicide, Dowry Deaths, Counterfeiting. Sequence of Crime heads is matching in both the data.

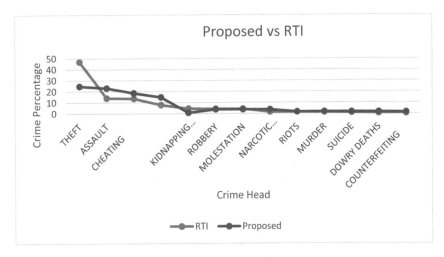

Fig. 13 Validation of proposed model crime data with RTI crime data

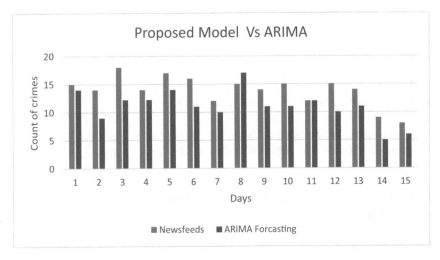

Fig. 14 Proposed model vs. ARIMA model

4.4.4 Findings in the Validation Work

Based on the validation of newsfeed data with RTI data the following point's identified. The density of Theft is high in both Newsfeed data and RTI data. The density
of Counterfeiting is low in both Newsfeed data and RTI data. The density crime
sequence is matching with more than 95% accuracy.

4.4.5   Hot Spot Validation

According to the news feeds data's highest crime news rate identified in Corporation Circle, Kempegowda and Bellandur are more density of crime identified. Based on RTI Data highest FIRs filed, City Market Police station, Whitefield, Commercial Street, K R Puram is in the highest rate of crime rate hand as been identified was almost equal geospatial values with the research work proposed.

4.4.6   Validation of Proposed Model Forecasting with ARIMA Model

Figure 14 Proposed model crime count with ARIMA forecasting values. The results from the model data have been compared with that of actual news feed data. It is found that the model accuracy is about 77.49. This is a fairly accurate model of crime prediction.

5   Conclusion

In this research, 1-year crime data has been used under the context of Indian and Bangalore crimes. Total 68 types of crime keywords are identified, and they are classified into 6 groups. The quality of the input newsfeed data has been compared and validated with that of RTI data. KDE is used for density analysis and compared with the ARIMA model. Proposed model predicts the possible crimes for the time span of 6 h, 1 day, 3 Days, 1 Week, 1 Month to help the crime authority take preventive measures well in advance. Our results have shown that the news feed data can be used for extracting spatiotemporal information about the prediction performance of 68 types of crime. We have achieved a prediction accuracy of 77.49% with our crime prediction models. We have validated our crime prediction model with that of ARIMA model and found equivalent prediction performance. In the future, this work can be extended to topic modeling in text analysis to reduce the false acceptance ratio. Additional features such as socio-economic characteristics of the population can be included in the news feed analysis. With the help of our system, police authorities in Bangalore can deploy their resources effectively. This application will result in a reduction of effort and improvement in crime response rates.

References

1. Algahtany, M., Kumar, L.: A method for exploring the link between urban area expansion over time and the opportunity for crime in Saudi Arabia. Remote Sen. **8**(10), 863 (2016). https://doi.org/10.3390/rs8100863

2. Mafumbabete, C., Chivhenge, E., Museva, T., Zingi, G.K., Ndongwe, M.R.: Mapping the spatial variations in crime in rural Zimbabwe using geographic information systems. Cogent Soc. Sci. **5**(1), 1661606 (2019). https://doi.org/10.1080/23311886.2019.1661606

3. Buczak, A.L., Gifford, C.M.: Fuzzy association rule mining for community crime pattern discovery. In: ACM SIGKDD Workshop on Intelligence and Security Informatics (ISI-KDD 2010), pp. 1–10. Association for Computing Machinery, New York, NY, USA (2010). https://doi.org/10.1145/1938606. Article 2.

4. Prathap, B.R., Ramesha, K.: Twitter sentiment for analysing different types of crimes. In: 2018 International Conference on Communication, Computing and Internet of Things (IC3IoT), Chennai, India, pp. 483–488 (2018). https://doi.org/10.1109/IC3IoT.2018.8668140

5. Prathap, B.R., Ramesha, K.: Geospatial crime analysis to determine crime density using Kernel density estimation for the Indian context. J. Comput. Theor. Nanosci. **17**(1), 74–86 (2020). https://doi.org/10.1166/jctn.2020.8632

6. Angers, J., Biswas, A., Maiti, R.: Bayesian forecasting for time series of categorical data. J. Forecast. **36**(3), 217–229 (2016). https://doi.org/10.1002/for.2426

7. Prathap, B.R., Rameha, K.: A pragmatic study on heuristic algorithms for prediction and analysis of crime using social media data. J. Adv. Res. Dyn. Control Syst. **11**(2), 30–36 (2019)

8. Catlett, C., Cesario, E., Talia, D., Vinci, A.: Spatio-temporal crime predictions in smart cities: a data-driven approach and experiments. Pervasive Mob. Comput. **53**, 62–74 (2019). https://doi.org/10.1016/j.pmcj.2019.01.003

9. Chae, J., Thom, D., Bosch, H., Jang, Y., Maciejewski, R., Ebert, D., Ertl, T.: Spatiotemporal social media analytics for abnormal event detection and examination using seasonal-trend decomposition. In: 2012 IEEE Conference on Visual Analytics Science and Technology (VAST), 143-152 (2012). https://doi.org/10.1109/VAST.2012.6400557

10. Chainey, S., Tompson, L., Uhlig, S.: The utility of hotspot mapping for predicting spatial patterns of crime. Secur. J. **21**, 4–28 (2008). https://doi.org/10.1057/palgrave.sj.8350066

11. Clancey, G., Kent, J., Lyons, A., et al.: Crime and crime prevention in an Australian growth centre. Crime Prev. Community Saf **19**, 17–30 (2017). https://doi.org/10.1057/s41300-016-0012-1

12. Manna, M., Das, P., Das, A.K.: Application of deep learning techniques on document classification. In: Nguyen, N.T., Chbeir, R., Exposito, E., Aniorté, P., Trawiński, B. (eds.) ICCCI 2019. LNCS (LNAI), vol. 11683, pp. 181–192. Springer, Cham (2019). https://doi.org/10.1007/978-3-030-28377-3_15

13. Kamalov, F.: Kernel density estimation based sampling for imbalanced class distribution. Inf. Sci. **512**, 1192–1201 (2020). https://doi.org/10.1016/j.ins.2019.10.017

14. Mohler, G.O., Short, M.B., Brantingham, P.J., Schoenberg, F.P., Tita, G.E.: Self-exciting point process modeling of crime. J. Am. Stat. Assoc. **106**(493), 100–108 (2011). https://doi.org/10.1198/jasa.2011.ap09546

15. Gerber, M.S.: Predicting crime using Twitter and kernel density estimation. Decis. Support Syst. **61**, 115–125 (2014). https://doi.org/10.1016/j.dss.2014.02.003

16. Hiropoulos, A., Porter, J.: Visualising property crime in Gauteng: applying GIS to crime pattern theory. S. Afr. Crime Q. **47**(1), 17 (2014). https://doi.org/10.4314/sacq.v47i1.2

17. Hu, T., Zhu, X., Duan, L., Guo, W.: Urban crime prediction based on spatio-temporal Bayesian model. PLoS ONE **13**(10), (2018). https://doi.org/10.1371/journal.pone.0206215

18. Jameson, L., Toole, N.E., Plotkin, J.B.: Spatiotemporal correlations in criminal offense records. ACM Trans. Intell. Syst. Technol. **2**, 4, 18 (2011). https://doi.org/10.1145/1989734.1989742

19. Karnataka Crime statistics, 10 August 2018. https://ksp.gov.in/Page.aspx?page=Crime+Statistics+of+Karnataka. Accessed 15 August 2019

20. Saravanakuma, S.: Crime mapping analysis: a GIS implementation in Madurai city. Int. J. Sci.. Res. (IJSR) **5**(3), 1894–1897 (2016). https://doi.org/10.21275/v5i3.nov162301

21. Marzan, C. S., Baculo, M.J., Bulos, R.D., Ruiz, C.: Time series analysis and crime pattern forecasting of city crime data. In: Proceedings of the International Conference on Algorithms, Computing and Systems - ICACS 2017 (2017). https://doi.org/10.1145/3127942.3127959

22. Mburu, L.: Spatiotemporal Interaction of Urban Crime in Nairobi, Kenya. GI_Forum 2014 – Geospatial Innovation for Society (2015). https://doi.org/10.1553/giscience2014s175

23. Mcclendon, L., Meghanathan, N.: Using machine learning algorithms to analyze crime data. Mach. Learn. Appl. Int. J. **2**(1), 1–12 (2015). https://doi.org/10.5121/mlaij.2015.2101

24. Algahtany, M., Kumar, L., Barclay, E., Khormi, H.M.: The spatial distribution of crime and population density in Saudi Arabia. Crime Prev. Community Saf. **20**(1), 30–46 (2017). https://doi.org/10.1057/s41300-017-0034-3

25. Boppuru, P.R., Ramesha, K.: Geo-spatial crime analysis using newsfeed data in Indian context. IJWLTT **14**(4), 49–64 (2019). https://doi.org/10.4018/IJWLTT.2019100103

26. Jayaweera, I., Sajeewa, C., Liyanage, S., Wijewardane, T., Perera, I., Wijayasiri, A.: Crime analytics: analysis of crimes through newspaper articles. In: 2015 Moratuwa Engineering Research Conference (MERCon), Moratuwa, pp. 277–282 (2015). https://doi.org/10.1109/MERCon.2015.7112359

27. Ohlan, R.: Are regional crime rates in India natural? Crime, Law Soc. Change **73**(1), 93–110 (2019). https://doi.org/10.1007/s10611-019-09851-8

28. Wang, X., Brown, D.E.: The spatio-temporal modeling for criminal incidents. Secur. Inf. **1**(1) (2012). https://doi.org/10.1186/2190-8532-1-2

29. Box, G.E., Jenkins, G.M., Reinsel, G.C., Ljung, G.M.: Time Series Analysis: Forecasting and Control. Wiley, New York (2015)

30. Nau, R.: The mathematical structure of ARIMA models, 3rd edn. Duke University (2017). https://people.duke.edu/~rnau/411home.htm

31. Irvin-Erickson, Y., La Vigne, N.: A spatio-temporal analysis of crime at Washington, DC metro rail: stations' crime-generating and crime-attracting characteristics as transportation nodes and places. Crime Sci. **4**(1), 1–13 (2015). https://doi.org/10.1186/s40163-015-0026-5

32. Zhou, B., Chen, L., Zhao, S., et al.: Spatio-temporal analysis of urban crime leveraging multi-source crowdsensed data. Pers. Ubiquit. Comput. (2021). https://doi.org/10.1007/s00779-020-01456-6

33. Rummens, A., Hardyns, W.: The effect of spatiotemporal resolution on predictive policing model performance. Int. J. Forecast. **37**(1), 125–133 (2021). https://doi.org/10.1016/j.ijforecast.2020.03.006

Cybersecurity Analysis: Investigating the Data Integrity and Privacy in AWS and Azure Cloud Platforms

Sivaranjith Galiveeti, Lo'ai Tawalbeh, Mais Tawalbeh, and Ahmed A. Abd El-Latif

Abstract The information technology field remains dominant in terms of adopting modern technologies. Additionally, there is increased adoption of the technologies in diverse realms and industries. One such rapidly emerging technology stands with the advancement of cloud computing. Today, cloud platforms are being sought after by a significant number of users and organizations to leverage their operations and productivity. The technology continues to gain more attention in the IT-Business arena. Cloud platforms offer greater flexibility in supporting real-time computation and arises as a more robust framework delivering offerings over the Internet. Amazon Windows Services (AWS) and Microsoft Azure are two key cloud platforms that allow users to utilize the cloud as a source of data storage, access, and retrieval. In the modern period of rapid global technological change, AWS and Azure cloud solutions are broadly adopted as public storage platforms for bulk information systems. Both server platforms offer private, public, hybrid, and community offerings to distinct organizations.

Further, the two cloud technologies can be upgraded to mitigate against information imposition. Different security features allow diverse users to allocate a mutual foundation for storing and accessing data. Systematic appraisals among various characteristics, including platform independence, bulk employment, client requirements, security, size of data size, and other associated assets, are considered.

This study is geared towards examining the infrastructure, platform, and data security issues that occur in cloud technologies, specifically with the application of

S. Galiveeti
University of the Cumberland's, Williamsburg, KY, USA
e-mail: sgaliveeti6276@ucumberlands.edu

L. Tawalbeh (✉)
Department of Computing and Cyber Security, Texas A&M University, San Antonio, TX, USA
e-mail: ltawalbeh@tamusa.edu

M. Tawalbeh
Computer Engineering Department, Jordan University of Science and Technology, Irbid, Jordan
e-mail: matawalbeh18@cit.just.edu.jo

A. A. A. El-Latif
Menoufia University, Shibin El Kom, Egypt
e-mail: aabdellatif@nu.edu.eg

© The Author(s), under exclusive license to Springer Nature Switzerland AG 2021   329
Y. Maleh et al. (eds.), *Artificial Intelligence and Blockchain for Future Cybersecurity Applications*, Studies in Big Data 90,
https://doi.org/10.1007/978-3-030-74575-2_17

Azure and AWS. Individual users and entities desire to experience the agility and scalability attributed to cloud technology platforms. These clients seek to develop and execute novice applications faster with the reduced cost through migration to the cloud platform. Amazon AWS and Microsoft Azure support distinct database management systems. The cloud providers portray different architectures, patterns of resource management, and degrees of complexity of the information systems' enhancers, impacting scalability, performance, and price. While the cloud platforms continue to provide immense data security solutions, major challenges limit their growth and application. Future research needs to focus on more robust solutions and best practices to maintain cloud data security and integrity.

1 Introduction

A rapidly emerging technology in the IT arena occurs with the use of cloud platforms. Currently, the cloud is being used by a considerable number of individuals and organizations when undertaking their everyday tasks. Common examples of cloud platforms include Gmail and Microsoft Office 365 [18]. Declaration of cloud technologies comes with immense benefits, including improved coverage of location, reduced costs system setup and better accessibility. Nonetheless, these benefits are prone to significant limitations that affect cloud platforms' applications, for example, limited resources, lack of skilled staff, and security issues. Given low maintenance costs, coupled with increased adaptability of cloud tools, there is prevalent utilization of these technologies. Different vendors offer different products.

For the cloud technology framework, sensitive information is generated from a vast collection of spheres [15]. An elaborate example is health data, which demands cloud computing environments that are highly secure. Given the growth of cloud technology in recent times, confidentiality and information safeguard needs continued to shift, thereby protecting users against disclosure of information and surveillance. Protective regulations require the sustenance of confidentiality when dealing with personally identifiable data [15]. In the last few years, considerable efforts have been directed towards utilizing various approaches to improve the privacy of data and enable more secure cloud solutions. Examples of strategies applied in advancing these platforms' greater security include multi-party computing, anonymization, genuine solution module, and encryption [27]. Nonetheless, there remains a key challenge regarding how to appropriately develop usable privacy-sustenance cloud platforms to manage sensitive data in a secure manner. Two main aspects leading to the arising concerns are existing privacy and data protection legal pressures and limited or lack of acquaintance with different security resolutions necessary for establishing robust cloud technologies.

The information technology sector remains dominant in terms of applying the latest technologies within a rapidly evolving business setting. Today, the internet arises as an important tool that enables different users to share resources in a rapid manner. With the increased demand of the Internet, a new networking and computing period has erupted and currently becomes clear within the cloud computing domain [28]. These technologies have advanced the capacities of organizations as well as the control mechanisms for storing data in various devices. As a service-targeted framework, cloud computing includes a number of capabilities, for instance, web 2.0 and distributed ledgers. These capabilities allow cloud computing to provide users with scalable capacities for data storage and greater accessibility to resources from any geographic location. The study focuses on data security and integrity management for cloud technology service providers, particularly Microsoft Azure and Amazon Web Services (AWS).

Cloud technologies arise as the future form of innovation based on the internet. Cloud computing offers customizable and simplified services for clients to access documents and integrate various cloud solutions [25]. With cloud technology, users find a means of storing and accessing information within cloud environments from any geographic location by simply linking to related applications via the Internet [13]. Users decide on the service providers to select for information storage and the type of information to store. Recently, alluring characteristics of cloud technology have accelerated the integration of cloud settings in the information technology sector. There have been extensive reviews of associated innovations by both the IT sector and the scholarly field. With a simplified payment system, coupled with an on-demand process, the organizational computing model is being transformed [12]. Enterprises have shifted from on-premise to off-premise databases, which are available via the internet and controlled by cloud service providers. While cloud computing plays a critical role in future systems due to offer immense advantages, security challenges are generated, and it is considered a developing concern [34]. Various studies have focused on related security issues.

Over 33% of companies worldwide store information in cloud landscapes, either on private or public platforms [9]. As corporate adopt multi-cloud designs, where are malicious attackers who take advantage of existing susceptibilities resulting in misconfiguration of infrastructure. An all-inclusive protection approach can help address the threats within these multi-cloud systems and allow firms to realize technology-related benefits. Cloud environments, including Microsoft Azure and AWS, continue to offer users greater rewards than other data storage environments [9]. The number of multi-cloud users is forecasted to expand rapidly in the long-run as more individuals and organizations continue to adopt cloud technologies.

Interestingly, over 20% of documentation available in cloud environments includes sensitive data, for instance, intellectual property [29]. Given an insecure cloud platform, this implies the accessibility of such confidential data to hackers. Today, entities operate under set guidelines and regulations that govern the management and control of information. In adherence to such legal frameworks, organizations must understand how to access their data, who accesses it, and what actions are taken following such access [29]. Further, in situations where internal stakeholders such as staff utilize data in an unacceptable manner, companies may face malicious attacks leading to mistrust and possible legal penalties.

The consensus among business stakeholders usually constraints how data is used and the parties allowed access. When inside users transfer private data within cloud environments without authentication, such agreements become violated or disregarded, resulting in legal processes. There is reduced trust on the part of the client as more data is breached. Adoption of cloud technology meets users' needs for IT resources, including services, applications and networks. As an internet-based framework, cloud computing provides the needed convenience and bulk resources for retrieval and use [35]. Information security is on the rise as more and more cloud platforms become adopted and integrated within systems. These challenges have remained the key inhibitors for the utilization of cloud technologies. There is a dire need for future research to focus on potential solutions to mitigate against arising risks and architectural challenges.

This chapter will explore cloud technology infrastructure, features, security, pricing, and governing adherence impact on the choice to accept and use cloud technology service platforms by individual users. Also, we investigate the connection between cloud service infrastructure and acceptance of cloud technology service platforms. More precisely, in this chapter, we are aiming:

- To establish the influence of cloud service features on acceptance of cloud technology service platforms.
- To uncover the correlation between cloud service security and the acceptance of cloud technology service platforms.
- To reveal the influence of cloud service pricing on the adoption of cloud computing.
- To determine whether or not cloud service regulatory framework impacts on adoption of cloud computing.

The study's significance comes from the fact that it includes privacy, confidentiality, technical infrastructure, regulatory frameworks, and solutions. Key contributions of the study include: the development of a clear framework to identify security patterns existing in cloud technology; identification of security-related issues, as well as privacy and confidentiality concerns faced by clients of cloud tools and resources; and providing the existing gaps in this field of study, and the potential solutions aimed at mitigating security risks in cloud platforms. First, a description of the concept of cloud computing and security issues is presented. Next, compliance and regulatory framework are developed depicting the situation for cloud technologies at the moment.

Given the limited research on key solutions, this study also focuses on the existing gaps across different studies related to data security in the cloud. Proposed solutions are given on the basis of the features of Azure and AWS cloud platforms. Other modern strategies, including the integration of machine learning, are recommended for future applicability. Finally, the paper is concluded, with future implications being presented. The significance of this research is to shape decisions pertaining to the choice of efficient cloud technologies by individuals, businesses, and organizations. Such decisions are grounded on typical features that meet the needs of potential users or clients in terms of performance and cost-efficiency, and the potential to assist cloud providers in acknowledging the existing limitations relative to other competitive solutions.

The rest of this chapter is organized as follows: Sect. 2 presents a synopsis of cloud computing technology, includes its definition, history, and infrastructure. Besides, it discusses previous researches gaps. Section 3 investigates several theories and frameworks that allow users insight into how the adoption of new technologies. In Sect. 4, we discuss the chosen two popular cloud service platforms for this research, Amazon Web Services, and Microsoft Windows Azure, then illustrate the applications and benefits of these platforms. Section 5 discusses the security matter in cloud computing platforms includes its issues and patterns. Best data security and integrity solutions provided by AWS and Windows Azure are presented in Sect. 6. Finally, Sect. 7 concludes this chapter and provides some recommendations depends on this study.

2   Literature Review

This second segment offers a synopsis of cloud computing technology. Its definition, history, infrastructure, key features, delivery and deployment models, security patterns, and the two popular cloud service platforms chosen for this research are discussed.

2.1   Cloud Computing

Globally, technology adoption is changing individuals' and organizations' lives in different ways as new developments become available to facilitate the manner in which individuals conduct their activities regularly [24]. Over the years, various entities have devised better and less costly innovations to manage information storage and dependability problems to clients, a term presently referred to as cloud computing. In mid-2000, various U.S. organizations embraced cloud technology to access services as demanded, with the innovation being embraced in other nations [24]. Clients store information and access, it through the web, making it possible for users to access the cloud storage area. Cloud technology has changed the IT arena due to its fast growth and demand. The rapid expansion in conveyed cloud computing has prompted the development of bulk data centers, which entail complex servers.

2.1.1   Definition

There are various definitions of cloud computing as identified from diverse studies. However, the generally accepted definition is attributed to NIST, which refers to cloud computing as "the framework for facilitating simple, urgent network connection to a separate system of PC utilization resources that can be distributed and released swiftly with restricted managerial intervention or client interference" [24].

The primary features which differentiate cloud computing from traditional computing options have been recognized and normally include: pattern on scalability and responsive facilities, buy on-demand delivery of service, payment for utilization of cloud system resources without the open loyalty of cloud clients, mutual and multi-tenancy, and accessibility of all devices via the Internet [24].

2.1.2   History of Cloud Technology

Cloud technology has experienced a fast change in history from the 1960's to the current day and potentially in the future [24]. In the last part of the 1960's, J. Lick-lider, the man credited with encouraging the headway of APRANET, concocted the idea of Intergalactic PC Network, which is comparable to the web today [24]. Later in 1970, virtualization was introduced, which involves running different working frameworks simultaneously in a restricted setting with programming, such as VMware. Subsequently, the development led to the introduction of virtual machines. By 1990, telecoms organizations started offering VPN services by offering clients shared availability to existing architecture [24]. In the mid 2000's, Amazon became the leader of cloud computing, delivering services through Elastic process cloud and less complex storage offering. Further, the company launched the 'pay as use' framework for individuals and organizations. In the late 2000's, Google turned into a major rival in the area of internet business, and by 2006, the organization had delivered its first cloud-based solution known as Google Docs; the tool permits a client to save and share documents accurately with different clients [24].

2.2   Infrastructure of Cloud Technology

The National Institute of Standards and Technology (NIST) arises as a well-known organization globally, given its extensive research within the IT domain. The organization exemplified the five basic features, three services, and four ways clouds are deployed in cloud technology's infrastructure design [24].

Within the cloud computing arena, five key actors or players exist. These include the consumer, the service provider, the auditor, the carrier, and the broker [18]. A cloud consumer is an individual or entity that utilizes offerings from cloud service providers in a business connection landscape. The cloud service provider ensures that related cloud services are made available to potential clients or users. The auditor performs independent evaluations of cloud offerings, security and processes associated with cloud deployment. The connectivity and delivery of cloud services from cloud providers to clients through the fundamental network is the cloud carrier's responsibility [15]. The broker administers the consumption, execution and delivery of cloud services, while promoting positive relationships between cloud providers and consumers. All these parties are involved in the creation and use of data in cloud platforms. For deployment, the categories include private, public, mix and group platforms.

2.2.1 Cloud Computing Delivery Models

In distributing offerings, primary platforms include software-as-a-service, infrastructure-as-a-service, and platform-as-a-service.

2.2.1.1 Software as a Service (SaaS)
In software-as-a-service platforms, all offerings generate form the service supplier. These models of utilization are prone to security risks. SaaS offers services whereby users are not required to manage any operating system installation and setup. The service offeror implements these roles. Examples of SaaS are Email, Customer Relationship Management (CRM), and Games (Fig. 1).

2.2.1.2 Platform as a Service (PaaS)
The platform-as-a-service tool is the service provider's responsibility, with the service consumer only focusing on the data and application. The PaaS offering enables clients to develop an application utilizing the cloud platform supported tools and settings. The user also manages the installation and configuration activity. Examples are web server and Decision support (Fig. 2).

2.2.1.3 Infrastructure as a Service (IaaS)
In view of infrastructure-as-a-service, the service provider is responsible for enabling storage, networking, and server configuration. Further, the provider controls the data, the operating systems and application. IaaS provides organization access to important web design, such as servers, without purchasing and controlling the internet-setting facilities itself. A primary advantage is that users would only have to pay for the

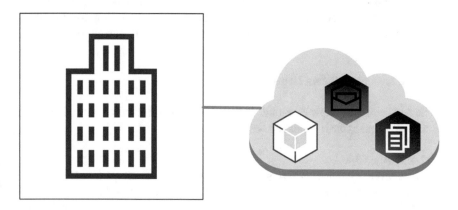

Fig. 1 SaaS model

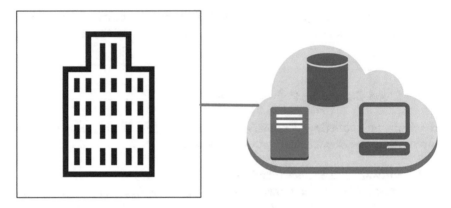

Fig. 2 PaaS model

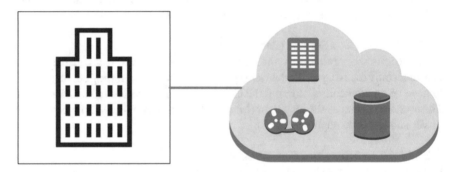

Fig. 3 IaaS model

period of time the offering is being consumed. The platform can be used to prevent buying, storage, and administration of basic operating systems service parts, speedily measuring back and forth to meet demand. Examples include Servers, and Virtual Machine. The five features of the technology are measured offering, asset pooling, wide network access, on-demand self-service, and prompt resistance (Fig. 3).

2.2.2 Cloud Computing Deployment Models

In this section, the various cloud utilization models, including public, private, community, and hybrid clouds are described.

2.2.2.1 Public Cloud
It is the conventional and most regular method for cloud offerings' provision, where a merchant or organization gives different cloud services, specifically SaaS, IaaS and PaaS through the web architecture. Every potential client can access their cloud services by making a conventional application on the web and experiencing the cloud

service provider's enlistment process. For the deployment paradigm, the services are noticeable to all the web clients and open to numerous clients simultaneously. The public cloud foundation crosses public and local geological limits. The administration and control are the obligation of the organization that gives or sells services. Examples of freely accessible cloud solutions are Google AppEngine from Google, Amazon Elastic Compute Cloud (EC2), and Windows Azure Services platform. The administrations present a few preferences to the client as the client just pays for what they use. It can undoubtedly scale to address the client's issues, the application, and related support costs are met by the cloud provider (Fig. 4).

2.2.2.2 Private Cloud
This depicts a restrictive cloud design that gives facilitated services to several users. It is separated from the web or public organizations by a firewall and is accessed by staff of the specific entity. It is assembled and managed by a solitary organization that possesses the cloud framework. There are advantages of actualizing private cloud, which include architecture and software that are custom-fitted to the firm's requirements, the security plan and usage is implemented by the organization, thereby, allowing a sense control (Fig. 5).

2.2.2.3 Community Cloud
In this setup, the existing framework becomes shared by a few associations that together form the network. The administration of the framework might be shared between the organizations through arrangement of a shared management guideline. In some instances, the control might be conducted by an external party on behalf

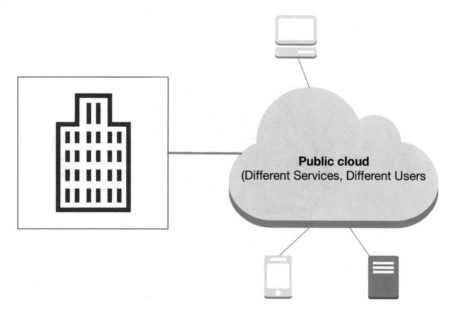

Fig. 4 Public cloud model

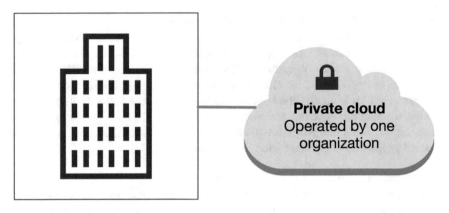

Fig. 5 Private cloud model

of the enterprise that form the community. The community cloud model depicts merits in cost sharing between the concerned entities forming the community and guarantees access to similar data, making joint effort simpler (Fig. 6).

2.2.2.4 Hybrid Cloud
The model encompasses both private and public clouds. It is generally utilized where an entity constructs a private cloud for the most confidential and fundamental services. Further, the hybrid model redistributes cloud offerings for the less-basic

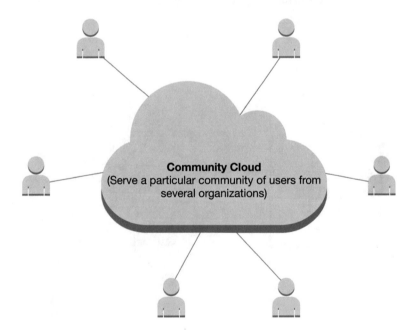

Fig. 6 Community cloud model

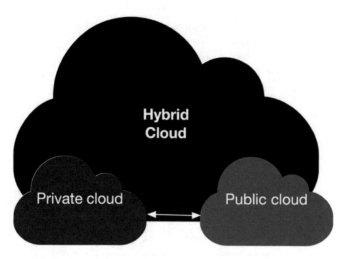

Fig. 7 Hybrid cloud model

services from a public cloud service provider. The model also allows entities to create harmony between making core services and the associated costs. Accordingly, the usage of the hybrid cloud assumes a significant function in minimizing capital costs on the firm's IT architecture execution because a segment of the services needed by the organization is re-appropriated from public cloud suppliers (Fig. 7).

With regard to the deployment frameworks, serious security concerns are evident in the public cloud. This is due to its prevalent internet as a linkage means for connection and its open feature. This implies that policies and guidelines established for security need to be distinct for individual models of utilization. Overall, service providers have a responsibility to manage the infrastructure of cloud and the stored data within these platforms. Given these different functions, security becomes a common role for both the service providers and consumers (Fig. 8).

2.3 Gaps Analysis in Previous Researches

Several past research have directed attention to practical perspectives of information technology by evaluating the implementation of software-as-a-service frameworks to leverage on improved business processes instead of examining the technology infrastructure itself. The concept of enterprise resource planning for applications is a new idea deployed within the landscape of cloud computing [6]. An ERP application allows distinct business processes to be migrated to cloud platforms and customizing solutions to the Independent tasks in a visible man. In this view, a study by Gerhardter and Ortner [14] was geared towards assessing the aspects of success for adopting a software-as-a-service framework. To identify the relevant factors, the researchers examined the implementation of ERP from a traditional and modern viewpoint the

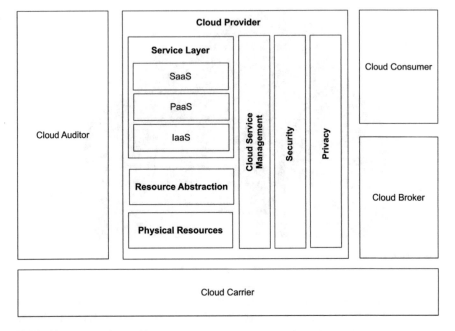

Fig. 8 Cloud computing architecture

latter scenario involved the adoption of cloud platforms for implementing Enterprise resource planning strategies.

In deciding whether to utilize cloud technologies or not, a number of models exist. The models investigate the connection between IT offerings and the right business framework to ensure successful cloud solutions implementation [7]. In implementing cloud solutions primary elements to consider include corporate sustainability classification of organizational model linkage and offerings' portability. Other studies have focused on determining a specific strategy based on a single business process, for example, a supply chain framework that involves the management of supply chain activities within an organization.

There is limited knowledge on clarifying the factors that motivate individuals or organizations to adopt cloud technologies. A major factor that has been considered a significant driving force for adopting cloud technologies is the demand for scalability and flexibility of information technology resources, coupled with demands for maximization of available resources [6]. While cost is an important consideration before implementing cloud technologies, various studies have concluded that it is not a basic requirement any decision-making. With multiple reasons being provided for the criteria taken to adopt cloud technologies, the study has stressed on the cost vs value component. For SMEs, who may have limited funds to invest in cloud technologies, an understanding of the exact value derived from the adoption will be a critical factor in deciding whether or not to utilize the technology [7]. From the study, it is clear that the main factors for acceptance of cloud technology are related to the technical advantages or benefits that are supposed to the savings made on costs. A business viewpoint is necessary before implementing cloud technologies. In the beginning, certain fields portray a potential of success by adopting cloud technologies while other areas may take a considerable amount of time to prosper. According to Bildosola et al. [6] study, key stakeholders need to be assisted by applying a model that showcases cloud technology's evolution pathway. Additionally, each organization needs to acknowledge its pathway within cloud platforms.

According to a study by Nemade et al. [23], cloud users will persistently have greater expectations for the technologies' performance. Given the complex and ever-increasing chain of data distribution in the IT domain, there are many service providers around the world. The concept of cloud computing entails diverse tools utilized to promote capability within a utility estimate software framework. Microsoft's Windows Azure is an elaborate illustration of a cloud platform tool that provides immense advantages to its users. In examining the idea of client utilization of cloud technology the concern of information quality is critical. In particular, sustaining data quality demand that new users to cloud technology have a positive experience. Today, beginning users comprise of a significant proportion of the entire population of cloud users.

In ETAM

Windows Azure is prone to various problems, which calls for effective remedies aimed at ensuring that these platforms are highly secure. A survey conducted on chief information officers and IT executives by the IDC showed that the major issue with cloud computing technology is security [1]. Furthermore, Singh et al. [30] concluded that the controls utilized in cloud platforms are less secure. Service providers in cloud technologies have a huge responsibility of ensuring that data is protected.

The use of cloud tools is on a rapid rise. According to a recent research by Alam [2], an individual user operates at least four applications in a given time with over 40% of business entities operating key applications on public domains. Nonetheless, even with the advent of cloud computing, there are significant security issues requiring providers to identify and evaluate uses and resources for reliable service delivery. In this regard, the Alam [2] study considers misapplication as a major concern for data security and integrity. An example is the use of botnets by malicious attackers, thereby leading to the creation of malware. Several suggestions arise from this study, including a thorough self-examination of user network, as well as a strong process for authentication. Alam [2] study also focuses on the vulnerability of network interfaces. Service consumers use the boundaries to engage with available cloud resources, which implies a need for secure authentication, confirmation, and monitoring of data processing.

Malicious insiders are another major concern for organizations. On this note an entity is prone to malevolent attacks by its staff due to the constraints existing in hiring verification and accessibility two primary assets of the company [2, 19]. As a result, the potential for risks arises due to lack of or poor monitoring and management practices. Possible remedies to prevent such problems include reliable supply chain management utilizing a more transparent process of information security and control and reporting compliance practices. Moreover, the use of cloud technologies is faced with susceptibilities because of shared infrastructure.

Specifically, vendors of Internet as a service platform use single elements which are not readily harmonious with the entire cloud tool or infrastructure [2, 19]. This implies that providers must check and enhance compartmentalization to ensure secure configuration of cloud applications. Further, there is a need for an effective confirmation and access control process complemented with evaluating potential vulnerabilities. Lastly, a key challenge for cloud platforms rests with loss of data. Such situation occurs due to poor data backup and unverified user access, which results in stolen information or unauthorized retrieval. This is a major concern for entities, given that their reputation is affected if proactive measures are not put in place. For this concern, possible remedies would include sustained integrity of information during its transmission, the utilization of strong password keys, and robust control of interface access.

In their study, Kofahi and Al-Rabadi [17] stated that a major problem for enterprises in adopting cloud technologies is the processes and needs demanded for while delivering the model necessary for maintaining the whole system, rather than the technology itself. In this view, the hardware experts and employees' services are not required following the adoption of cloud computing, the utilization team can provide a brief outline of the hardware by taking advantage of information centers. In cloud computing, there is a general inconsistency about data security issues coupled with best practices to completely control associated threats. Future research pertaining to cloud technology susceptibilities as well as an investigation of clear strategies to prevent such issues as critical.

On another note, a recent research by Das et al. [8] has examined the idea of edge computing AS an efficient technical model to address the issues in cloud technology. Given the framework a new layer known as the edge computing layer is included to complement the conventional computing model. Only actual information processing is delivered to the layer with other complex processes being implemented on the cloud platforms [8]. Overall, an edge IT system includes the consumer devices, the cloud server, and the edge computing layers.

Given the distributed nature of information and the integration of tasks on distinct stratums, there are significant issues regarding security performance and confidentiality. Artificial intelligence tools are in rampant use with edge computing events. These techniques are utilized to supplement the three layers and adding to data storage and networking capacities. The edge framework is normally applied for autonomous automotive, surveillance cameras, and smart City [8]. The events are complex as there are diverse client or customer devices, greater heterogeneity of data, and privacy and security concerns.

Further, the events present significant needs for network bandwidth and latency. The edge computing system is presently in its initial phases and lacks a universal standard for such events. This demands an all-inclusive computing benchmark collection that can be used to measure and enhance the applications. There is also a lack of testing capacity with edge frameworks and a lack of inducement to share information due to privacy concerns. These complexities mean that future studies need to develop comprehensive end-to-end application events that authenticate or confirm the system's architectures and algorithms applied in particular environments.

3   Theories of Technology Adoption

It becomes critical for decision-makers to comprehend prevailing problems that affect users' judgment to accept a certain technology system [32]. Such understanding can assist developers in focusing on the development stage of the system. In this regard a primary question for both researchers and practitioners is why do individuals all organizations adopt new technologies? By responding to this question, decision-makers can identify improved design assessment approaches and forecast users' response regarding the fresh innovations. Technology acceptance paradigms have been utilized in various fields to gain a clearer understanding of uses and their behavior. Several frameworks have been established to allow an insight into adoption of new technologies by users.

3.1   Theory of Reasoned Action (TRA)

Initially, TRA was formulated by Fishbeine and Azjen in 1975 for sociology and psychology works. Recently the model has assisted in examining uses information technology adoption behavior. In this view, inhuman conduct becomes forecasted and describes three primary cognitive elements: attitudes, social values, and intentions. Such individual conduct occurs voluntarily and in a rational manner. The strength between attitude and volition can be enhanced by utilizing various approaches such as context and action. A key limitation of this model is the need for voluntary user participation to validate the paradigm's application. Further, the model fails to respond to the rule of human conduct survey bias and moral aspects.

3.2   Technology Acceptance Model (TAM)

TAM generates from TRA. In TAM, the elements of user norms and preferences are eliminated. In describing users' motivation, the model considers three primary components, including perceived usefulness, perceived ease of use, and attitude toward adoption, with the first two being dominant values. The former value poses a significant influence on an individual's or organization's attitude towards technology adoption. Therefore, the two principal elements can be considered as the positiveness or negativeness toward the technology system. Besides, external factors are also considered in this model, including user knowledge and experience, the features of the user's system engagement in system design, and the execution activity's characteristics. The model has been widely applied in the IT domain and gained considerable empirical support.

Nonetheless, the technology acceptance model is prone to certain constraints, given the disregard of technology adoption's social impacts. Besides, there is still needed to add external aspects to the model to offer a more consistent IT system adoption forecast. The internal motivating forces are disregarded in the TAM, which implies the model's limited capability to be utilized within a client-oriented setting where the acceptance and utilization of IT lead to the realization of tasks and meeting users' emotional needs.

3.3   Expansion of TAM (ETAM)

In ETAM, additional variables within the TAM2 enhance understanding detailed description and particularity of the original TAM. Two main research studies have recommended the utilization of ETAM. In one, the causes of behavioral intention and perceived usefulness were examined. For this model, known as TAM2, two classifications of concepts were proposed, namely cognitive and social impacts to TAM. Resultantly, the forecasting robustness of perceived usefulness became improved. The second research determined concepts that affect perceived ease-of-use. The causes of perceived ease of use are categorized into modifications and anchors. For the former, values established from direct user experience to the given IT system are considered, while the latter pertains to general norms of IT system adoption.

3.4   Theory of Planned Behavior (TPB)

TPB adds perceived behavioral control as a fresh aspect extending the TRA paradigm. This standard is identified by the availability of skills assets opportunities perceived importance of the available assets and the capability to realize results. While the TRA and TPB both focus on an individual's behavioral volition, TPB model considers perceived behavioral control for a user's actions that cannot be considered voluntary. Thus, the theory of planned behavior model adds an element of self-efficiency as well as eliminating limitations. The focus on professed behavioral management poses a direct impact on actual conduct and indirectly affects behavioral violations.

Simply put the theory of planned behavior is geared towards three key aspects that impact on behavioral intention. These include perceived behavioral control, attitude, and subjective belief. Two primary constraints appear with the application of TPB. One rests with the attitudes towards it which may be highly relevant if an IT system is inaccessible. Two, a revised version of the theory of planned behavior may appear more feasible as a framework for concept development, and, in particular, impacting the degree of an individual's voluntary nature in selecting the adoption of IT in the work settings.

3.5   Model of PC Utilization

This model arises with the information security view to predicting user acceptance and PC use. Behavioral intention becomes disregarded within the model of PC utilization, given the focus on evaluating real behavior. Moreover, individual habits are disregarded even the total logical connection with prevalent use in the environment of PC use. The PC utilization model is particularly focused on assessing the direct effect of adoption, perceived effects, social factors, complexity, and job alignment on user behavior. Overall, employment fit, social aspects, long-term effects, and complexity pose significant impacts on personal computers usage. Nonetheless, reinforcing conditions and effects lack a major impact on related use of IT.

3.6   Motivational Model

Internal and external motivating factors determine IT system adoption. Extrinsic or external motivation entails the users' viewpoint regarding the realization of value adding results, which are different from the particular tasks, such as enhanced job performance. In this regard, perceived usefulness is considered as a type of external motivation. On the other hand, internal motivation refers to the view that individuals will want to implement an activity for no obvious support other than conducting the task. In this case, perceived pleasure is considered a form of internal motivation. Overall, the perceived ease of use and equality of outcomes affects perceived pleasure and usefulness. Additionally, the value of a given activity appears as a moderating variable of the perceived ease of use and quality of outcome, which impacts perceived usefulness. Behavioral intention is affected by the worth of outcome and perceived ease of use.

3.7   Unified Theory of Acceptance and Use of Technology (UTAUT)

Venkatesh and Morris formulated the UTAUT framework as they contrasted eight previous paradigms attributed to the IT system domain. The eight models focused on three key fields, including communications psychology and sociology. They include TAM, TPB, integrated TAM and TPB, TRA, diffusion of innovation, PC use theory, motivational model, and social cognitive paradigm. For the UTAUT paradigm, forecasted factors were identified regarding the adoption of IT structures. These factors are generated from matching the original concepts derived from prior theoretical bases, including effort anticipation, performance anticipation, communal impact, and enabling contacts. Sex, age, familiarity and deliberateness of adoption are four additional moderating factors identified within the UTAUT model.

3.8 Compatibility UTAUT (C-UTAUT)

Initially, Agarwal and Karahanna established compatibility values, which were incorporated into the UTAUT model by Bouten. The compatibility UTAUT model seeks to enhance the paradigm's informative power and offer deeper insight into how the model's cognitive experiences are established by recognizing and examining new boundary contexts [32]. The research aimed at exploring the connection between behavioral viewpoints and compatibility norms, therefore, it was unnecessary to determine actual conduct of adoption. Moreover, the study was cross-sectional, which implies that a focus on behavioral intention would depart from the key question of retrospective assessment [32].

3.9 Diffusion of Innovation (DOI) Theory

The DOI model focuses on various technological advancements by considering four primary aspects that impact the distribution of a new concept: time, communication channels, innovation, and social system. DOI is not solely applied at individual or organizational, but additionally allows a hypothetical framework to examine acceptance at an international scale. The diffusion of innovation theory integrates three major elements: adopter features, technology features, and innovation-decision activity. There are five main phases in the technology choice-making activity, including validation, insight, execution, decision, and influence. These stages occur through successive routes of communication for a given communal structure's members, and within a stipulated period. There are five vital factors concerning innovation features, including relative merit, compatibility, complexity, trialability, and observability, all which affect acceptance of technology [32]. There are five adopter features: early embracers, innovators, stragglers, early majority, and the late majority. Overall, diffusion of innovation model places greater emphasis on the IT system features, firm qualities, and environmental factors. Further, it poses reduced robustness in explanatory and less logical predictions of results relative to other technology use theories.

3.10 Social Cognitive Theory

The social cognitive model is derived from social psychology and grounded on three main factors: behavior, temperament, and setting. These three aspects become interrelated to allow a forecast of both individual and group behavior. From this paradigm, user conduct can be modified by determining a clear approach. The theory considers behavior a major aspect, and focuses on acceptance, utilization, and performance as core elements of such behavior [32]. Social cognitive model is incorporated to assess the IT usage by utilizing several concepts, such as self-efficacy, performance of results expectations, affect, apprehension, and personal outcome expectations.

4 Cloud Computing Service Platforms and Examples

There exist diverse cloud computing solutions having immense benefits. This study considers two popularly adopted platforms, including Amazon Web Services, and Microsoft Windows Azure.

4.1 *Amazon Web Services (AWS)*

Given the positive image and long years of experience of Amazon, the organization continues to gain trust in information technology, and, in particular, cloud technology.

To build on trust, the organization has largely focused on adhering to set security standards [28]. Therefore, the company stands in a favorable position to provide cloud related offerings. Amazon takes several steps to ensure that its cloud services provide clients with positive experiences. In this regard, a key system is the identity access management (IAM) framework, which the company utilizes to manage access to its vast resources [33]. The model is applied in identifying, validating and confirming groups or individual users to permit accessibility to its resources. There is a common interface with the framework to control users' access keys, guidelines, and passwords. Such framework allows for clear definition of which individual is allowed access to a particular resource.

The IAM model works under several steps. First, an individual user registers for an account by using their email address and password. Such authentication gives users complete access to the available resources and offerings within the AWS platform [28]. Following the creation of a user account, a definite password, access key, and username is provided to provide permissions to a given user account. IAM user functions are recommended so that no user can alter the resources within the AWS platform. Given the degree of access or everyday uses, a group for the framework can be established [28]. Such groups can be established through distinct guidelines for successful management. This means that the group's level should determine permissions as opposed to an individual's degree. Resultantly, users experience improved access. The use of the least access privilege is also a critical approach to ensure better access control. In addition to the least privilege, IAM roles' use is effective since these tools reinforce the utilization of temporary security validations. A clear example is Amazon S3 which provides access guidelines alternatives including user-oriented guidelines and resource directed policies [28]. This means that a user can override other policies for the other or take advantages of both policies to ensure enhanced access rights to available resources.

Replication Versioning is another tool that users can utilize to ensure that data is readily available [28]. A secure linkage between the user browser and console endpoint can be established with the AWS console management system. Further, Amazon includes an integrity assessment to validate a client or user request while maintaining data integrity through utilizing several techniques, for instance, digital signatures [28]. The company also maintains multi-variable verification as a means of ensuring best practices in setting security standards. The multi-factor authentication tool can also be combined with the IAM user account to successfully except user credentials which involves sending digit codes to a virtual or hardware tool.

4.2 Microsoft Windows Azure

Similarly, Microsoft is a proud leader in the arena of information technology, given its prominence in providing cloud-based solutions. The organization is known for its high data standards since it complies with many guidelines pertaining to data security [28]. Specifically, the firm provides an individual sign-on offering that allows access to vast amounts of applications. This offering arises as a cloud oriented directory identified as the Azure active directory or Azure AD. The directory enables application developers to establish a common authentication control by establishing settings for guidelines and policies. The application of Azure AD comes with access control based on rules which involves allowing permission to individuals, groups, or applications to view the available resources on the Azure platform. The rule-based access controls are provided by an administrator who oversees resource management activities' storage and authorization [28]. Following the assignation of access by an administrator, a user account is given a function. While the administrator can manage accessibility to a user account's activities, they cannot alter the information objects. For such allowance, the administrator can provide permission to view the access keys of the storage platform.

Clients can utilize a common access signature to gain permission to certain information objects within stipulated amounts of time. The signatures arise as strings of security symbols linked to a website to enable access to storage accounts and outline the limits, for example, the date of access [28]. The Shared Key is another strategy that users can use to encrypt a storage account's access keys and other limits that require a user to sign in. For this approach, the storage of access keys is only possible in the application of the client. Anonymous access is also possible in the storage account if the cloud platform is public. The Azure AD is recommended to validate permissions and access to Azure resources [28]. Given a multi-factor authentication process, a one-time password is provided and sent to the user in form of a phone call, a mobile application or SMS.

4.3 Applications and Benefits of Cloud Technology Platforms

Today, organizations focus on different investments to meet the return on investments true profitability maximization improve decision-making and lower costs. A key form of investment for organizations rests with developing robust ERP frameworks [11]. Given an enterprise resource planning system, a firm can gather record, manage, and deliver information across all operation units. Further, an ERP framework helps farms break down data existing in production stock planning engineering human resources sales and marketing production and all company divisions. Resultantly applying an ERP system means improved quality for organizations, better communications, reduced costs, and increased productivity [11]. With lower costs, a company stands in a better position to provide customer value and raise its share in the industry translating to higher profits.

For current ERP systems, the Internet arises as a means of connection for involved parties [11]. The ERP frameworks are integrated with e-commerce platforms to ensure greater collaboration with partners, suppliers, and other key stakeholders. This implies improved monitoring of incoming and outgoing inventory thereby developing greater visibility and control of its activities. Organizations need to consider the related budget before implementing such a system, including hardware maintenance fees training costs and licenses. Simply put, an ERP system is used to combine functional units within a company to deliver collaborated tasks or processes [11].

Furthermore, third parties external to the organization can also be included within this framework for instance suppliers and clients. As a primary component of an infrastructure an ERP system provides business solutions. The framework arises as an all-inclusive application that integrates an entity's entire units resulting in a corporation's holistic IT perspective.

Over the years, ERP systems' application has not grown and scrubbed due to increased evolution and upgrade activities to promote functionality and collaboration features. ERP service providers, for example, Oracle, have established distinct models supporting the functional units of a company [11]. Conventionally there are two main classes of Enterprise resource planning framework, which include hosted and on-premise systems. The former refers to an offering offered to a user or group by a service provider hosting the physical service and managing it remotely. In this case, a direct network linkage is applied meaning that the Internet will be disregarded. The latter involves running the ERP system within the organization's infrastructure, such as computers, networks and servers. In this regard, the organization is solely responsible for the system's operations and management and according to the licensing model.

In cloud computing, there exists a vast number of applications and solutions. As ERP systems are geared towards connecting the universe players within and external to an organization integration is easier. With a subscription model of payment there is more transparency in costs within cloud platforms than traditional ERP platforms [11]. This means that organizations only pay for what they require. The automation of procurement processes means that service providers can deliver solutions to consumers all over the world full stop cloud platforms allow users from different geographic locations to access resources.

Moreover, service providers execute guidelines for encryption thereby setting protection standards that safeguard data from consumers. Before implementing an ERP system client server choice of trying out the cloud systems in their three forms [11]. The free trials enable consumers to have confidence in the cloud ERP solutions being provided.

Presently, information technology hardware has become more available, inexpensive, and powerful due to the augmented development in storage and processing tools, coupled with the internet's growth [10]. The trend has resulted in the establishment of fresh frameworks of applying computers referred to as cloud technology. Cloud computing resources for information technology are considered functions, which can be rented and issued by clients via the web. Gartner's survey documentation revealed that public cloud solutions were anticipated to reach over $400 billion by 2020 [10]. Given the increased awareness and adoption of IT frameworks, there is a need to opt for a cloud service provider having the right solutions to ensure long-term success. Nonetheless the challenge dance with making the right decision among various alternatives for example Azure and AWS. These two platforms are considered global leaders in the field of cloud computing.

The software-as-a-service framework is dominant in North America, representing a leading and mature sector for the new technology. Additionally, Gartner's research indicates that takes these unexpected difference between the U.S. and the European region in terms of utilizing cloud tools. In this regard, the U.S. is anticipated to lead in adopting the technology, given the EU privacy rules, the multiple regional business transactions, and an impending recession. With the introduction to cloud solutions, different organizations adopt the technology in distinct ways. In this case, many organizations in Spain are expected to utilize cloud computing by 2015 with only 69% of SMEs considering the technology. For SMEs, there is a clear rise in the adoption of cloud technology from over 13% in 2011 to 69% in 2015 [6]. According to expert statistics, major firms are expected to utilize the mix or hybrid system in cloud computing, and complement them with other features, including IaaS, SaaS, and PaaS. Conversely, small and medium-sized enterprises are expected to direct attention from the public cloud due to higher SaaS applications' demand.

A keynote for executives and organizations is that the application of cloud technologies is complemented with other rapidly emerging tools including social networks mobile computing and big data analytics. The technologies or tools are interdependent and can be used or combined to leverage on performance. Moreover, there is limited knowledge regarding the concept of cloud technology coupled with its value for users or organizations. According to the study, over 54% of SMEs cite the lack of knowledge about cloud computing, which implies a loss in competitive advantage. On the other hand, the limited understanding of cloud technologies' advantages occurs as a hindrance to cloud platforms' adoption.

The decision to move to cloud platforms is challenging as there exists no broadly accepted guidelines for cloud offerings. On this note, it is important to develop regulations and a clear basis to assist organizations comprehend their needs. Additionally, decision makers and policy makers should have support tools to swiftly transition into IT systems to run on the cloud. Since it is an emerging technology cloud computing presents management innovation as leadership interventions in various corporate processes will be required to address traditional practices' inefficiency when handling or dealing with new technology. Adopting such technology also means organizational change made by staff resistance full stop. In this case, the top-level management within an organization that prepares it's tough by providing training opportunities and communicating about changes in time to allow for successful change towards implementing cloud technologies. Simply put key stakeholders within an organization should be made aware of the needed changes while gaining more insight into the benefits of adopting cloud tools.

Cloud technology solutions are highly complex with unfamiliar infrastructure. A clear example is the Azure Compute structure, which controls virtual and physical assets within Microsoft's data centers. The management systems in this case are container scheduling, virtual machine, and other control roles. Moreover, the application of cloud tools is prone to high costs in terms of development and running. Thus, developers need to maximize on such platforms to obtain value. A promising approach is taking advantage of machine learning as a means of managing resources in the cloud.

5 Security Patterns in Cloud Computing

Given cloud computing, the security pattern issue arises as a vital component of software or application security. In recent times, the security pattern in cloud platforms has greatly evolved due to the increasing demands for protection against higher data attacks and weaknesses [26]. Today, the notion of data security becomes expanded from the traditional viewpoint of a simple threat and becomes a requirement for all organizations and users to ensure data protection. The software-as-a-service platform in cloud computing demands structured data regarding suggested solutions and knowledge to assist the development of secure cloud platforms [21]. Currently there is ongoing research on several security patterns with data being directed to specific domains of security such as web development. Furthermore, there is poor organization and standardization of data about best practices and data security training. This means that software-as-a-service developers lack clear guidelines pertaining to best practices and security information needed to develop cloud applications.

Rath et al. [26] study is focused on an approach applied in defining and categorizing existing security patterns in cloud computing. Five main phases are identified, beginning from the security needs determination to the classification of security patterns. A key step rests with security needs in cloud platforms. This entails a clear definition of the security needs in cloud computing. Ultimately the key objective is outlining all potential security needs required for establishing protection trust and adherence to legal guidelines. Different data processing needs are considered in terms of regulatory frameworks, such as responsibility for private information control.

In the end, a security data checklist is provided for optimum application of software-as-a-service. The next phase includes evaluating risk, which entails checking existing susceptibilities and risks or uncertainties in cloud platforms. On this note, it is important to review the necessary security and identity and huntsmen to ensure safeguarded systems and proper data security controls within the design and execution of the software-as-a-service initiatives in cloud platforms. Ultimately, the activity results in a well-managed safety analysis report which lists all possible security threats that can impact cloud projects. With such data, the security and threat evaluation documentation can assess and retrieve security features in the software-as-a-service platform.

Third, it is critical to identify the existing security properties to ensure effective safeguard against malicious attacks. Since there are distinct types of attacks directed towards particular users, resources or systems, there is a need for diverse security features to handle such attacks. Cloud technologies are prone to significant security risks, for example denial-of-service attacks. This implies that it is vital to consider key issues like privacy integrity and trust. The fourth stage entails defining the security pattern determined by the security features identified in the previous phase. In this step, it is critical to investigate the conventional frameworks' potential security patterns to determine whether or not the existing patterns in other models can be applied in the cloud setting. For proper definition of security patterns, three major security components are considered, including the security of data, the system's security, and effective communication or protected dialogue [26]. The last stage entails the categorization of security patterns. This means that the identified patterns are classified together whenever there is a similar issue or context.

5.1 Security Issues in Cloud Technology

For entities, data storage is primarily the responsibility of the organization at large. On the other hand, cloud platforms require that data is stored in an external point from the user or client and in the service provider's control. This implies that more security measures are necessary in cloud computing to complement the existing conventional measures to ensure data protection [18]. Following the creation of information, distribution occurs freely between the phases. At each stage of the data lifecycle process, there is a need for proper protection of data, beginning from its creation to its erasure. Encryption is one technique used to safeguard data during its transmission phase [4].

A critical step in cloud computing rests with auditing. In this regard, it is necessary to trace information pathway, particularly when dealing with the public cloud. The CIA triangle features three key data qualities, including confidentiality, integrity and availability. The former involves information privacy, with data belonging to an individual user being protected from unauthorized access. On the other hand, integrity entails the guarantee that the cloud data is free from malicious attacks. The latter refers to a commitment that the service consumer receives access to their data when needed without any denial. In the utilization of the public cloud these three fundamental features are examined before application.

Confirmation and verification control aim to determine a user's identity in accessing and utilizing a cloud platform's resources. In organizations, the idea of computing requires that validations are stored in a server directory. Authentication within a private cloud is conducted through a virtual network [3]. For public cloud, the Internet is used as a means of connecting the service providers and consumers. This implies that public cloud platforms are most susceptible to risks than other platforms.

Additionally, the use of passwords may not be effective in ensuring data protection in public cloud. Service providers must establish tight controls aimed at ensuring data protection through highly secure techniques for user validation. The validation process in cloud technologies extends to machines, which require the authorization of particular actions such as system updates. A robust method of validation should be used in cloud applications since a variety of devices are utilized. A recent research by Alsmadi and Prybutok [3] examined the behavior of distributing and storing information, and the existing incongruence between scholarly literature and market reports regarding cloud technology.

5.2 Compliance and Regulatory Framework

There are distinct regulations and compliance requirements in different nations around the world. In this view, it is critical to ensure strong and strict standards of adherence to set guidelines. As a result, various geographic locations can access information when needed [26]. An important consideration when examining the issue of regulatory frameworks is the patterns of security.

A key focus is on information citizenship, which demands for legal penalties when private data is manipulated. When users enter and store data in cloud platforms, developers and service providers are accountable if such information is improperly used or handled and can accrue legal penalties. This implies that a provider needs to develop a solution which complies with the established regulatory frameworks. The geographic location where the offerings are provided is also an important considera-tion. Another component rests with data erasure a process conducted through the use of photography. Developers need to find a strategy in which information is securely and practically erased following its storage in cloud platforms. Resultantly search approach ensures that the legal adherence needs are fully met. Third, a common role framework arises. It is important to identify the individual accountable for data loss modification or any other form of alteration. Cloud service consumers must efficiently control their legal and regulatory adherence in cloud applications [26]. In contrast to typical information, confidential data processing is prone to robust controls and its retention, which is governed by regulatory frameworks.

Different countries have distinct policies regarding the retention of information occurring in cloud platforms. This demands for relevant adaptability from service providers. Another vital aspect of the life cycle of information entails data processing stages since its inception to its erasure [26]. The effective control of data in particular private data is a significant role for service providers operating in the cloud setting. A major issue in data security rests with unintended deletion of information. In this regard, service providers need to understand best practices in recovering information that is maliciously deleted and develop preventative measures to eliminate any opportunities for malicious attackers to access and erase existing information.

6 Data Security and Data Integrity Best Practices and Solutions

Today, many entities are considering cloud platforms as storage sources for their massive amounts of data. With the rising trend in adopting cloud tools across organizations, there is a direct rise in attacks from internal and external parties. These hackers identify existing susceptibilities in cloud networks, and trigger an unauthorized access, leading to data loss, sharing and disclosure of private information [22]. An all-inclusive cyber security approach can help address the potential risks in a multi-cloud setting and provide the firm a chance to realize the benefits of the cloud technology. AWS and Microsoft Azure are the main cloud platforms of data storage used by companies across the globe.

6.1 AWS

For data integrity compliance, the use of AWS offers several tools to ensure that the entity adheres to set regulatory frameworks and data management needs. Given information citizenship, the application of AWS provides geographic tags services, identified as location blocking tools, which are utilized to control access. A major strategy rests with CloudFront application, which constraints access to data based on the nation of origin [26]. Additionally, the tool provides access to content if request is made from acceptable locations, while disregarding those made from blacklisted nations. More robust measures would include a combination of CloudFront and other services to provide greater access management to information based on a variety of limits, including latitude and postal code. In terms of data erasure in cryptography, AWS offers an offering referred to as KMS to control the applied keys. The service enables individuals to establish and administer keys and manage encryption in different applications [20].

In meeting the framework of common responsibility, the use of AWS guarantees diverse offerings to safeguard information and systems [26]. The user has a choice to utilize either the free or purchased types. A key focus of AWS stands with ensuring the availability and protection of data within cloud platforms. The geo limiting tool can manage information across regional boundaries [26]. The AWS includes distinct forms of tools for both backup and storage of data to retain information, thereby preventing malicious attacks. The AWS information lifecycle function is responsible for enabling clients to control the process for related resources and application data.

6.2   Windows Azure

Adopting a sewer comes with diverse offerings and tools for enforcing the legal adherence processing and data control. For information citizenship, the use of Azure Front Door occurs as a viable solution that can be applied in constraining access to information and application concerning the geographic location [26]. Additionally, the Front Door tool comprises a web application firewall or WAF, allowing users to operate on a predefined custom access policy for particular pathways on endpoints to permit access from specified regions. In terms of information erasure through psychographic means, the Azure key vault includes other offerings such as certificate management [26]. Within a common role framework, using distinct tools to ensure that data and the entire system has safeguarded us necessary. While not all offerings are freely provided, the user can choose either the free or paid tools depending on their requirements.

Windows Azure is limited by legal needs of ensuring the fundamental security and availability of cloud platforms. The Front Door offering can also be applied in controlling the transmission of information across different locations [26]. In terms of information retention, the use of Azure provides distinct data storage and backup techniques that can be utilized in safeguarding data from unintended deletion or malicious attacks. A clear example is the Azure cosmos DB, which arises from a multi framework directory service [26]. Given the life cycle of information, applying a sewer globe storage life cycle and she was a guideline-based policy that clients can apply to manage data from its inception, consumption, and destruction. Finally, the Azure backup tool can be utilized for data storage in such a position that data is retained even when attackers attempt to remove the content [26]. The tool provides backup services for all resources in Windows Azure.

Another form of attack in cloud data is the denial of service problem. To eliminate related attacks techniques such as the signature-directed strategy, the filter-based strategy and firewalls are necessary [31]. Using the filter approach a flow label filter is applied in identifying denial-of-service attacks occurring and a low rates. A denial-of-service attack gradually raises the rate of traffic thereby affecting the network. By employing a signature strategy, a cloud's network traffic is assessed with the pattern of attacks being compared [31]. Given the signature directory, several predefined signatures are developed, leading to the blocking of possible attacks matching the database.

Another type of attack is malware injection. Under typical situations, a service consumer creates an account in the platform, and the service provider develops a copy of the client's computer-generated account in the directory structure [31]. Service providers measure the activities of the user with high integrity and effectiveness. It is recommended that data integrity in the hardware position is highly maintained since any attacker will face clear challenges when trying to interfere with the infrastructure as a service cloud platform. The final location table or if it is utilized to control virtual content within operating systems. Given the tool, service providers can identify the application that our client will implement [31].

Additionally, the providers can verify with the earlier situations obtained from the client's device to ensure the integrity and validity of subsequent events. Therefore, it is important to install a hypervisor which service provider can use to measure the most protected portion of the cloud platform that any attacker cannot hamper. The Hypervisor tool assists in documenting all events that affect data integrity in the file allocation table appearing in virtual machines within cloud platforms [31]. An additional remedy entails storing the client's operating system in the first or initial stage upon signing up for a new account. Here, cross assessment becomes conducted with the functioning system.

Because of the CIA triangle, various steps can be taken to ensure that a user to cloud platforms has the best experience. Following data creation, its classification, determination of the sensitive type, a clear definition of guidelines, and creation of access strategies for diverse data types are necessary steps. Further, it is important to develop policies for the destruction and archiving of information [18]. Secondly, it is vital to ensure that data storage is accompanied by efficient logical and physical safeguard, including a plan for recovery and backup of information. Third, it is crucial to understand the type of information to be shared with whom and the process of sharing and defining policies related to common data frameworks. Service level agreements or SLA is a collective term used to describe the various policies applicable to cloud computing.

Another step is establishing a corrective action strategy to ensure data security when it is hacked or maliciously altered due to vulnerabilities existing in either the network or the communication devices. This means that data integrity she will be assessed bored at the level of competition and the information level. In this view, competition integrity entails the verified applications that are allowed access two available information. It is important to prevent any actions that go against the typical computing process. With a proper identity and access management or I am system issues of integrity confidentiality and availability are addressed.

Before purchasing a cloud solution, a user has their priorities regarding the utilization of such services. Given the diverse properties of cloud platforms like Azure and AWS, the service consumer may find it hard to understand the necessary features that they may be needing for their projects. Past studies have failed to make valid conclusions regarding the actual features required from an IT viewpoint and the systems' infrastructure. Given the limited data on the most applicable features, clients must first assess their needs to gain more insight into the kind of solutions they would need. Kamal et al. [16] study compared the features of AWS and Microsoft azure. In this regard, AWS is recommended for internet as a service frameworks.

Nonetheless, the platform is relatively expensive and less secure compared to Microsoft Azure. Azure, on the other hand, presents noble features and remains dominant in SaaS and PaaS frameworks. Additionally, the solution is attributed to Microsoft Company, which is a global brand.

Service providers within cloud platforms make huge investments to develop robust hardware and software systems. A critical consideration for this providers is optimizing the application of the high investments on resources and ensuring improved performance and availability. A potential approach to facilitate maximizing the adoption of these resources is integrating machine learning into cloud tools. On this note, Bianchini et al. [5] study has focused on examining the opportunities and designs for including machine learning into cloud platforms' resource management planning. The case of a supercomputer framework is provided as an example showing how the combination of machine learning into the Azure platform leverages forecasts of behaviors in service and containers. The established prediction system in the study leads to such transformation of cloud platforms. While the study portrays that machine learning models can help managers make more informed decisions on planning resources, there is still limited research on the topic.

7 Conclusions and Recommendations

In cloud computing, key issues stand with the confidentiality and protection of the data available in cloud tools. Given the cloud setting of sharing resources, virtualization, mobile computing, service level agreements, and heterogeneity, cloud platforms are highly susceptible to attacks. The study has focused on presenting the data security concerns and remedies to mitigate against arising issues. Nonetheless, there are emerging challenges that lack clear mitigation approaches, presenting long-term risk for prospective clients of cloud technology. There are fresh developments in cloud technology, such as the internet of things and software-oriented networking, which imply greater capabilities and storage provisions to address arising issues in cloud platforms. With these developments comes related challenges in cloud technology, and which demand for proactive measures and solutions. Given the rampant changes and dynamics of technology, there is a need for clear policies and regulations regarding data security and the irrelevant update to maintain data integrity in cloud computing.

Further examination should be actualized to ensure credibility among clients and entities seeking to embrace cloud technologies. It is indispensable to design effective frameworks that can identify expected unapproved admittance to information and assure customers that the cloud computing solution is secure. Throughout the long term, and since the advent of the Internet, there have been huge concerns about adulterated gadgets, virus attacks, and information loss. Currently, more people and organizations use the Internet by utilizing different devices, with decreased instances of potential malware attacks. A few programming organizations began contributing vigorously to assemble innovation to recognize infection assaults and keep them from getting to a gadget or information. Presently, limited attacks happen because of developed advancements and best practices.

Likewise, cloud computing specialist suppliers and IT associations must contribute more to develop checking frameworks to distinguish unapproved access, prevention, and best practices to actualize cloud technology. More examinations should be executed to assess whether progress has been made to forestall unapproved admittance to data. As portrayed for the situation examines, enormous associations have lost income because of a few robbery issues. Instead of re-authorizing clients to utilize authentic items, programming organizations slowly manufactured another innovation to advise clients to utilize pilfered programming and the expected danger to utilizing robbery programming.

More examination is expected to decide the advancement made on teaching clients to use distributed computing safely. Further, it is basic to recognize the best direction for end-client or organizations to use distributed computing to expand their trust in distributed computing. Additionally, greater responsibility for the cloud specialist co-op is required where information gets traded off. Restricted advancement has been made to make cloud specialist co-ops responsible for information misfortune or unavailable. Today, organizations are more receptive to end-clients for utilizing their administrations if there is some trade off in information misfortune or unapproved access.

References

1. Ahmad, I., Bakht, H., Mohan, U.: Cloud computing- threats and challenges. J. Comput. Manage. Stud. **1**(1), 1–12 (2017). https://www.researchgate.net/publication/319725257_Cloud_Comput ing__Threats_and_Challenges
2. Alam, S.B.: Cloud computing – architecture, platform and security issues: a survey. World Sci. News **86**(3), 253–264 (2017). http://www.worldscientificnews.com/wpcontent/uploads/2017/08/WSN-863-2017-253-264-1.pdf
3. Alsmadi, D., Prybutok, V.: Sharing and storage behavior via cloud computing: security and privacy in research and practice. Comput. Hum. Behav. **85**, 218–226 (2018). https://doi.org/10.1016/j.chb.2018.04.003
4. Al-Haija, Q.A., Tawalbeh, L.A.: Efficient algorithms and architectures for elliptic curve cryptoprocessor over GF (P) using new projective coordinates systems. J. Inf. Assur. Securi. (JIAS), **7**, 063–072 (2010)
5. Bianchini, R., Fontoura, M., Cortez, E., Bonde, A., Muzio, A., Constantin, A., Moscibroda, T., Magalhaes, G., Bablani, G., Russinovich, M.: Toward ML-centric cloud platforms. Commun. ACM **63**(2), 50–59 (2020). https://doi.org/10.1145/3364684
6. Bildosola, I., Rio-Bélver, R., Cilleruelo, E.: Forecasting the big services era: novel approach combining statistical methods, expertise and technology roadmapping. In: Cortés, P., Maeso, E., Escudero, A. (eds.) Enhancing Synergies in a Collaborative Environment. Lecture Notes in Management and Industrial Engineering. Springer, Cham (2015a)
7. Bildosola, I., Río-Bélver, R., Cilleruelo, E., Garechana, G.: Design and implementation of a cloud computing adoption decision tool: generating a cloud road. PLoS ONE **10**(7), e0134563 (2015b). https://doi.org/10.1371/journal.pone.0134563
8. Das, A., Patterson, S., Wittie, M.: EdgeBench: benchmarking Edge computing platforms. In: 2018 IEEE/ACM International Conference on Utility and Cloud Computing Companion (UCC Companion), Zurich, pp. 175–180 (2018). https://doi.org/10.1109/UCC-Companion.2018.00053

9. Duncan, R.: A multi-cloud world requires a multi-cloud security approach. Comput. Fraud Secur. **2020**(5), 11–12 (2020). https://doi.org/10.1016/S1361-3723(20)30052-X
10. Dutta, P., Dutta, P.: Comparative study of cloud services offered by Amazon, Microsoft and Google. Int. J. Trend Sci. Res. Dev. (IJTSRD) **3**(3), 981–985 (2019). https://www.ijtsrd.com/papers/ijtsrd23.170.pdf
11. Elmonem, M.A., Nasr, E.S., Geith, M.H.: Benefits and challenges of cloud ERP systems – a systematic literature review. Future Comput. Inf. J. **1**(1–2), 1–9 (2016). https://doi.org/10.1016/j.fcij.2017.03.003
12. Fernandes, D.A.B., Soares, L.F.B., Gomes, J.V., Freire, M.: Security issues in cloud environments: a survey. Int. J Inf. Secur. **13**(2), 113–170 (2013). https://doi.org/10.1007/s10207-013-0208-7
13. Hughes, R., Muheidat, F., Lee, M., Lo'ai, A.T.: Floor based sensors walk identification system using dynamic time warping with cloudlet support. In: 2019 IEEE 13th International Conference on Semantic Computing (ICSC), pp. 440–444. IEEE, January 2019
14. Gerhardter, A., Ortner, W.: Flexibility and improved resource utilization through cloud based ERP systems: critical success factors of SaaS solutions in SME. In: Felderer, M., Piazolo, F., (eds.) Innovation and Future of Enterprise Information Systems, pp. 171–182. Springer, Heidelberg (2013)
15. Gholami, A.: Security and privacy of sensitive data in cloud computing. Doctoral Thesis, Stockholm, Sweden (2016). https://www.diva-portal.org/smash/get/diva2:925669/FULLTEXT01.pdf
16. Kamal, M.A., Raza, H.W., Alam, M.M., Su'ud, M.M.: Highlight the features of AWS, GCP and Microsoft Azure that have an impact when choosing a cloud service provider. Int. J. Recent Technol. Eng. (IJRTE) **8**(5), 4124–4132 (2020). https://doi.org/10.35940/ijrte.D8573.018520
17. Kofahi, N., Al-Rabadi, A.: Identifying the top threats in cloud computing and its suggested solutions: a survey. Adv. Netw **6**(1), 1–13 (2018). https://doi.org/10.11648/j.net.20180601.11
18. Kumar, P.R., Raj, P.H., Jelciana, P.: Exploring data security issues and solutions in cloud computing. Procedia Comput. Sci. **125**, 691–697 (2018). https://doi.org/10.1016/j.procs.2017.12.089
19. Kushwah, V.S., Bajpai, A.: Cloud computing: a future e-learning environment. Int. J Res. Electron. Comput. Eng. **5**(4), 63–67 (2017). https://www.researchgate.net/publication/321016275_Cloud_Computing_A_Future_eLearning_Environment
20. Lo'ai, A.T., Tenca, A.F.: An algorithm and hardware architecture for integrated modular division and multiplication in GF (p) and GF (2n). In: Proceedings of the Application Specific Systems, Architectures and Processors, 15th IEEE International Conference, pp. 247–257, September 2004
21. Lo'ai, A.T., Saldamli, G.: Reconsidering big data security and privacy in cloud and mobile cloud systems. J. King Saud Univ. Comput. Inf. Sci. (2019). https://doi.org/10.1016/j.jksuci.2019.05.007
22. Muheidat, F., Tawalbeh, L.: Mobile and cloud computing security. In: Maleh, Y., Shojafar, M., Alazab, M., Baddi, Y. (eds.) Machine Intelligence and Big Data Analytics for Cybersecurity Applications. SCI, vol. 919, pp. 461–483. Springer, Cham (2021). https://doi.org/10.1007/978-3-030-57024-8_21
23. Nemade, B., Moorthy, S., Kadam, O.: Cloud computing: Windows Azure platform. In: ICWET 2011: Proceedings of the International Conference and Workshop on Emerging Trends in Technology, pp. 1361–1362, February 2011. https://doi.org/10.1145/1980022.1980341
24. Opara, C.M.: Cloud computing in Amazon Web Services, Microsoft Windows Azure, Google App Engine and IBM cloud platforms: A comparative study. A Thesis Submitted to the Graduate School of Applied Sciences of Near East University (2019). https://docs.neu.edu.tr/library/6842203396.pdf
25. Rao, R., Selvamani, K.: Data security challenges and its solutions in cloud computing. Procedia Comput. Sci. **48**, 204–209 (2015). https://doi.org/10.1016/j.procs.2015.04.171
26. Rath, A., Spasic, B., Boucart, N., Thiran, P.: Security pattern for cloud SaaS: from system and data security to privacy case study in AWS and Azure. Computers **8**(34), 1–8 (2019). https://doi.org/10.3390/computers8020034

27. Tawalbeh, L.A., Jararweh, Y., Mohammad, A.: An integrated radix-4 modular divider/multiplier hardware architecture for cryptographic applications. Int. Arab J. Inf. Technol. (IAJIT) **9**(3) (2012)
28. Saeed, I., Baras, S., Hajjdiab, H.: Security and privacy of AWS S3 and Azure Blob storage services. In: 2019 IEEE 4th International Conference on Computer and Communication Systems (ICCCS), Singapore, pp. 388–394 (2019). https://doi.org/10.1109/CCOMS.2019.882 1735
29. Sharif, H.U., Datta, R.: Cloud data transfer and secure data storage. Int. J. Eng. Appl. Sci. (IJEAS) **7**(6), 11–15 (2020). https://doi.org/10.31873/IJEAS.7.06.04
30. Singh, I., Mishra, K.N., Alberti, A.M., Singh, D., Jara, A.: A novel privacy and security framework for the cloud network services. In: 2015 9th International Conference on Innovative Mobile and Internet Services in Ubiquitous Computing (IMIS) (2015). https://doi.org/10.1109/IMIS.2015.93
31. Subramaniam, T.K., Deepa, B.: Security attack issues and mitigation techniques in cloud computing environments. Int. J. UbiComp (IJU) **7**(1), 1–11 (2016). https://doi.org/10.5121/iju.2016.7101
32. Taherdoost, H.: A review of technology acceptance and adoption models and theories. Proceedia Manufact. **22**, 960–967 (2018). https://doi.org/10.1016/j.promfg.2018.03.137
33. Tawalbeh, L.A., Muheidat, F., Tawalbeh, M., Quwaider, M.: IoT Privacy and security: challenges and solutions. Appl. Sci. **10**(12), 4102 (2020)
34. Tawalbeh, M., Quwaider, M., Lo'ai, A.T.: Authorization model for IoT healthcare systems: case study. In: 2020 11th International Conference on Information and Communication Systems (ICICS), pp. 337–342. IEEE, April 2020
35. Jararweh, Y., Al-Ayyoub, M., Song, H.: Software-defined systems support for secure cloud computing based on data classification. Ann. Telecommun. **72**(5), 335–345 (2017)

Blockchain-Based IoT Forensics: Challenges and State-of-the-Art Frameworks

Md Azam Hossain and Baseem Al-Athwari

Abstract There is no doubt that the recent emergence of Internet of Things (IoT) paradigm brings significant changes in many aspects of our life including smart homes, smart cities, healthcare, farming, etc. Despite the unlimited advantages of IoTs, the tremendous increase of interconnected smart devices attracts more threats and hence introduces many challenges related to the security and digital forensics of the IoT environment. Although the IoT forensics is relatively new domain of research, several IoT forensics frameworks have been proposed recently to investigate cybercrimes. However, blockchain-based IoT forensics is the most promising approach. This chapter introduces the digital forensics from the point of view of IoT environment. It also discusses recent IoT forensics challenges and presents the most recently developed blockchain-based frameworks for the IoT forensic.

1 Introduction

In recent years, new technologies have become an integral part of everyday life, such as Internet of Things (IoT), fifth generation of telecommunications (5G), social networks, distributed blockchain technology, etc. Emerging technologies make our daily life faster, easier, and more enjoyable by developing fascinating devices, apps, and resources and such technologies bring the most valuable services at our fingertips. The exponential growth of the IoT devices, like smartphones, smart watches, CCTV cameras, washing machines, and medical implants pave the transformation to the smart world [6].

M. A. Hossain (✉) · B. Al-Athwari
Department of Computer Engineering, Kyungdong University,
Gangwon-do, Republic of Korea
e-mail: azam1708@kduniv.ac.kr

B. Al-Athwari
e-mail: baseem_cs@kduniv.ac.kr

Generally, IoT devices generate and process confidential information, and hence they are becoming a rich source of information for cybercriminals. The underlying IoT infrastructures also become an ideal target for intruders and cyber-attackers due to its unique characteristics [16]. Hence, we need to conduct digital forensics investigation process in order to prosecute the malicious activity in IoT environment known as IoT forensics.

The existing digital forensic tools and procedures do not fit with the IoT environment due to many factors including high connectivity, heterogeneity, wide distribution and openness of IoT systems. The huge number of heterogeneous interconnected IoT devices generates huge amount of data which creates a major challenge for the IoT forensics professionals to identify, acquire, examine, analyze, and present the evidences. The diverse data format used by the IoT devices would also pose concerns in data analysis. Current digital forensic methodologies are centralized in nature which create much doubt about investigation transparency and reliability. Moreover, malicious actors can tamper the evidence because most of the data are stored in the IoT devices (e.g., wearable, phones) which in turn raise the question about evidence integrity and trustworthiness [12]. In addition, every day exponential increased number of IoT devices becomes part of the IoT systems which demands scalable distributed infrastructure for conducting forensics process.

Despite the great efforts by many researchers on digital forensics, the IoT forensic is still in its early stage and there is a lack in the literature regarding the approaches that can be used during investigation [5]. One of the most promising technique is utilizing the blockchain. Considering its unique features, blockchain attracts many applications in diverse domains such as healthcare, supply-chain business, insurance, etc. with regard to the IoT, blockchain technology offers a unique set of functionalities which highly suitable for IoT forensic. That is, blockchain is a distributed, decentralized digital ledger that maintains the growing list of blocks in the peer-to-peer network. These features open the door to apply blockchain technology in IoT forensics investigation process as it can ensure evidence integrity, availability, traceability, accountability and system scalability.

In the context of IoT, a block is a collection of transactions and a transaction refers to the exchanged data among various devices in the IoT environment. Distributed ledger stores time-stamped blocks connected in a chain, providing an immutable, publicly accessible and verifiable by a consensus algorithm to ensure evidence trustworthiness. Since ledger is publicly available and distributed among participated stakeholders of an IoT environment, it eliminates the control of central authority on the data. This makes also impossible to insert, delete, modify the transaction data and ensures evidence integrity and availability.

This chapter introduces the digital forensics from the point of view of IoT environment. It also discusses recent IoT forensics challenges and presents the most recently developed blockchain-based frameworks for the IoT forensic.

The structure of this chapter is defined as follows: Sect. 1 is introductory. Section 2 defines digital forensics(DF) and presents the widely accepted stages of DF investigation process such as evidence identification, acquisition, examination, analysis, and finally presentation. Section 3 introduces the IoT forensics concepts and discusses the unique characteristics of IoT environments.

Section 4 focuses on various data/evidence sources of the IoT environment and their challenges faced by forensics professional. Section 5 deals with blockchain definition, features, and types of blockchain. A comprehensive review of blockchain-based IoT forensics approaches and their complexity is discussed in Sect. 6. Finally, Sect. 7 summarizes the discussion and highlight some future work direction.

2   What Is Digital Forensics?

The Digital Forensics (DF) discipline is a subset of conventional forensic science. It is described as a legally acceptable procedure to collect, inspect, analyze, record the evidence and finally produce the digital evidence to the court for persecution [10]. Digital forensics involves the study of data collected from digital devices such as wearable, medical devices, smart home appliances, smart vehicles, aerial drones, security systems, and sensor network. In 2006, US Federal Rules of Civil Procedure (FRCP) extended the scope for using electronically stored information (ESI) as evidence in civil cases [9]. FRCP defines the discoverable artefacts such as electronically stored information including writings, drawings, graphs, charts, photographs, sound recordings, images, and other data or data compilations stored in any medium from which information can be obtained either directly or, into a reasonably usable form for forensics investigation.

The National Institute of Standards and Technology (NIST) defines the digital forensic as "an applied science to identify an incident, collection, examination, and analysis of evidence data" [13]. Widely accepted digital forensics process [15] comprises five main phases as shown in Fig. 1.

Identification: DF investigation process starts with identification of an incident and evidence. In this phase, computer forensics examiners meticulously identify evidence, analysis the legal framework, prepare the tools required for DF process and correlate with other incidents.

Acquisition: In acquisition process, forensics examiner extracts digital evidence from various media such as hard disk, RAM, operating systems registry file, log file, USB, cell phone, e-mail, etc. labels, packages and preserves the integrity of the evidence.

Examination: At this stage, forensics examiner extracts and examines artefacts collected from the crime scene and appropriately preserves the evidence.

Analysis: This is the most crucial phase in the DF process. Forensics expert analyzes the artefacts, interprets and correlates with evidence to reach a conclusion, which can serve to prove or disprove at court.

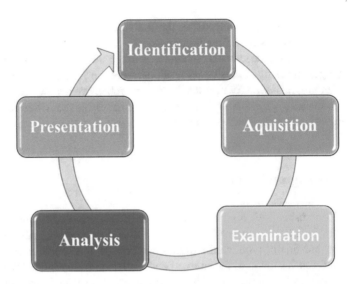

Fig. 1 Digital forensic process

Presentation: In the final phase, forensics investigator presents the results of the investigation and makes a report to affirm his or her findings about the case. This report should be appropriate for admissibility of the evidence.

In digital forensics process, preserving integrity of the digital evidence and following strict chain of custody for the information is compulsory. Although there are shuttle differences in the investigation cycle into phases, but the whole cycle should be completed using certify tools and scientifically proven methodology.

3   IoT Forensics

IoT Forensics is an emerging branch of digital forensics, where forensics activities deal with more complicated and heterogeneous IoT infrastructures (e.g., Cloud, network, etc.) and devices or sensors such as wearable, smart homes, cars, aerial drones, and medical implants, to name a few. It is a comparatively new and novel field and it has similar goal of digital forensics with respect to the way of investigation carried out in legal and scientific manner including digital evidence collection to establish the facts about an incident.

In traditional digital forensics, evidence sources are usually limited where investigators mainly collect the evidence from PC, laptops, usb, flash drives, smartphones, tablets, server, network gateway, etc. On the other hand, in IoT forensics, evidence sources are generally vast and divers.

3.1 Characteristics of IoT Environment

IoT forensics differs from conventional digital forensics because it needs to deals with numerous unique characteristics of IoT environment such as [27]

- Devices in IoT-enabled environments are diverse and resource-constrained (e.g., energy, computing power, and storage capacity).
- IoT devices generate a huge amount of data called "Big IoT Data".
- Various data formats are used to store and process data by IoT devices.
- Digital evidence of IoT devices has limited visibility and short survival period.
- In IoT environment, evidences are mostly spread across multiple platforms, e.g., on the edge devices, cloud, and data centers, which makes it one of the major difficulties to get access for forensics investigation.
- IoT devices have inherently different hardware architectures and heterogeneous operating systems.
- IoT devices are manufactured using proprietary hardware, software and multiple standards by various vendors.

3.2 Type of IoT Forensics

IoT forensics is broadly categorized as cloud forensics. Network forensics and IoT device forensics as shown in Fig. 2 [25].

Cloud Forensics: IoT devices inherently resource-constrained in terms of processing capability, storage capacity and energy, and for this reason they are connected to virtualized data center to process and store data. Cloud forensics deals with the IoT data stored in the cloud in order to conduct forensics investigation.

Network Forensic: IoT devices communicate with each other through some networks. In IoT environment, various types of networks are formed, including personal area networks (PAN), local area networks (LAN), wide area networks (WAN),

Fig. 2 General type of IoT forensic

metropolitan area networks (MAN), etc. Network traffic data and abnormal behavior log contain very useful evidence to perform forensics investigation process.

IoT Device Forensics: Digital evidences are collected from the devices used in IoT environments. Forensic experts gather evidence data primarily from local storage of physical devices where data are stored in IoT devices.

4 IoT Forensics Data Sources Challenges

It is well recognized, in near future, IoT will touch every aspect of our life including homes, cities, health, industries, etc. Even though IoT will make our life more comfort and easy, security and privacy are still the most critical challenging issues in the IoT environment. Considering its unique features, including interconnectivity of massive number of heterogeneous devices, dynamic changes, and the complicated architecture, IoT environment is exposed to the possibility of being attacked easily by different types of attacks including hardware, operating systems, applications, data, and communication protocols [25]. Unfortunately, to the best of our knowledge, there is no standard forensics procedure that can handle all of attacks. Instead, each attack is handled separately. Therefore, there is a tremendous need for a common forensics process which can help to ensure best practices of cyber-security that consider all the security issues related to the IoT environment. The effectiveness of IoT forensics process is highly depends on identifying the source of the forensic evidence.

Identification of the source of the evidence in the IoT environment is the first and one of the most challenging tasks in the digital forensic process. Considering the complicated infrastructure of the IoT environment in terms of the huge number of interconnected heterogenous devices, variety of forms of networks, different operating systems, and different applications supported by the devices, digital forensics professionals face difficulties to locate the source of the evidence. Unlike the traditional digital forensics where the sources of the of the evidence are usually restricted to a limited type of devices such PCs, servers, or even mobile devices, the forensic data sources in the IoT context are heterogeneous and of wide range including:

End User Devices: include computers, servers, printers, scanners, laptops, mobiles, etc. that provide services directly to the users. These devices allow users to create, share and obtain information. Despite the different sizes and specifications of these divices in terms of their computing resources (CPUs, RAM, and storage), these devices can be considered an easy target for the attackers to obtain, alter, or even delete the sensitive information stored in these devices. Although these device can provide a vital amount of data, however, due to the sensitivity of the stored data and the privacy-related issues of the end users, the digital forensic professional might face difficulties to extract the evidences from these devices.

Network Devices: include all the devices that provide connectivity between the IoT devices to allow them to communicate and share the resources. Some of these devices provide extension and concentration of connection between the IoT devices at the Local Area Networks (LANs) level such as switches and wireless router, and some of them such as router, provides Wide Area Networks (WAN) connection and responsible for routing the data between the source and the destination. Therefore, in the case of any attack, it might be helpful to check the network logs to identify the source of evidence. However, considering the variety networking infrastructure of the IoT environment in terms of communication media (red and wireless) and the area covered by each network (PAN, BAN, LAN, MAN, and WAN) digital forensics professional need training on how to trace the network devices and extract the evidences without disturbing the network performance including other users who are sharing the same network infrastructure.

Sensors: Sensors are the most essential components for IoT and play a great role during IoT forensics. The majority of IoT devices are equipped with one or more sensors. Sensors basically detect external information around them according to their purposes. There are different types of sensors, including environmental, chemical, medical, and phone-based sensors [19]. They are also manufactured in different shapes and sizes. Considering their small sizes, some sensors could be hard to locate them by the IoT forensics professionals. Moreover, due to their location, most of the sensors can not be accessed easily or because they couldn't be distinguished from other home appliances. Sensors also have a limited battery life and computing resources (memory, processor, storage) which can not support them to store significant evidences to the IoT forensics professional.

Controller: Controllers play a vital role in the IoT environment. That is, controllers are responsible for collecting data gathered by the sensors and providing network or Internet connectivity. Controllers may have the ability to process the data received from the sensors and make immediate decisions. Considering the vital role of the controllers in the IoT environment, they might be one of the most targeted devices by the attackers and hence provide a significant evidence to the IoT forensics. However, due to their computing resource constraints, they may send data to a more powerful computer for analysis. This more powerful computer might be in the same LAN as the home gateway or might only be on the cloud and can be accessed through an Internet connection which makes it difficult for the IoT forensics to extract information regarding the attack.

Actuators: Actuators are often work together with the sensors and controllers. Actuators take electrical input and transform the input into physical action. For instance, in smart home, sensor might detect excess heat in a room, the sensor sends the temperature reading to the controller. The controller can send the data to an actuator which would then turn on the air conditioner. Similar to the sensors and controllers, the actuators are running continuously. Therefore, data could be easily overwritten as they have limited memory and as a result retrieving evidence from them is a challenge for the IoT forensics.

Smart Devices: Smart devices are the core of the IoT environment. Day by day, there is exponential increase of smart devices connected to the internet. In our world today, the number of smart devices exceeds the number of people on the planet. These might include home appliances, medical implants, cars, and embedded systems. Considering the increasing number of smart devices, and the diversity of the vendors, IoT forensics professional might face considerable challenges to collect the evidences from these devices. Considering the privacy, owners/users of smart devices should be informed to get their permissions to access the data stored in their smart devices. In addition, although some data can be stored in local memory of smart devices, some devices, due to their memory and processing constraints, send the data to another nearby devices or even to the cloud for the processing which makes it difficult to be retrieved and collect the evidences.

Cloud: Cloud helps to provide high quality computing services to the IoT devices. As mentioned earlier, due to their computing resources constraints [3], IoT devices send their data to be processed and stored in the cloud. Despite the numerous benefits brought by the cloud to the IoT [1], collecting crime-related data is a big challenge for the IoT forensics. That is, investigators have to gain access to the cloud and that requires the involvement of the service provider who may be hesitant to share information or providing investigators with access to their cloud-based environment [2].

5 Introduction to Blockchain

Blockchain technology has been foreseen as a disruptive technology by industry and scientist community [20]. It is predicted that blockchain technology could play vital role in managing and securing IoT environments. Due to immutability and distributed nature of blockchain could be highly suitable solution for IoT forensics. The section starts with an introductory background about blockchain, and then describes the key features of blockchain.

5.1 Blockchain

Blockchain concept was first surfaced in 1991 by Stuart Haber and W. Scott Stornetta, who implemented a cryptographically secured chain of blocks (document) system where document timestamps could not be tampered with. Almost after two decades, in 2008, Satoshi Nakamoto introduced the Bitcoin built on blockchain; a new electronic virtual cash system on a peer-to-peer network without trusted third party [21]. Since then, blockchain technology has evolved as a disruptive technology and has swept across many industries. Recently, the application of blockchain technology has expanded rapidly beyond the financial and banking world such as cloud storage, cybersecurity, payment processing, content distribution, reals estate, tourism sector, energy industry, health care, etc.

5.1.1 Blockchain Definition

Blockchain is a shared, immutable and distributed ledger system in which a record of transactions known as block is maintained and blocks are linked in a peer-to-peer network without trusted third party [23]. A typical blockchain has several basic features such as:

Timestamps: A timestamps defines the time and data when a record is created in the chain.

Immutability: It defines that data cannot be modified or tempered by any malicious attack and guarantees that it is impossible to create a counterfeit version of data.

Decentralization: Blockchain network is decentralized and which means that there is no centralized authority to govern the network. This feature of blockchain makes it more popular because it can avoid single point of failure, less prone to breakdown, fully user controlled, and offers transparency to every participant.

Consensus: The decision making process in the blockchain architecture is consensus algorithm-based. This allows the participated active nodes to take part in the decision making process.

5.1.2 Blockchain Structure

Figure 3 illustrates the basic structure of a blockchain which consists growing number of blocks. The description of each field in a blockchain is given as follows:

Block: Block in a chain is timestamped and validated record by participated miners using the consensus algorithm which ensures the data integrity and authenticity. Blocks are broken into two parts: body and header.

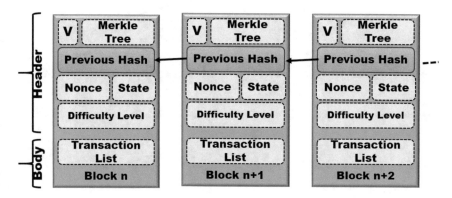

Fig. 3 Blockchain structure

Body: Body of a block stores a list of transactions or data.

Header: Header of a block contains several fields such as blockchain version, merkle tree root, previous hash, nonce, difficulty level, and state. Each field in a header is described as follows:

Version (V): Version field indicates the protocol/software upgrades.

Merkle Tree: A merkle tree is data structure which stores the hash value of the transaction in hierarchical fashion and hashing is performed from the bottom to top starting from individual transaction [18] as depicted in Fig. 4. As shown in Fig. 4 **Hash(1)** stores the hash value of transaction **Tx 1**, and similarly, **Hash(2)** to **Hash(8)** store hash value of transactions **Tx(2)** to **Tx8)** respectively. h_{12} stores the hash of **Hash(1)** and **Hash(2)**, h_{34} stores the hash of **Hash(3)** and **Hash(4)**, h_{56} records the hash of **Hash(5)** and **Hash(6)**, and so on. h_{1234} is the hash of h_{12} and h_{34}, and in the same fashion the nodes reach to the root also known as merkle root. Finally, root stores the hash of h_{1234} and h_{5678} as shown in Fig. 4.

Investigator and other participants in the blockchain can easily verify and locate the transaction by using the merkle root. This tree structure provides an efficient and secure verification of content consistency. It generates a digital fingerprint of the entire transactions set by accumulating the data in the tree, which allows easy verification whether a node is added in the root. Merkle tree structure is similar to binary tree which has even number of leaf nodes. If the number of transaction in leaf nodes is odd, then simply last transaction is duplicated to yield even number of leaf nodes.

In a merkle tree, branches can be fetched separately, which allows to verify the integrity of each branch independently. As a result, in a verification process it significantly reduces the amount of data need to be examined.

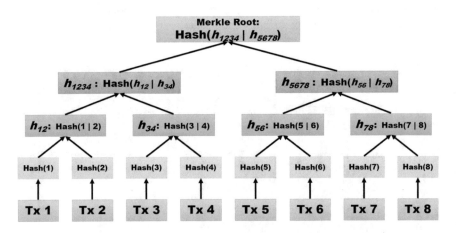

Fig. 4 Merkle tree structure

Nonce: The Nonce is a random number of length 4 bytes which is used only once for proof of work consensus algorithm. Blockchain miners first need to solve and find a valid nonce when competing for a new block to be added into the blockchain. In return, miners are awarded incentive for validating a node into the chain [23].

Difficulty Target: The difficulty target is a number that dictates how long it takes for miners to add new blocks into blockchain. Difficulty target increased or decreased based on the previous 2016 blocks took less or longer time respectively [22].

5.2 Type of Blockchain

Blockchain network can be divided into two categories such as:

Public or Permissionless Blockchain: it is open for everyone to join the network and maintain the transaction. Examples of such kind public blockchain are Bitcoin, Ethereum, etc. [26].

Private or Permissioned Blockchain: In this blockchain network permission is restricted to a specific group of participants or certain organization and it is not open for everyone. Private blockchain offers the opportunity to get the full benefit of blockchain while controlling the access right of network. It is relatively small blockchain network which allows to customize the consensus algorithm in order to improve efficiency. Hyperledger Fabric [4] is an example of permissioned blockchain.

6 Blockchain-Based Framework for IoT Forensics

Moving towards the decentralized solution for IoT forensics investigation process, and managing explosive amount of cyberattacks incidents is the key to success. Blockchain technology could be a suitable enabling technology which meets the demand and requirement of IoT forensics such evidence integrity, distribution and secure verification. Digital evidence can be easily added and collected from the blockchain network and the immutability feature of the blockchain will protect its legitimacy and consistency. Investigation authority can reliably access the forensically relevant and important evidence from any node of the chain.

In IoT-based ecosystem, IoT users, device manufacturers, IoT service providers, law enforcement office, forensics experts, and other participants in blockchain could maintain a copy of the ledger. Therefore, the evidence could not be removed or counterfeited by a single control entity, and the issue and risk of the "single point of failure" is eradicated.

Very recently, blockchain-based IoT forensic frameworks have been proposed in order to deal with the dynamic challenge pose by the IoT paradigm. In this section, the most recent proposed blockchain-based IoT forensics framework is presented

Table 1 Summary of Blockchain-based IoT forensics framework

Blockchain-based IoT forensics framework	Implementation	Category	Author/Year
A blockchain-based decentralized efficient investigation framework for IoT digital forensics	Ethereum	Public	Ray et al. 2019
FIF-IoT: A forensic investigation framework for IoT using a public digital ledger	Not available	Public	Hossain et al. [11]
Blockchain-based digital forensics investigation framework in the Internet of Things and social systems	Not available	Public	Li et al. [15]
Biff: A blockchain-based iot forensics framework with identity privacy	Not available	Private	Phong et al. [14]
A Cost-efficient IoT forensics framework with Blockchain	EOS, Stellar, Ethereum	Public	Mercan et al. [17]

since 2018 to 2020. Table 1 illustrates the summary in brief of blockchain-based IoT forensics approaches.

Ryu et al. [23] proposed blockchain-based decentralized framework to conduct the IoT forensics investigation. The framework is divided into three main layers such as participants layer (top layer), blockchain (middle layer) and devices (bottom layer). During the interaction between IoT devices, they usually generate data and each interaction is called a transaction. In their proposed architecture, a transaction has five fields such as source device identity (SID), destination device identity (DID), exchange data (D), digital signature (S) and transaction id. Digital signature of a transaction was generated using the private key and source deceive identity (SID). Then, transaction id is produced by hashing twice of SID, DID, D and S using the SHA-256 hash function. Transactions are added one after another continuously into a block until the block size is exceeded. Once the block is completed, it is linked into the blockchain layer (middle layer). Participants from the top layer such as device users, manufacturers, service providers, and forensic investigators can verify the integrity of the blockchain. The proposed framework was simulated using Ethereum platform and smart contract interface is constructed using the Mist [7] browser to carry out the evidence generation, collection and report presentation.

A public digital ledger based IoT forensic framework named as FIF-IoT is presented by Hossain et al. [11] in order to discover facts in cyberattack incidents within various IoT environments. FIF-IoT model aggregates all the communication happening among the various entities in the IoT environment such as IoT devices to IoT devices, IoT device to cloud, and users to IoT devices in the form of transaction. The transactions are sent to blockchain network where the miners from different stockholders obtain transactions and mine the new block by combining the relevance transactions. Finally, the blocks are glued into the public, distributed, and decentralized blockchain network. FIF-IoT framework is capable of catering integrity, confidentiality, anonymity, and non-repudiation of the publicly-stored evidence. In addition, FIF-IoT framework proposed a scheme how to authenticate and verify the collected evidence during the investigation process.

IoTFC, a blockchain-based digital forensics investigation framework in IoT and social system environment was proposed by Li et al. [15], which can provide evidences traceability, provenance of data, reliability between IoT entities and forensic investigators. Building blocks of this architecture are users, IoT devices, Merkle tree, block, and smart contract. This framework collects the evidences only from the devices that are relevant and involved in a particular case. IoTFC method first gathers all the evident items and creates a distributed ledger in order to store and record the transactional evidents (TEs). Then these evidents (TEs) are shared and distributed to the legitimates participants through the blockchain network. To support the tamper proof environment, IoTFC builds a public timestamped log mechanism ensuring the full provenance of each evidence for all investigators without the existence of a trusted third party. This framework also graded the evidence into five types according to difficulty level such as g1 (easy to identify e.g., plain text, unencrypted image, QR), g2 (deliberate attempt to hide e.g., renamed extension), g3 (hard to identify), g4 (difficult to identify e.g., encrypted data, password) and g5 (very difficult to identify e.g., steganography).

BIFF is a private blockchain-based IoT framework proposed by Phong Le et al. [14] to store all the events during digital forensic process. This model offers a cryptographic-based technique to eliminate the identity privacy problem. BIFF framework has three entities such as digital witness (DW), digital custodian (DC), and law enforcement agency (LEA). Each entity has different roles and rights in the IoT forensic process. LEA is the most important entity in the proposed framework who is the most trusted entity and responsible for evidence gathering, examining, evaluating, and archiving from DW and DC. Framework also defines each entity access right which includes read, write and verify right. All participant entities have read access but write and verify access rights are given to selective entities. BIFF framework has four main components such as transactions, smart contract, block, and consensus protocol. In order to ensure the privacy of an entity, BIFF framework combined the digital certificate techniques into the merkle signature.

Mercan et al. [17] proposed a cost-efficient IoT forensics framework leveraging multiple blockchain in two layers. This framework uses the multiple low-cost blockchain platforms which provide the multi-factor integrity (MFI). MFI feature of the model allows to withstand against any kind of malicious attack because attackers still need to break at least one more obstacle in order to breach the integrity of evidence. The proposed approach tries to reduce the data size to be written in public blockchain network by deploying hash function and merkle tree. In the very first stage, hash values of relevant IoT data are sorted into the 1st level EOS [8] and Stellar [24] blockchain network. In the second step, data center collects all confirmed transactions those are stored in 1st level blockchain network and builds a merkle tree. Finally, merkle root is computed and hash of all hashes are submitted to the 2nd level Ethereum blockchain. By delineating multi-level blockchain, framework significantly reduces the cost.

From the above discussion, we can argue that blockchain enabling IoT forensics solution is a promising emerging field and it is growing attention among the forensic scientists because it offers evident integrity, provenance, traceability and decentralized management. Yet effective mechanism need to be defined in order to ensure the data privacy and avoid race attack.

7 Conclusion

Rapidly growing IoT environment is creating plethora of challenges for conducting IoT forensics. Therefore, there is an essential need to develop innovative IoT digital forensic techniques that can handle the challenges encountered by IoT forensic professional. Since IoT-based attacks escalate, it may become more impossible to convict perpetrators effectively with the existing traditional digital forensics mechanisms. Current proposed blockchain based frameworks lay the foundation for future practical forensic investigation work. Law enforcement agencies, IoT service providers, and device manufactures should join hands to withstands against challenges of IoT security and work together to provide a standard mechanism to deal with the cybercrimes in legitimate and standard manner securing the forensics evidence life-cycle.

References

1. Al-athwari, B., Azam, H.M.: Resource allocation in the integration of IoT, Fog, and Cloud computing: state-of-the-art and open challenges. In: International Conference on Smart Computing and Cyber Security: Strategic Foresight, Security Challenges and Innovation, pp. 247–257. Springer, Cham (2020)
2. Alenezi, A., Atlam, H., Alsagri, R., Alassafi, M., Wills, G.: IoT forensics: a state-of-the-art review, challenges and future directions (2019)

3. Altmann, J., Al-Athwari, B., Carlini, E., Coppola, M., Dazzi, P., Ferrer, A.J., Haile, N., Jung, Y.W., Marshall, J., Pages, E., et al.: BASMATI: an architecture for managing cloud and edge resources for mobile users. In: International Conference on the Economics of Grids, Clouds, Systems, and Services, pp. 56–66. Springer, Cham (2017)
4. Androulaki, E., Barger, A., Bortnikov, V., Cachin, C., Christidis, K., De Caro, A., Enyeart, D., Ferris, C., Laventman, G., Manevich, Y., et al.: Hyperledger fabric: a distributed operating system for permissioned blockchains. In: Proceedings of the Thirteenth EuroSys Conference, pp. 1–15 (2018)
5. Atlam, H.F., Alenezi, A., Alassafi, M.O., Alshdadi, A.A., Wills, G.B.: Security, cybercrime and digital forensics for IoT. In: Principles of Internet of Things (IoT) Ecosystem: Insight Paradigm, pp. 551–577. Springer, Cham (2020)
6. Bhushan, B., Sahoo, C., Sinha, P., Khamparia, A.: Unification of blockchain and internet of things (BIoT): requirements, working model, challenges and future directions. Wirel. Netw. **27**, 55–90 (2020)
7. Dannen, C.: The mist browser. In: Introducing Ethereum and Solidity, pp. 21–46. Springer, Cham (2017)
8. EOSIO: next-generation, open-source blockchain protocol. https://eos.io/. Accessed 20 Dec 2020
9. Federal Rules of Civil Procedure. Rule 34. http://goo.gl/NfL61. Accessed 20 Dec 2020
10. Horsman, G.: Raiders of the lost artefacts: championing the need for digital forensics research. Forensic Sci. Int. Rep. **1**, 100003 (2019)
11. Hossain, M., Karim, Y., Hasan, R.: FIF-IoT: a forensic investigation framework for IoT using a public digital ledger. In: 2018 IEEE International Congress on Internet of Things (ICIOT), pp. 33–40. IEEE (2018)
12. Janarthanan, T., Bagheri, M., Zargari, S.: IoT forensics: an overview of the current issues and challenges. In: Digital Forensic Investigation of Internet of Things (IoT) Devices, pp. 223–254 (2021)
13. Kent, K., Chevalier, S., Grance, T., Dang, H.: Guide to integrating forensic techniques into incident response. NIST Spec. Publ. **10**(14), 800–86 (2006)
14. Le, D.P., Meng, H., Su, L., Yeo, S.L., Thing, V.: BIFF: a blockchain-based IoT forensics framework with identity privacy. In: TENCON 2018-2018 IEEE Region 10 Conference, pp. 2372–2377. IEEE (2018)
15. Li, S., Qin, T., Min, G.: Blockchain-based digital forensics investigation framework in the internet of things and social systems. IEEE Trans. Comput. Soc. Syst. **6**(6), 1433–1441 (2019)
16. Li, W., Wang, Y., Li, J., Au, M.H.: Toward a blockchain-based framework for challenge-based collaborative intrusion detection. Int. J. Inf. Secur. **20**, 127–139 (2020)
17. Mercan, S., Cebe, M., Tekiner, E., Akkaya, K., Chang, M., Uluagac, S.: A cost-efficient IoT forensics framework with blockchain. In: 2020 IEEE International Conference on Blockchain and Cryptocurrency (ICBC), pp. 1–5. IEEE (2020)
18. Merkle, R.C.: A digital signature based on a conventional encryption function. In: Conference on the Theory and Application of Cryptographic Techniques, pp. 369–378. Springer, Heidelberg (1987)
19. Mohamed, K.S.: Iot physical layer: sensors, actuators, controllers and programming. In: The Era of Internet of Things, pp. 21–47. Springer, Cham (2019)
20. Mufti, T., Saleem, N., Sohail, S.: Blockchain: a detailed survey to explore innovative implementation of disruptive technology. EAI Endorsed Trans. Smart Cities **4**(10), 164858 (2020)
21. Nakamoto, S.: Bitcoin: a peer-to-peer electronic cash system. Bitcoin, vol. 4 (2008). https://bitcoin.org/bitcoin.pdf
22. Omote, K., Yano, M.: Bitcoin and blockchain technology. Blockchain and Crypt Currency, p. 129 (2020)
23. Ryu, J.H., Sharma, P.K., Jo, J.H., Park, J.H.: A blockchain-based decentralized efficient investigation framework for IoT digital forensics. J. Supercomput. **75**(8), 4372–4387 (2019)
24. Steller: Blochchain Network. https://www.stellar.org/. Accessed 20 Dec 2020

25. Stoyanova, M., Nikoloudakis, Y., Panagiotakis, S., Pallis, E., Markakis, E.K.: A survey on the internet of things (IoT) forensics: challenges, approaches, and open issues. IEEE Commun. Surv. Tutor. **22**(2), 1191–1221 (2020)
26. Vujičić, D., Jagodić, D., Ranđić, S.: Blockchain technology, bitcoin, and ethereum: a brief overview. In: 2018 17th International Symposium INFOTEH-JAHORINA (INFOTEH), pp. 1–6. IEEE (2018)
27. Yaqoob, I., Hashem, I.A.T., Ahmed, A., Kazmi, S.A., Hong, C.S.: Internet of things forensics: recent advances, taxonomy, requirements, and open challenges. Future Gener. Comput. Syst. **92**, 265–275 (2019)

Printed in the United States
by Baker & Taylor Publisher Services